WHY TEACH MATHEMATICS?

Mathematics Education Library

VOLUME 33

The titles published in this series are listed at the end of this volume.

WHY TEACH MATHEMATICS?

A Focus on General Education

by

HANS WERNER HEYMANN
Universitat Siegen, Germany

KLUWER ACADEMIC PUBLISHERS
DORDRECHT / BOSTON / LONDON

Library of Congress Cataloging-in-Publication Data

ISBN 1-4020-1786-3

Published by Kluwer Academic Publishers,
P.O. Box 17, 3300 AA Dordrecht, The Netherlands.

Sold and distributed in North, Central and South America
by Kluwer Academic Publishers,
101 Philip Drive, Norwell, MA 02061, U.S.A.

In all other countries, sold and distributed
by Kluwer Academic Publishers,
P.O. Box 322, 3300 AH Dordrecht, The Netherlands.

Printed on acid-free paper

Printed in the Netherlands.

CONTENTS

PREFACE TO THE ENGLISH EDITION

Seven years after its publication in Germany, this book is now available to English readers. In the German-speaking countries, the publication led to intense controversies on the significance, the goals, and the future development of mathematics instruction in schools.

Currently, there is much debate in many countries on how to evaluate the findings of such international comparative studies as TIMSS and PISA, debate on which consequences should be taken for designing national curricula and on which reforms of mathematics instruction in schools should be initiated. I hope that this book might provide substantial suggestions beneficial to resolving these and similar issues, for its major concern is twofold: first, to determine the essential characteristics of general education in schools, and, secondly, on the basis of this, to specify the status of mathematics instruction *within the framework* of general education in schools.

Due to the fundamental nature of my argument, I did not consider it necessary to 'update' the book superficially for the English edition. Thus, I am presenting the book to the English-speaking public in its original form, the single exception being that I have omitted the original second chapter since that chapter ("*Bildung* und *Allgemeinbildung*") seems to be relevant only to those readers especially interested in German educational debates. Most of those readers, in turn, will belong to the German-speaking public. The omission of this chapter necessitated a few insignificant alterations in the (otherwise unchanged) introduction and at the beginning of the present second (originally, third) chapter.

CHAPTER 1

INTRODUCTION

> *This liability to reconditeness is the characteristic evil which is apt to destroy the utility of mathematics in liberal education. So far as it clings to the educational use of the subject, so far we must acquiesce in a miserably low level of mathematical attainment among cultivated people in general.*[1]
>
> Alfred N. Whitehead, 1913

> *For me, mathematics was more of a burden that was to be gotten rid of as quickly as possible. What always bothered me about mathematics was the usually abstract material, the formulas and numbers or letters whose meaning and context I simply could not understand.*[2]
>
> An educational science student, 1988

Mathematics is a major and mandatory subject in schools worldwide. Apparently, there is an international consensus that schools should impart basic mathematical cognitive abilities and procedural skills to all children and adolescents – regardless of the respective culture or political system.

But this major significance of mathematics for the educational canon (hardly a matter of fundamental debate) contrasts strikingly with a considerable number of problems. At first glance, these present themselves as follows:

A great number of children, adolescents, and adults encounter enormous difficulties with mathematics. For these people, the difficulties are intrinsic in the distinctive characteristics of this subject matter. In many cases, the mathematics which they are obliged to learn in school only attains the status of knowledge required for examinations – learned superficially and, correspondingly, quickly forgotten again. Many students experience the 'abstract' nature of mathematics as a handicap. They feel that they are encountering a baffling, practically incomprehensible system of concepts and rules. And although no other subject except perhaps language arts is allotted more classroom time, even the knowledge and skills related to elementary mathematical content remain deficient.[3]

Often, this situation is complemented by a negative attitude toward the subject, which, then, enjoys a marked depreciation as a whole. It is not at all rare that precisely academics flaunt their ignorance of mathematics (cf., e.g., Fölsch 1975, pp. 155-156).

For the majority of adolescents, then, we can assume that the ostensible educating effects of mathematics instruction are correspondingly slight. The institutionalization of mathematics teaching was and still is linked to the hope that more could be learned than the mathematical content on the lesson plan. Even before the introduction of a general school system, Plato, Humboldt, and many others acclaimed mathematics as an excellent means of training the intellect, quite apart from any considerations to utility. The reality of the mathematics classroom seems to be far removed from such effects on the individual's education and personal development. At any rate, there is hardly any empirical evidence available to support these claims. Mathematics instruction seems to have promoted abilities related to systematical and critical thinking, problem solving, mathematical modeling, and rational argumentation only for a minority of students.

Moreover, it is becoming increasingly difficult to bridge the gulf between conventional school mathematics and the mathematical applications that have made our life circumstances possible – circumstances largely determined by technology and automation. The gap between the objective and the subjective significance of mathematics is constantly expanding. On the one hand, our environment is increasingly becoming dominated by developments and permeated by products that would be inconceivable without the astute mathematical knowledge and skills of a relatively small group of highly qualified experts. On the other hand, mathematics itself tends to disappear behind its applications. The mathematics that is the foundation of our standard of living is, as it were, invisibly built into the technological devices we use. From the perspective of the technology user, mathematics makes itself superfluous: operating a computerized cash register in a supermarket requires less 'mathematical' ability than computing the costs with paper and pencil. Neither students nor adults, as long as the latter are not engaged in mathematics-related professions, are now in a position to experience any practical usefulness of the mathematical knowledge they learned in school, apart from a few very elementary skills.

In view of the problems mentioned above, the following premise, which I am consciously formulating pointedly, deserves careful consideration: *Conventional mathematics instruction in schools does justice neither to foreseeable societal demands nor to the individual needs and qualification interests of a majority of adolescents.*

With the outlined situation in mind, convincing arguments will certainly be necessary to justify the decision to use state authority to compel all children and adolescents to learn mathematics during the entire course of their schooling – unless, of course, one is content to take recourse to the indisputable function of mathematics instruction as a subject matter pertinent to selection and as a means for recruiting the next generation of professionals for mathematical and mathematics-related vocations. The usual arguments that are continually put forth in favor of a generally mandatory mathematics instruction can be found in the preambles of both earlier and more recent syllabi and guidelines – with certain variations corresponding to the respective spirit of the times. Yet, in view of the problems mentioned above, the rationale behind such arguments must appear extremely dubious. One reference to the objective significance of mathematics, e.g., is particularly widespread and is related both to practicality and to intellectual history: the stipulation that only mathematics, especially as mediated by science and technology, has enabled and continues to enable the (further) development of our modern civilization, and the stipulation that mathematics is, in and of itself, regardless of its practical applications, a cultural achievement of the highest order. But by no means can either one of these stipulations lead to a solution of the problem of how much and which mathematics children and adolescents should learn – to their own benefit and to that of society. It seems that in the conventional perspective the indisputability of the premise is confused with the soundness of the entire argument.

To put it succinctly, there is no adequate differentiation between the *value and significance of mathematics as such* and the *necessity of compulsory mathematics instruction for all students*. This would require a soundly tenable standpoint external to the discipline, for neither the mathematical discipline itself nor an analysis of the objective application of mathematics in our society can solely provide criteria that allow a prudent judgment on the issue of which and how much mathematics all the students in our society should learn in which way.

This book undertakes an attempt at such a clarification on the basis of a modern concept of general education. Concepts of general education seek to clarify the idea of general education as a basic education for all members of a society and to concretize this idea with respect to teaching practice. Despite a considerable number of competing concepts of general education that have been developed in the last two and a half decades with widely varied emphases, there is still an alarming lack of orientation in our society whenever issues of general education in school arise. For this reason, I will develop and substantiate my own conception in detail. This concept is at the basis of other considerations in this book and is the actual pedagogical core of the present study.

Once this concept of general education has been developed, a "criterion" will then be available that can be referred to for judging and critiquing the existing form of mathematics instruction. Using this criterion, we can systematically determine elements of an alternative mathematics instruction that can justifiably claim to aim at a concrete realization of the idea of general education.

Thus, our major concern within the scope of this book is, beginning from a standpoint of sound educational theory, to systematically arrive at a better understanding of the deficits outlined above and to provide a perspective for a profile of mathematics teaching as part of a general education. This book differs from many other earlier and more recent attempts to more exactly define the problems of mathematics instruction and to search for solutions in that all of the arguments consistently refer back to the modern concept of general education developed here. And it differs from many other earlier and more recent attempts to outline general education in school on the whole in that the consequences of the more general reflections are carefully considered with respect to a particular subject matter, mathematics instruction. For the fact that instruction in schools is primarily specialized instruction is arguably one of the inevitable preconditions of institutionalized general education. If general education is to succeed, then it must succeed *in* and *through* specialized subject-matter instruction.

Apart from occasional overviews and references to corresponding endeavors in other countries, my discussion is limited to current issues under debate in German-speaking countries. *Systematical* consideration of the relevant approaches developed in other countries[4] would have exceeded both the scope of the book and the capacity of a single author. Nevertheless, the concept of general education developed here and the consequences for school mathematics instruction are not exclusively designed for the conditions prevalent in today's Germany. The arguments presented here certainly do also offer suggestions for reassessing mathematics instruction in other countries with a comparable social structure.

This study is committed to a hermeneutics informed by the social sciences and, in the broadest sense, is methodologically situated within the tradition of humanist pedagogy. Perhaps in opposition to the major trends in the discipline, I still consider Wilhelm Flitner's (1966, p. 18) appeal to pursuing pedagogy as a "réflexion engagée" to be inspiring, innovative, and necessary. The major function of pedagogy – as one of the humanities and as a discipline concerned with human action – still seems to be to provide objective and linguistic clarifications of its subject area. Public affairs ('res publicae') can then be considered and negotiated more rationally and with greater benefit to all of those involved. Public education and personal development, especially 'general education' within the context of compulsory education, is a 'res publica' par excellence. This, then, also applies to mathematics instruction as one component of general education.

Thus, to a great extent, the arguments presented here have the quality of a systematical 'search'. What is sought after are reasonable goals for limited spheres of social action: for teaching and learning in school on the whole and, in particular, for teaching and learning mathematics in school. The suggestions in this book develop from reflections on the limited sphere of school mathematics in the light of goals our society generally considers reasonable. The following statement also applies to other school subjects and areas of learning: If in our society there are ideas about what sort of effects the institution of schooling should have on the next generation, ideas that are explicable and likely to find a consensus, then it would make no sense to design the branch of mathematical learning according to totally different criteria and to set it apart from more general goals that for the most part are considered appropriate.

The appropriateness of global and abstract objectives is often only demonstrated by their effectiveness in more specific contexts: If they lead to undesirable or even intolerable results in everyday practical activity, then they must be revised or countermanded. But in order to substantiate such inappropriateness, the objectives must, at least, first be clarified to the extent that their practical consequences are also evident. Time and again, we can observe the inclination to engage in reflection on the subject of mathematics largely independently of reflection on schooling with its social and interdisciplinary contexts. The uncommunicative division of labor among mathematics educators and general educationalists that becomes established in this way and that can extend to the point of mutual ignorance obstructs important learning processes for both factions:[5]

The educationalists forgo the opportunity of exposing their more global considerations to a practical test involving specific subject matter and, thus, to possibilities for revision.

Undisturbed by more general reflections on the purpose and objectives of schooling, mathematics educators can devote their time to detailed, subject-matter specific, or even specialist, problems – up to the point of totally losing sight of any considerations that go beyond their own discipline.

This book is my attempt to mediate between these two separate perspectives, to increase the mutual perceptibility and recognition of both positions. By explicating the overarching pedagogical goal of schools in the form of a concept of general education, a criterion for judging the adequacy of mathematics instruction will be made available. And by considering the contribution of mathematics teaching to general education in detail, the limitations involved in defining the more specialized subject area by the more general will also become apparent. It will become evident that no concept of general education can automatically produce good mathematics instruction – it must be complemented by mathematical competence and subject-related creativity. The more specialized school subject makes more sense and gains a better-defined profile in the light of the general idea; it acquires a corrective against becoming too involved with specialist concerns. In view of the more specialized subject matter, the more general idea acquires concrete meaning and relevance to human action.

The main argument of my book is as follows: If we bring the idea of general education and personal development into a sufficiently concrete and up-to-date form, it can, indeed, have substantial consequences for the problems involved in teaching school subjects (in this case: mathematics) and can thus function as a critique, a clarification, and an orientation. Beyond the intentions more strictly concerned with the relevant academic disciplines, the book, which critically examines a number of traditional pedagogical ideas in a new light, also has a practical aim that is of a more conservative nature – though not in the sense of conventional political polarization: The goal is neither a fundamental revolution of ideas on general education nor a radical revision of mathematics instruction or of schooling in general; instead, suggestions for altering the current unsatisfactory instructional practice result from connecting what is usually separated, from calling to mind neglected and forgotten contexts, from a dialectical, integrated overview.[6]

The Specific Structure of the Book

The first part of the book presents detailed arguments for seven distinguishable, interconnected objectives of a school system aiming at general education. With these seven objectives, a specific concept of general education and personal development is then expounded and made applicable as a criterion for the projected pedagogical critique of mathematics instruction (chapter 2). Then, the objectives are systematically used to identify deficits of the current widespread form of mathematics instruction and to determine which innovations at which levels (curricular and related to lesson planning) are necessary in order to do more justice to the claim of realizing general education. To this end, existing concepts of mathematics teaching are assessed anew and related to one another – in the light of the idea of general education and personal development (chapter 3). The most important results of the study are summarized in the final chapter, which attempts to answer the following question: What must we stress in order to realize a mathematics instruction that is truly a part of a general education? (chapter 4).

Acknowledgments

More people than I could list here have contributed to this book – by engaging in stimulating discussions, by expressing concurring or opposing views, by responding critically to earlier versions, and by sharing experiences or providing references to works that would have remained unnoticed. Of those persons not referred to by name in the main sections of the book, I would at least like to mention Rainer Opitz and Erwin Steinhoff. I also wish to acknowledge my thanks to Heinrich Bauersfeld for his support. It is my pleasure to express my gratitude to these people and to many others – colleagues at school or at the university, friends and relatives. This book is for my wife, Karin.

CHAPTER 2

AN OUTLINE OF THE CONCEPT OF GENERAL EDUCATION AS A FRAME OF REFERENCE

Since we cannot be universal and know all that is to be known of everything, we ought to know a little about everything.[7]
Blaise Pascal, around 1660

Your general education and everything designed to further it are simply nonsense. What really matters is that a person understands something conclusively, is able to attain a level of excellence that cannot readily be achieved by the others in his vicinity.[8]
Johann Wolfgang v. Goethe, around 1820

One of the great things that education can and should give is the power of seeing the general in the particular, the power of feeling that this, although it is happening to **me**, *is very like what happens to others...*[9]
Bertrand Russell, 1961

The 'idea' of general education and personal development is confronted with the 'reality' of schools aiming at general education. Any concept of general education attempts to mediate between the 'idea' and the 'reality'. It concretizes the idea of general education and personal development. It elaborates the idea of general education with respect to existing practices and it provides this idea with an interpretation related to the prevailing historical conditions. From the perspective of an individual author or of a group of collaborators, a concept of general education hypothetically anticipates a conceivable social consensus on the overarching pedagogic goals of schooling. In the final analysis, competing explications of general education in the form of various conceptions are prerequisites of a rational social discourse aimed at establishing a consensus on new syllabi, curricular guidelines, school organizational procedures, etc.

At the core of the concept of general education and personal development to be discussed here are seven objectives of schools:

- preparing for later life
- promoting cultural competence
- developing an understanding of the world
- promoting critical thinking
- developing a sense of responsibility
- practicing communication and cooperation

- enhancing students' self-esteem.

With reference to these seven objectives, it is possible to pedagogically evaluate both curricular guidelines of instruction and concrete lessons – more precisely, it is possible to asses their potential for furthering general education. It will become apparent that this cannot be accomplished by applying mechanical or additive operations to the objectives listed above, but only by means of rational arguments that provide a frame of reference for the objectives.

In what follows I will be discussing and proposing arguments for these seven objectives of schools. To do this, I will employ certain strands and elements both of the theoretical tradition concerned with education and schooling and of more recent educational debates, without assigning a privileged position to any one of these elements. Considered by itself, none of the objectives discussed below is particularly original. For each one, theorists can be named who have focused their own educational theories on this objective, thereby neglecting other aspects or assigning to them a derivative character. In my opinion, such attempts at derivation are questionable. My priority is to go beyond the specific views elaborated within the context of these objectives thus far and to work out the interconnections among them, their dialectical cross-references, their mutual need to complement one another. I am interested in demonstrating how these objectives mutually accentuate, but also restrict one another, how considering any one of these objectives to be absolute necessarily results in undesirable excesses, side-effects, or disregard for other considerations, all of which are unacceptable for classroom practice and in a political sense if the idea of general education is to be taken seriously. Only by tracing these interconnections can the claim of presenting an up-to-date and feasible concept of general education with reference to the entire ensemble of these objectives be upheld. Otherwise, it would have made more sense to exhaustively discuss seven existing educational or curricular theories – one for each of these objectives – and, indicating the particular perspective of each of these approaches, to compile them synoptically.[10]

Naturally, the selection of objectives for schools proposed here is open to criticism. Why should there be exactly these seven, why not others, why not more or less? In a very strict sense, there is no conclusive theoretical justification for this selection. Nonetheless, a pragmatic justification that is not to be underestimated does exist. In my opinion, practically all of the significant considerations which figure in current debates on general education can be discussed in relation to these objectives. They reflect these considerations and can help to put them into a systematic context.

The objectives listed here are neither devoid of overlap nor do they all exhibit the same level of theoretical abstraction. And, of course, it is inevitable that the subjective experience and values of the author find their way into the formulation of such a concept. I hope to make these influences sufficiently clear.

2.1 Preparation for Later Life

Hardly any other specification of an objective of schooling would meet with such general public approval as the following: Schools aiming at general education should prepare adolescents for life as adults. From the perspective of the students and of their parents, the directly concerned 'customers' of classroom instruction, the demand for preparation for later life corresponds to the demand for the personal benefit of what is to be learned, for its practical utility in the context of everyday life. On the other hand, seen from the perspective of society, in committing itself to the preparation for later life the school safeguards the continued existence of society: The structural correlate of the objective of preparing for later life is the reproductive function of the school.

But just how adequate is such a demand as a criterion for decisions on how and which instruction is to take place in schools? Certain experiences, most recently those attempts to reform curricula in Germany during the 1960s and 70s, have demonstrated that the superficial consensus is quickly in danger of dissipating whenever possible concrete realizations are about to be put into practice. The stipulation that schooling should provide preparation for later life can be all too easily propagated from widely divergent standpoints on social theory and educational policy.

The aim of the present section is twofold:

- I want to point out that the basic rationale behind preparation for later life as an objective of public schools is an indispensable element of a modern conception of general education.
- On the other hand, I want to demonstrate the limitations that apply to this objective as a criterion for general education. I will argue in favor of its augmentation by further criteria by demonstrating the constrictions involved in overtaxing the principle that schooling should provide preparation for later life.

Especially with regard to this second purpose, headway can be made by examining certain problematical aspects of Saul B. Robinsohn's curriculum theory, a theory in which for the first time preparation for later life became the major systematic starting point for a curricular revision. A short discussion of the key qualifications currently under debate will provide additional considerations. These, then, facilitate a more precise specification of the objective of preparing for later life.

2.1.1 Preparation for Later Life as the Starting Point for Designing a Utilitarian Curriculum: Saul B. Robinsohn

Robinsohn's approach – very influential in educational science, but with hardly any repercussions in the actual classroom – has merited widespread reception and frequent critiques. Yet, repeatedly, Robinsohn's intentions have been misunderstood and he has been undeservedly reprimanded for epitomizing a technocratic aberration. Here, we are not concerned with contributing to a very general sort of critique. On the contrary, confronting Robinsohn's model helps to determine the specific significance of preparing for later life for the concept of general education to be expounded in this study.

Robinsohn demonstrates a profound familiarity with the German tradition of educational theory and his critique of this tradition is carefully considered (1973 [1967], pp. 13ff). He assesses the weaknesses of, for example, Erich Weniger's syllabus theory with regard to a constructive conception of syllabi in much the same way as Blankertz, whose critique he explicitly endorses (ibid., p. 25). Many of his remarks on the problems surrounding education can be interpreted as the attempt to reconcile German educational theory to Anglo-Saxon pragmatism:

> "For the individual, the process of education means becoming prepared to behave in the world around us. The fact that the educational process is oriented to existing culture, that the interpretation of reality is carried out with the help of socially transmitted forms and content, does not contradict this definition; on the contrary, it is implicitly contained within it." (ibid., p. 13)

If he then designates "preparing the individual to cope with situations of everyday life" as the "general goal of education" (p. 79), then this is by no means only to be understood as being *technically* capable of dealing with such situations or as having capacities of some pragmatic immediacy. Robinsohn emphasizes that "attention should be increasingly paid to qualities such as cooperativeness, empathy, self-confidence, and imagination" (ibid., p. XX); particularly significant are "abilities such as mental flexibility, problem awareness, greater sensitivity" (p. XIX). Thus, preparation for later life, which Robinsohn declares to be the main objective of any formal education, is initially to be understood as *preparation for later life in the broadest possible sense of the word.*

Yet, with his efforts to proceed from this starting point and plan curricula as rationally as possible, a peculiar limitation appears in Robinsohn's conception. This limitation results from that connection between life circumstances, qualifications, and elements of curricula which seemed so persuasive at first glance. In designating the task of curricular research as follows:

> "finding and employing methods by means of which these situations and the functions required by them – the qualifications necessary to cope with them, the content of instruction, and the objects effecting this further qualification – can be identified with an optimum objectification" (ibid., p. 45),

Robinsohn unwittingly invalidates his broad definition of preparation for later life. In order to demonstrate this, I will concentrate on the following three points:

- the use of the concept of qualification;
- the problem of undecided life situations;
- the implications of the utilitarian claim.

The Concept of Qualification

Originally, the concept of qualification stems from research on professions and professionalism. Qualification can be defined as the "capacity to work, as the entirety of the particular subjective, individual abilities, knowledge, and skills which allow the individual to carry out a specific working function" (Baethge 1974, p.

479). The corresponding inference for utilitarian vocational education is the predominant goal of imparting the required qualifications as effectively as possible.

Toward the end of the 1960s, the concept of qualification was increasingly broadened to include activities beyond those of professional life. In part, this occurred with the explicit intention of replacing the classical concept of *Bildung*. In this respect, Robinsohn's procedure was simply logical: Just as the qualification for a job enables me to cope with the situations typical of this job, there could be qualifications for life situations not relevant to my profession which would enable me to deal with precisely such situations.

By contrast to the concept of education and personal development, the concept of qualification is inextricably linked to a rigorous means-ends line of thought. Thus, the qualification is the *means* which helps me to attain the *end*, the mastery of a certain situation. In his comparison of the concepts of education and qualification, Kade (1983) emphasizes the fact that qualification is associated with utilization, application, and adaptation. He correctly notes that even in ordinary language, a connection between certain realms of activity and "qualification" can hardly be maintained, that, e.g., it "seems to be inappropriate if someone talks about qualifying himself or herself for love or for some artistic activity" (ibid., p. 860). The conception of qualifications as means to attain certain ends (whether determined by an employer, by society, or by the individual himself or herself), is in marked contrast to the idea usually associated with education and personal development, namely, that it serves its own purpose. In addition, the concept of qualification exhibits a further peculiarity that limits its possible use in this context: a qualification must be operational, subject to precise description and production. The pragmatist way of thinking is oriented to feasibility. Along these lines, the primary concern is with qualification elements such as abilities, skills, knowledge and behavior that can be sufficiently atomized, analytically distinguished from other elements, and standardized. There have been attempts to broaden the content of the concept of qualification in order to release it from such constrictions (Lenhardt 1974, p. 36-37); with some of the ideas he puts forward, Robinsohn himself also seems to be using a very broad concept of qualification as a starting point. But the question arises if, then, the conceptual precision (presumably one of the advantages in comparison to the concept of *Bildung*, education and personal development) does not become so diluted that it would be better to totally abandon it. In any case, for the realization of his conception of curriculum, Robinsohn is obliged to take recourse to the properties of the concept of qualification in the more restricted sense mentioned above – its status as a means in a means-ends scheme, its ability to be operationalized and to be atomized.

Problems Involved in Undecided Life Situations

To a certain extent, establishing a correspondence between qualifications and their applications is possible for a considerable number of professions and fields of work, especially with reference to craftsmanship and industrial work. Work situations are often such that they themselves can be interpreted as being linked to the means-ends schema, e.g., if a particular product is to be manufactured or a certain service is to be performed. However, this correspondence causes considerable problems whenever it is to be applied to the entirety of human life situations. Many situations are undecided and without any codified set of requirements to be met, situations which we encounter, for instance, when we are in love, when we engage in artistic activities, become acquainted with others, or provide help in emergencies, etc. In these and many other situations, what is required is more the readiness to become involved with another person or a certain circumstance rather than applying accumulated knowledge or abilities (cf. Kade 1983, p. 867). On the one hand, the general concept "situation" includes cases that are not covered by predictive classifications – there are situations that are practically incomprehensible to us, situations that we would never have expected –; on the other hand, a considerable number of situations do not immediately pertain to some purpose in such a way that they could be dealt with by a course of action that can be planned in advance. Instead, they allow for unforeseen developments and require a correspondingly unrestricted attitude on the part of the persons involved.

The problem with Robinsohn's approach is that by linking qualifications to life situations, he excludes in advance, probably unintentionally, a large number of the conceivable life situations: all of those which are open, not standardized, and not to be dealt with by pragmatic action. The general conception of situations applies to such a large sphere that it can hardly be used at the same level of abstraction as the (relatively limited) concept of qualification. If this difference is ignored, the concept of situation becomes much too reductive. The concept of qualification is the offspring of a means-ends line of thought, whereas the general concept of situation goes far beyond this.

For Robinsohn's concept, an additional difficulty arises with regard to the future aspect of the life situations to be identified. If schooling is not only to provide qualifications for coping with the present, but also, at a later date, to enable adults to take action, then the future life situations must be predicted as accurately as possible. Yet, which social and political, economic and ecological prognoses should serve as the basis for identifying future life situations? Furthermore, how distant a future is to be taken into account? Are there more and less desirable futures among those conceivable? Who makes this decision? To what degree does the future – intentionally or not – depend on qualifications imparted today?[11]

Implications of the Utilitarian Claim

Thus, in Robinsohn's conception, qualifications are to be thought of as means to achieve ends. As such, they are related to the general purpose of coping with those situations with which the individual is confronted during the course of his or her life in a particular society – to his or her benefit or detriment. For this reason, qualifications manifest no value in their own right – in contrast to the classical conception of education and personal development. If values can be spoken of at all, they are subsumed under the situations to which the qualifications are oriented. Thus, constructing a curriculum can be understood as an optimizing process not bound to any values or simply committed to pragmatism as its supreme value.

Of course, the means-end scheme also has significant repercussions for the constellation "curriculum elements – qualifications", which has not yet been examined here. For, in Robinsohn's conception, the elements of the curriculum are, in turn, means to producing the desired qualifications. The need to establish correspondences among these three levels promotes the tendency to atomize and to overlook more complex interconnections at all three levels. From this perspective, an ability without any clearly demonstrable relevance to a certain situation must appear questionable and thus the further development of such an ability in school must seem irrational. Conversely, more complex and undecided situations for which the structures and procedures involved are less foreseeable resist the required decomposition into individual qualifications. But this means that from the outset, constructing curricula involves a substantial risk that a considerable portion of the conceivable situational complexes, of human abilities, and of potential content of instruction is simply disregarded. I would like to illustrate each of these three levels with an example that resists being operationalized in the sense discussed above and thus does not conform to Robinsohn's means-ends framework. Seeking to establish a romantic relationship with someone is certainly a life situation that cannot easily be reduced to simple elements; musical creativity (e.g., in composition) is certainly an ability that can be promoted if sufficient talent exists, but it can hardly be taught with a view to a specific goal as an array of exactly defined individual qualifications; finally, a novel of some literary worth might be considered as a many-faceted content for instruction open to a wide variety of meaningful contexts, yet, a precise list of which qualifications for the mastery of which life situations can be produced by an intensive study of such a novel is hardly possible.

At all three levels, those elements that can be so operationalized that they are grasped by the methodical rationality of Robinsohn's procedure and can be identified empirically and analytically delineate a concept of preparation for later life that, in the final analysis, is oriented to pragmatic usefulness. To express this in a more constructive manner: with Robinsohn's conception, certain elements of a curriculum imparting a general education can indeed be designed reasonably well; insofar as preparation for later life in a utilitarian sense is aimed at, his procedure seems to warrant more success than the still widespread perpetuation of traditional syllabi modified by ad hoc considerations. His concept seems inadequate for determining elements of general education which transcend a perspective focused on usefulness.

Before I attempt to gain some benefit from the rational core of Robinsohn's objective of preparing for later life for my own concept of general education, I would like to digress briefly and examine vocational education. With a critical evaluation of the concept of *key qualifications*, we can argue that the model of pragmatic curricula construction proves inadequate. We can also find certain indications of which content is to be focused on in schools if they are concerned with preparation for later life in a modern world.

2.1.2 *Preparation for Later Life and Key Qualifications*

Originally, the discussion centering on key qualifications was primarily confined to professional training and, in an even narrower sense, to training within a company. In the meantime, however, the term "key qualifications" has become a catchword of educational policy. As often occurs in such cases, the political monopolization of the concept was accompanied by both an extension and a dilution of the concept. The debate on key qualifications developed almost concurrently with a new debate on general education, with a major focus at the end of the 1980s and the beginning of the 1990s. Remarkably, those people who engaged in the debate on general education or on key qualifications, respectively, hardly ever referred to the other strand of debate at this time. Nonetheless, there are certainly points of intersection between the two. – What effects did the increasing interest in key qualifications have?

For some time now, profound changes in the qualifications required for a significant number of professions have begun to emerge. These changes are a manifestation of increasingly rapid social and technological transformations. Especially within the last two decades, since the intrusion of the computer and communication technologies into practically all areas of social life, this development has become even more conspicuously dynamic and explosive. The profiles of many professions, some of which had a centuries-old tradition (e.g., typesetter, printer, draftsmanship), have had to undergo radical transformation within only a few years. Among the characteristics of this transformation applicable to a large variety of professions, the following are particularly conspicuous:

- The knowledge and abilities acquired during the original training for a specific profession becomes outdated at an increasingly rapid rate. More and more people change professions during the course of their working life; at the same time, the profiles of professional performance often undergo profound alterations within a few years. 'Lifelong learning' seems to be inevitable. Often, the ability to learn new and different skills and to cope with new requirements flexibly is decisive for success in any particular field.
- Extremely specialized, exactly functional process tasks (for instance, on the assembly line) are being replaced by less restricted, more flexible teamwork which, in addition to technical knowledge and abilities, requires social competencies.
- In industrial production, narrowly defined operational activities are increasingly becoming relegated to 'intelligent' technological systems and, for this reason, tend to become superfluous. As a result, there is an increasing need for control and service activities which require a greater amount of independence, self-responsibility, and creativity related to problem solving than traditional jobs did. There are comparable developments within the entire service sector.

Elements formerly more typical of executive positions have now begun to spread into a much broader spectrum of work activity.

Advocates of the concept of key qualifications often refer to developments like those mentioned above. The starting point for their argumentation is the assumption that in light of the changes outlined above, a certain qualification gap arises between traditional vocational training and the actual requirements – a gap that should be filled by teaching "key qualifications". Such teaching, according to this view, can enhance, above all, the flexibility and mobility of employees.

Mertens (1974) introduced the term "key qualification" as the main concept of his scheme of "training for a modern society". He uses the term to designate qualifications by which "a broad spectrum of practical tasks can be made accessible by direct and rapid application transfer" (ibid., p. 40). Mertens does not claim to already know the relevant key qualifications in a definitive sense. He simply establishes the need, outlines the direction of further considerations, and provides examples that, in his opinion, deserve careful consideration.

After some delay, Mertens' ideas met with considerable response during the 1980s – both in educational science and among the designers of trainee and internal further training programs.[12] A number of authors published catalogs of possible key qualifications. The following is an example of such a catalog, in this case the one devised by the "Group Project: Key Qualifications in Vocational Training" (1992):

1. "Problem-solving abilities and creativity
2. Cognitive and learning abilities
3. Reasoning and assessment abilities
4. Cooperative and communicative abilities
5. Ability to assume responsibility
6. Independence and achievement potential" (ibid., p. 22).

This catalog goes beyond the limits of Mertens' original concept, which was primarily aimed at training intellectual abilities, insofar as it also mentions value-system related attitudes and opinions pertaining to one's personal working and social behavior ('social competence').

Critiques and Some Conclusions

The approach focused on key qualifications has been subject to intense criticism from various quarters – with sound arguments from an educational science perspective by, for example, Zabeck (1989) and Tillmann (1994). Zabeck finds fault with the lacking psychological underpinnings of the concept; he considers the desired transfer of qualifications described in very general terms to be illusory; in addition, he discovers a questionable tendency toward separation from practical occupational work. The leitmotif of his critique is his suspicion that "the key qualifications just might be a 'chimera'" (Zabeck 1989, p. 77). From the ranks of the trade unions, it had often been suspected that the concept of key qualifications was simply a further attempt, this time a very subtle one, to adapt the education and training of workers to the exploitative interests of businesses. Tillmann (1994) demonstrates that this allegation is by no means simply pure invention; and he lists a number of sound reasons why it would at least be naïve to assume a rather remarkable convergence between modern economic requirements and pedagogical aims oriented to emancipation.

Regardless of how far the concept of key qualifications can be realized within vocational training and education programs and of which political interests this concept serves, the concept places effective limitations on the idea of immediate qualifying procedures for exactly definable on-the-job applications. Keeping in mind the statements about the concept of qualification in the previous section, we can maintain that most of the key qualifications called for are actually not qualifications in a strict sense. They are not means to attain exactly defined goals. To express this more pointedly: The concept of key qualifications in its entirety proves to be a concept of general education oriented to occupational purposes. Thus, the pragmatic qualification model transcends itself.

Even if one concurs with Zabeck that the concept of key qualifications is more apt to direct attention to a certain problem than to offer a solution, the following conclusion can be drawn from the debate on this issue: For vocational education and further training programs, it is inadequate to simply make appropriate qualifications available for clearly definable purposes (namely, coping with foreseeable situations at the future workplace). And this is by no means a question of pedagogic goals, but, rather, a consequence of radical social and economic changes. Evidently, the structure of the modern work world with its enormous boosts of change and the resulting uncertainty of prognoses, with its demands for flexibility and mobility, necessitates at least considerable modification of the simple model of direct qualification for definable applications, if not indeed the abolishment of such a model. The pragmatic aspects of technical, economic, and scientific actions cannot be directly applied to the training necessary for precisely such actions. In my opinion, taking recourse to such characteristics as responsibility, independence, creativity, etc. – i.e., to properties more commonly associated with the classical idea of 'pure' individual education and personal development – demonstrates that the advanced complexity of technological and economic developments imposes certain inherent limits on the system of linking the individual to definite purposes. The simple adaptation of those participating in the work process to external specifications also involves the risk of employees no longer being able to cope with the tasks before them and of the businesses employing them becoming less competitive.

But if even vocational training must refrain from construing a too closely pragmatic relationship between training and professional practice, then how much more will this hold true for general education, which is to allow a considerable number of professional options to remain open for each individual student as long as possible. In other words, the fundamental didactic problem of content and methods of instruction at public schools cannot be solved in a strictly pragmatic way. Preparation for later life that can be understood solely in a pragmatic sense does not go far enough.

The results of this analysis should not be misunderstood in the sense that the connection between life situations and the qualifications applicable to them, as postulated in Robinsohn's concept, would be totally irrelevant. This would be throwing the baby out along with the bath-water. The failure of the claims to generalization implied by Robinsohn's concept simply indicate that schools cannot restrict themselves to imparting qualifications whose usefulness or even necessity is immediately apparent from their applicability in future life situations. Put a different way, Robinsohn's perceptive interpretation of the issue of preparation for later life can stimulate a critical examination of how schooling deprives the younger generation of useful knowledge and abilities relevant to their later lives – and can do this in a much more precise manner than the simple continuation of designing school curricula in a way largely dependent on tradition could.

2.1.3 *Specifying the Objective of Preparing for Later Life*

Examining Robinsohn's curriculum concept and the concept of key qualifications has already led to a number of general considerations and insights which can now be constructively combined: What is the rational core of the objective of preparing for later life? In which form can such preparation become a legitimate goal for schools and thus become a meaningful and convenient criterion for an up-to-date general education?

At the outset, we should differentiate between preparing for later life in a strict sense and in a broader one. A close look demonstrates that the dispute over the priorities of formal schooling is not concerned with whether or not preparation for later life is an actual concern: No one will seriously want to promote an inability to cope with life through education. Rather, the contradictory positions can be ideally reconstructed as diverging conceptions of preparation for later life, the first in a stricter, the second in a broader sense, approximately as follows:

(a) *Preparation for later life in a strict sense*: Formal schooling is to prepare children and adolescents for later life by relating to concretely identifiable, limited situations in which the actions required of the persons involved are based on clearly definable knowledge, skills, and abilities. Instruction, then, aims at imparting and training exactly these qualifications.

(b) *Preparation for later life in a broader sense*: By being confronted with intellectually challenging materials and topics, students are to be given the opportunity to *develop* their *individual* powers and abilities as far as possible. In that case, they will also be sufficiently prepared for the practical requirements of private and professional daily life.

For better clarification: Not in accordance with his intentions, but as a result of the operational effects involved, view (a) corresponds de facto to Robinsohn's concept. The concept of key qualifications can be understood as an attempt to enhance traditional vocational training programs with elements of view (b).

Preparation for later life in the strict sense aims at usefulness, at the immediate practicality of what is to be learned. If schooling confines itself to this sort of preparation for later life, then it can justifiably be accused of utilitarian reductionism.

Preparation for later life in a broader sense can be realized, among other possibilities, by means of "pure" education, not restricted to a specific purpose. But if schooling's remove from specific purposes is elevated to a guiding principle in the sense that anything 'useful' should be excluded at the outset, then this view of preparation for later life can justifiably be accused of being out of touch with the real world.

The following consequences result from this differentiation for the concept of general education under discussion here: Preparation for later life in the broader sense can be furthered by schools through their efforts to realize the other objectives to be discussed below: promoting cultural competence, developing an understanding of the world, promoting critical thinking, developing a sense of responsibility, training communication and cooperation, enhancing students' self-esteem. Thus, if preparation for later life is listed here as one distinguishable (but not necessarily uniquely definable) objective of public schooling, then the emphasis is on *preparation for later life in a strict sense*: preparation for everything that youths in all probability will now or at some later date essentially require for daily life in the society in which they reach maturity – and which they, for the most part, would not learn without formal schooling. This last consideration is particularly significant.

The affirmative character of such preparation for later life in a strict sense, which involves adapting to the 'given', cannot be denied. For the emancipatory claims of education, a problem would only arise if orientation to utility in an exclusively pragmatic sense would function as the guiding principle. Thus, preparation for later life as a principle of general education requires further elaboration: If the idea of general education were simply equated with preparation for later life in a strict sense, it would remain disturbingly unspecified.

Evidently, though, there are a considerable number of common types of situations with which almost every member of society is often confronted – both within and without his or her own particular professional work. Such situations, which can be generalized to varying extents, include the following: situations in which one must engage in reading, writing, or elementary mathematics, in which it is necessary to independently acquire certain information, or in which certain group working procedures must be agreed upon, etc. For many of these types of situations there are additional, easily identifiable qualifications which can relatively reliably be taught in institutionalized instruction – for the first three situations mentioned above, for instance, the "cultural techniques" of reading, writing, and doing arithmetic.

Thus, designating *preparation for later life in a strict sense* as a component of a modern general education is to be understood as an appeal to taking such connections between daily life situations and corresponding qualifications, wherever these can be demonstrated or are even obvious, much more consistently into consideration for designing curricula than in the past. Robinsohn's concept can be used to describe this essential preparation for later life in a strict sense more exactly – if only one is willing to relinquish the claim that the entirety of formal schooling can be reduced to imparting qualifications. In other words: The means-ends scheme, the utilitarian principle, behind Robinsohn's concept is not elevated to the guiding principle of curricular design, but, instead, is acknowledged as a legitimate instrument for constructing a particular segment of the curriculum – that segment in which the factors ensuring its proper functioning maintain. An extensive neglect or complete abandonment of this principle would be pedagogically irresponsible: If we know which situations (nearly) all students will encounter and, furthermore, what they will have to know and be able to do in order to cope with such situations, and, finally, if there are methods for imparting this knowledge and ability by means of classroom instruction, then it would be an act of cynicism not to do so.

On the basis of these considerations (if not strictly deducible from them), we can now establish a catalog of relevant criteria.

Schools aiming at a general education should impart qualifications which help to deal with existing and presumably common life situations in our society as long as the following stipulations apply:

(1) The qualifications should be necessary in the sense that a lack of such qualifications would noticeably restrict one's ability to lead a 'normal' life.

(2) They should be of such a nature that they could not normally be acquired by youths casually and without systematic instruction – for instance, from family members or from peers.

(3) Nor should they be such that they could readily be acquired through temporary special instruction courses.

(4) The structure of such qualifications should allow them to be adequately imparted by means of systematic, public instruction.

Basic cultural techniques such as reading, writing, and mathematics, but also abilities such as being able to articulate oneself with a certain sophistication and a substantial number of learning and working techniques evidently adhere to all four of the criteria mentioned above. I would now like to demonstrate the practicality of these four criteria by examining them in detail in the light of some examples and counterexamples.

Ad (1). Hardly any single type of situation exists which, if considered in isolation, cannot be avoided. Illiterate people, who are, indeed, still present in our society, demonstrate this point. Although situations in which one would have to be able to read and/or write provide perfect examples of real and omnipresent standard situations, the illiterate people in our society can and do get along. Nevertheless, they do pay a high price for this: They are constantly forced to put themselves into humiliating circumstances of dependence, to engage in a masquerade, to have feelings of belonging to an underprivileged minority group (cf. Kazis 1991). Thus, it is certainly reasonable to consider the vital importance of qualifications not in the strictest sense possible, but with respect to leading a socially accepted 'average' life. Then, such qualifications as being able to drive a car, to conduct a telephone conversation, or to dance, but certainly not the ability to write programs for a computer, would be covered by criterion (1). With a look to the future, one might maintain that the qualification of being able to use a word processor would also have to be listed here.

Ad (2). Children and adolescents acquire many of the qualifications that are practical in and vital to everyday life situations and adhere to criterion (1) just 'naturally', casually, and on the side, or by conscious, but not systematic, efforts on the part of their parents, friends, siblings, etc.: daily use of one's native language, fairly civilized rules of conduct, riding a bicycle, adhering to certain rules of play, carrying on a telephone conversation, using the television guide, and many other things. In principle, schooling does not have to be concerned with 'teaching' all of this. However, specific differences pertaining to different social strata should be kept in mind: Some of the things which are 'natural' for children from the upper and middle classes would remain deficits for children from the lower classes if schools would not secure their instruction (and the converse is also true).

Ad (3). Criterion (3) allows us to eliminate skills such as driving a car or dancing, that is, qualifications which represent a limited skill. Of course, this does not mean that there should be no dancing at school or that corresponding courses must not be offered by schools. Nonetheless, instruction promoting such qualifications cannot be included in an obligatory set of standards for general education.

Ad (4). There are common situations of a complex nature for which qualifications that adhere to all of the criteria mentioned above can prove very useful, whereas supplying these qualifications is hardly conceivable under the prevailing circumstances of instruction in public schools. As an example, we can consider affectionate behavior toward one's partner in a sexual relationship. School instruction can provide opportunities to 'talk about' this matter, to turn one's attention to important aspects, but not to actually teach the corresponding 'qualifications'. Love is certainly one of the major subjects in life, but not a school subject. The modern trend toward making all topics and areas of everyday life into matters for educational theory encourages an overestimation of the actual possibilities available to classroom instruction. Criterion (4) heightens our awareness of this danger.

2.1.4 Significant Trends: What is Vitally Important in Today's World?

To conclude this section, let us consider the following question: Are there obvious tendencies or trends according to which the preparation for later life necessary today must differ from that which was considered appropriate one or two generations ago? The four criteria of the preceding section are, in themselves, unspecific as far as content is concerned; they simply presume the existence of an industrial and service sector-oriented society based on the division of labor with compulsory school attendance. In this respect, it would have been unreasonable to apply these or similar criteria thirty, sixty, or a hundred years ago.

Criterion (1) implicitly requires that social transformations be taken into account for the design of school curricula. In our discussion of key qualifications, some tendencies have already been mentioned which reflect a long-term change in occupational circumstances. The qualifications I list below are related to such tendencies and to parallel trends in private life. This list is not the result of an exact analysis and certainly cannot replace such an analysis; it is simply a tentative estimate involving certain elements of speculation. My intention is to specify actual qualifications in a strict sense, i.e., abilities, skills, and knowledge that can be employed as means of dealing with life situations. I will refrain from mentioning vague personality traits such as 'creativity' or 'problem-solving abilities', for everyone is glad to notice these characteristics in students, but, up to now, no one has really discovered how to produce them as a general disposition through classroom instruction. In my opinion, the following qualifications largely satisfy the four criteria from the last section.

As formal qualifications, the following have become increasingly important:

- techniques for learning, working, and organizing;
- the ability to acquire information on one's own, to actively use media and systems for storing information;
- the ability to adequately use technological aids (e.g., handheld calculators);
- the ability to organize and participate in teamwork;
- the ability to interpret those symbolic and graphic representations that are becoming increasingly important in private and professional everyday life;
- the ability to present information graphically and symbolically;
- the ability to select individually required information from an abundant supply of information;
- the ability to express oneself verbally; the ability to argue for one's own point of view.

As material qualifications, the following have become increasingly important:

- adequate fluency in (not absolute mastery of) at least one foreign language;
- basic knowledge of relevant economic and ecological contexts;
- basic knowledge of relevant political and legal topics;
- the ability to adequately estimate physical quantities, dimensions, and proportions.

Although they have often been proclaimed obsolete, the traditional cultural techniques of reading, writing, and mathematics are still of considerable import, as are versatile skills in the active and passive, oral and literary use of one's native language. These are fundamental qualifications which provide the basis for all further qualifications and without which participation in everyday social life would be impossible. Also still essential are the knowledge and mastery of forms of socially accepted conduct, even though these have been temporarily underestimated and have been considerably modified by the radical social transformations of the last thirty years.

By contrast, the general significance of rote memorization of information and of all routine skills which can be more effectively executed with the use of technological aids (pocket calculator, computer), particularly, that of making extensive calculations, has diminished.

The greater part of what is taught at non-vocational schools today cannot be subsumed under the list of criteria provided above. Generally speaking, the task of preparing for later life, as elaborated by means of the four criteria mentioned, only outlines a sort of 'minimum curriculum'. Schools should not refrain from teaching qualifications that satisfy all four criteria without having compelling reasons for doing so. Thus, the demand for preparation for later life in the strict sense is much less suitable for determining what is to be taught at schools than for indicating what, in any case, must not be lacking. For more detailed decisions, further considerations relevant to other objectives of general education are necessary. These are to be discussed below.

2.2 Promoting Cultural Competence

Of the objectives for schools to be discussed here, 'promoting cultural competence' is certainly the one which is vaguest at first glance – simply due to the concepts involved. Thus, certain fundamental reflections would seem to be in order.

2.2.1 Culture and Cultural Identity, Cultural Continuity, and Cultural Competence

The Concept of Culture

What in ordinary language is designated as 'culture' is widely diverse and to a large extent dependent upon the context (cf. Löwisch 1989, p. 18ff). The literature on the concept of culture and its distinct variants is just as extensive as the literature on the concept of education and personal development (*Bildung*). I would like to briefly explain what is meant whenever 'culture', 'cultural', and compounds based on these words are mentioned in the following pages.

Generally, the starting point for my considerations is a broad, sociological concept of culture. A much-quoted, almost classical description of 'culture' was formulated by the anthropologist Edward Tylor (1871, p.1) as follows: "That complex whole which includes knowledge, belief, art, morals, law, custom, and any other capabilities and habits acquired by man as a member of society." Such a concept of culture is first and foremost of a descriptive, not of a normative nature: Every society in its entirety has its own specific culture. Accordingly, here – in contrast to the common usage of ordinary language – no particular social products are labeled 'culture' and played off against others which are branded as 'less cultivated' or less valuable. Thus, by no means are cultural products only identical to the 'cultural events' of literary journalism or to the 'cultural possessions' of middle-class intellectuals. They range from forms of everyday social conduct and skills to the measures of moral behavior, and to artistic and scientific artifacts of the highest intellectual caliber; and they even include culturally specific misconceptions.[13] Current social problems and issues (e.g., ecological issues), then, are also cultural products in this sense.

Thus, culture designates the entirety of socially produced, not innate, but (in principle) common features of a particular group along with their materially and symbolically communicable artifacts. This does not by any means already firmly establish what is to be understood, for example, as 'our culture' in the sense of the culture of contemporary German society. Culture is constantly undergoing transformation. Since we all participate in it and it is always already a prerequisite of all of our activities, it must constantly be interpreted and (also empirically) redefined within the hermeneutical circle of social self-reflection.

Moreover, attention can readily be directed toward more extensive groups, in which case it may, for instance, make sense to speak of *Western culture*, or toward subgroups, where the topic is one of *subcultures*, whether they be those of West or East Germans, of youth, of drug addicts, of intellectuals, or of academic mathematicians. Subcultures can become established across national, linguistic, and other cultural boundaries; in this sense the prefix 'sub', which implies a hierarchical structure, is not really appropriate. One could also further specify the concept of culture with respect to types of situations. In this sense, I will be speaking of *everyday mathematical culture* in our society and of the *culture of mathematical instruction*.

Any particular culture always represents a selection from the entire spectrum of human possibilities – a selection that is realized collectively and vitalized by social action. This selection provides a society – or a certain cultural sphere or a social subgroup – with a specific identity on the basis of which it distinguishes itself from other groups and is perceived by others as distinct.

From the point of view of those who feel that they belong to a certain culture – from the 'inner view' of a culture – a normative conception of culture gains significance alongside a descriptive one. The values represented in a culture suggest, indeed, they seem to demand, that one also perceive value in culture itself. The explicit nurturing ('*cultura*') and further development of the values represented in a culture and characteristic of it can be designated as *cultivation*. Naturally, in such a complex culture as our own it is difficult to come to a consensus on the assessment of cultural development. Is, for instance, the increasing plurality in our society an indication of cultural decline, as suspected by conservatives, or is this itself an important cultural achievement in agreement with that displacement of particular values by more universal ones so typical of our cultural sphere?

Cultural Identity

The cultural identity of the individual corresponds to that of society (or of some other social group). The cultural identity of the individual represents, as it were, the subjective side of culture. Anticipating some of the considerations to be discussed further on, we can presume the following at this point: A *reflective* cultural identity that does not simply demonstrate an unquestioned adoption of handed-down patterns requires education and personal development. Adorno's (1978, p. 90) characterization of education (*Bildung*) as "culture with reference to its subjective appropriation" points in this direction. – In what follows I will be using the term *cultural identity*, if nothing else is explicitly stated, in the sense of the cultural identity of the individual.

Since the individual belongs to various social groups, his or her cultural identity is by no means unequivocally predetermined. The more complex the society, the more diverse is the construct of cultural identity. In a multicultural society such as ours, divided into many disparate subgroups, the identities of subcultures can easily smother an all-embracing cultural identity, which, in turn, would also have to extend to common basic values such as tolerance toward those who have different appearances, beliefs, or value systems. Again, anticipating further systematic arguments: Within the context of schooling, the development of a reflective cultural identity that acknowledges the possibility of being different would have to be a priority; the focus would have to be on cultural achievements which are sufficiently universal in character. Then, the universality of these achievements and of the values and insights associated with them permits the acknowledgement of difference, rapprochement with the members of other societies. In other words, a reflective cultural identity must prove itself by demonstrating its coexistence with an understanding of other cultures and other cultural identities without denying the peculiarities of one's own culture. National identity is only a special form of cultural identity. If the individual is perceived as a member of various groups with their specific (sub)cultures, then the most comprehensive group is humanity itself. But the sole reference to this most comprehensive group will hardly suffice to impart some sort of cultural identity to the individual. The challenge to developing a reflective cultural identity consists in establishing a balance between, on the one hand, being rooted in the particularities of one's own culture (or subculture) and, on the other, an openness toward foreign cultures (or subcultures), perceiving them as equally legitimate possibilities of human existence.

Cultural Continuity and Cultural Competence

Up to this point, we have neglected the historical dimension of culture. Culture develops over time and cannot be imagined without social transmission from generation to generation. Every society requires a minimum of cultural continuity, the transmission of cultural achievements from one generation to the next, and the appropriation and further development of the existing culture by the younger generation. In our society, schools are indispensable for maintaining cultural continuity through the social transmission of cultural achievements.

There is a significant reason for designating the promotion of *cultural competence* – and not (simply) of cultural continuity – as the objective of schooling in our present context. In agreement with common usage, I restrict the concept of cultural continuity to the temporal, or more precisely, the *diachronic* aspect of cultural development. With the introduction of the concept of cultural competence, I wish to acknowledge the fact that the processes of developing cultural identity in a modern society cannot be adequately analyzed if they are confined to this diachronic aspect. There are a number of arguments in favor of expanding reflection on cultural development by including the *synchronic* aspect. In this sense, the focus is not solely on maintaining and transmitting a cultural heritage as an interconnection between 'former times' and 'today', but also on linking the various traditions and subcultures which appear in any society simultaneously. Cultural competence, then, refers to both the diachronic *and* the synchronic aspect. Demanding of schools that they should promote cultural competence, then, includes, on the one hand, securing cultural continuity. In addition, it includes the task of mediating between simultaneously existing subcultures and of resisting the decomposition of society into disparate cultures. Thus, cultural competence is the more comprehensive concept.

At the moment, there are, above all, two tendencies which contribute to an increase of cultural diversity in modern German society (there are similar developments in comparable industrial societies): on the one hand, the constant increase in social differentiation, which is a result of the increase in specialization and in the division of labor and promotes the unceasing development of new subcultures; on the other hand, the coexistence of various cultures as the result of more intense economic interconnections and of politically or economically motivated migration. Since a democratic society can only function if it maintains or constantly acquires anew a minimum of discursive capability; mutual understanding between the various subcultures and subtraditions is becoming increasingly significant. In principle, where matters of general welfare, future developments, and political decisions of long-lasting consequence are concerned, matter-of-fact communication between the engineer and the journalist, the musician and the farmer, the construction worker and the professional politician must be possible. To the same extent, the representatives of such occupational groups must also be able to communicate as colleagues and fellow citizens with Muslim foreign workers from Turkey, with the Mennonite emigrant from Kazakhstan, with students from Black Africa, or with refugees from Sri Lanka. This problem is acknowledged by the demand that schools should promote cultural competence.

Thus, taking both the diachronic and the synchronic aspects into consideration accounts for the fact that cultural development is characterized both by the transmission of cultural achievements from one generation to the next and by the confrontation between subcultures, by the suppression, fusion, integration, and assimilation of foreign cultural elements. In both respects, the schools of a society are important supports for cultural development.

2.2.2 *Cultural Continuity as the Result of Passing On Tradition in School and Society*

In which way can schools bring about the cultural continuity called for above? How can one avoid continuing to teach obsolete content under the guise of passing on tradition? How can the risk be minimized that the ‘promotion of cultural continuity’ simply leads to the development of an elitist cultural consciousness among the members of the upper middle classes? The answers to these and similar questions are more controversial in our society than is the demand that schooling prepare for later life, the sensibleness and necessity of which is affirmed by practically the entire population. These might be two of the reasons for the dominance of controversial assessments:

First, neither cultural continuity nor its particular forms on the part of the human subject – thus, above all, the *cultural identity* of the individual – correspond to any immediate and conscious personal need of children and adolescents. For the average citizen it is by no means obvious that schooling should be concerned with cultural continuity.

Secondly, this topic readily yields to a political polarization which impedes the establishment of a social consensus on the objectives attributed to schooling. A superficial view permits the demand for the promotion of cultural continuity to be interpreted as the materialization of a (neo-)conservative cultural ideology behind which lurk motives that are perhaps restorative, nationalistic, ethnocentric, or, in the most harmless case, Euro-centric.

This means that it is necessary to much more conscientiously explain and substantiate the claim that promoting cultural continuity is an indispensable objective – especially for schools facing the challenges of the present and of the future in the interests of the younger generation. For this reason, the entire complex of passing on tradition in school and society must be examined in more detail.

Cultural continuity is produced by means of passing on tradition socially. Within this context, handing down tradition through schooling is only one, albeit certainly an essential means of this social transmission of culture. For one of the most general characteristics of schooling is that in the course of it, cultural products, achievements, and distinctive features are imparted to subsequent generations.

Upon inspecting past critiques of formal schooling, one encounters time and again – in a number of variations – the allegation that schools are oriented to the past, that the content of the instruction they provide is to a large extent obsolete and, for this reason, otherworldly, too far removed from everyday life, and useless.[14] At times it seems that this critique refers to the school's function of passing on tradition as such. But a closer look reveals that it is usually directed against traditionalist interpretations of this function, against a one-sided and backward-oriented selection of those cultural achievements which are fixed in the syllabus and decisive for everyday reality at school. In any case, the real problem is to determine the relationship between more recent and more remote cultural manifestations within the educational canon. The issue is not *whether* schooling has the function of passing on cultural traditions – for there is no way to avoid this –, but, rather, *how* the school can fulfill this function within the context of its objective of imparting an up-to-date general education.

At the base of the idea of cultural continuity is a view of the social transmission of tradition as a *necessary counterweight* to an unrestricted aspiration to progress. In memorable phrases very much to the point, Leszek Kolakowski (1970, p. 1) has expressed the dialectics of tradition and progress:

> "First, if the latter generations had not constantly rebelled against their inherited tradition, we would still be living in caves; second, if the revolt against the inherited tradition became universal, we would be living in caves once again. The cult of tradition and the resistance to tradition are equally indispensable to social life. A society in which the cult of tradition becomes omnipotent is doomed to stagnation; a society in which resistance to tradition becomes universal is doomed to destruction."

The demand for cultural continuity takes relaying cultural tradition as a basic structural principle of any society seriously, without making it absolute and separating it totally from its dialectic opposite pole, the abandonment of tradition in favor of social progress. The idea of cultural continuity as a social goal to be aspired to encompasses the hope that with good reason the future can be formed "in harmony" with preceding developments, but also the apprehension that without this link to the past any sort of progress would be without direction.[15]

Imparting Tradition Unintentionally and Intentionally

A further explication of the concept of tradition is well-advised to determine the sphere of the concept's validity more exactly. Here, we should distinguish between imparting tradition unintentionally and intentionally. Formal schooling always contributes to the unintentional transmission of tradition, regardless of the specific curriculum and of the conscious goals and intentions of the teachers and institutions backing the school. Simply due to the fact that in school, as in other social spheres, people come together and are associated with one another, are compelled to discuss things and communicate, a process of becoming accustomed to the norms of living together, working, and adapting to existing organizations takes place. At the same time, one gains experience in resisting external demands, in resisting the demands of the institution, and in attaining social recognition – in harmony with the goals of the institution or in resistance to them. To put it briefly: Just as with any other form of social praxis, the school also always contributes to the *socialization* and *enculturation* of the students. Its 'hidden curriculum' functions preceding to and alongside any pedagogical intentions because school is part of society. And this is also true even if the norms and behavioral patterns transmitted by schools and the values implicitly in effect are of a specific institutional nature and thus differ from those in other social spheres (the school as a 'protective space', as a 'pedagogical province').

Imparting tradition intentionally refers to subject matter explicitly listed in syllabi, to what traditional educational theory has called the 'cultural heritage'. If the task of promoting cultural continuity is to be allocated to schools, then the focus must be on examining how this can be done by intentionally passing on tradition. Beforehand, though, a further differentiation must be discussed which enables us to describe more accurately the intentional transmission of tradition within the school context.

Three Modes of Imparting Cultural Tradition

On a social level, there are three major modes of imparting tradition that are noticeable in numerous social spheres and characterize various stages of development of individual strands of tradition (in craft professions, the arts, science, etc.):

- unchanged and unreflective adoption of what is passed on (e.g., language, craftsmanship techniques, religious customs; in general, typical for nearly all social spheres in postfigurative cultures[16]);
- productive further development of what is passed on (characteristic of Western art until around 1900, of modern science and technology; in general, typical of dynamic strands of development within cofigurative cultures);
- reflection on what is passed on, conscious critique of tradition from a detached standpoint, which can lead to both a productive further development of the tradition on a less naïve level or to a total break away from tradition (characteristic of numerous developments within Western art of the 20th century).

Especially the first and third modes of passing on tradition are also of considerable significance to the intentional transmission of tradition at school; the second mode is more likely to be found in social life outside of school.

For many contemporary zealous critics, unreflective adoption is objectionable. Nevertheless, even in today's schools, it is still indispensable in many contexts. In hindsight, wanting to get students to critically reflect on *all* of the content of instruction was one of the misconceptions of an emancipatory pedagogy which arose due to the prevailing circumstances. It is comprehensible as a reaction to a teaching practice that was characterized primarily by self-assured traditionalist instruction. Of course, the maxim of passing on certain content unreflectively should not apply to teachers and designers of curricula, but to dealing with these subjects during instruction. To which content can we justifiably apply this maxim?

In general, I think it appropriate to apply it to a wide range of elementary and universal subjects which simultaneously serve to prepare the individual for later life and to promote cultural continuity, for example: fluency in the use of one's native language in oral and written communication, dealing with elementary geometric forms and numbers, as well as proficiency in elementary operations with numbers and their application in everyday situations. Such subject matter can be considered to be relatively straightforward: cultural continuity is assured by introducing students to culturally significant practices, by their learning – not least through imitation and practicing – how to do something, how it 'works'.

The situation is quite different regarding much traditional curricular subject matter whose relevance for the students' everyday lives is not at all obvious. In this case, students are justified in asking why they should be compelled to devote their time to something if its benefits are not immediately apparent. To list a few examples from a wide range of diverse topics of instruction (primarily, but not solely, for secondary schools): the symphonic music of 'Viennese classicism', Pythagoras's theorem, differential calculus, Caesar's *De bello Gallico* (as *pars pro toto* for Latin instruction as a whole), Luther's reformation, the French Revolution, styles of sacred architecture during the Middle Ages, Kepler's laws of planetary motion, poetry of the 17th century, interpretations of Romantic poems. If students are to experience such subject matter as personally meaningful, then there is no way to avoid critical reflection on it, to avoid evaluation and appropriation of such knowledge from a standpoint of critical detachment. Passing on tradition as an end in itself produces lifeless knowledge that more than likely is a hindrance to responsibly taking charge of one's own life and becomes a dead weight. A reflective cultural identity is not promoted by passing on tradition blindly. On the contrary, this way of imparting tradition involves the risk that the students will, above all, begin to feel bored and resist any part of tradition.

'Traditional', 'New', and 'Current' Topics of Instruction

A further (three-part) differentiation is designed to provide more clarity, for up to this point we have still been using the terms 'traditional' and 'passing on tradition' in a naïve sense. With respect to our fundamental, broad concept of culture, this might lead to certain misunderstandings. We have seen that in a strict sense any topic of school instruction must be regarded as a cultural product and that schools pass on even the most *up-to-date* subject matter in this sense.

But a concept of passing on tradition in such a broad sense is not very helpful. From a sociological point of view, we were able at first to neglect the historical dimension of cultural manifestations. Nevertheless, in the present context, it is of major importance and must be taken into consideration.

In everyday speech, it is quite correct to only then talk of 'passing on tradition' if the matter involved is characterized by having been in existence for a certain length of time. With reference to schooling, an appropriate rule of thumb would be to equate the duration required for new subject matter to become 'traditional' to a generation, i.e., to a period of about thirty years.[17] This would mean that everything that children and adolescents learn at school today can be designated as 'traditional' subject matter if it was already in the syllabus when their parents went to school.[18]

For the present context, any other subject matter should be considered to be 'new' or 'current'. 'New' subject matter is subject matter that has actually only become part of the canon for general education in the course of the last generation, although as a cultural manifestation it is already of a certain age. Set theory, which has been instilled into the curricula within the framework of 'New Mathematics' is one example – it originated as a mathematical subdiscipline during the 19th century. – By contrast, 'current' subject matter has to do with cultural manifestations which have only come into existence for schooling within the last thirty years, for example: working with computer software, ecological issues, contemporary history. Naturally, it is possible that certain 'traditional' topics have continuous relevance, while some 'new' or 'current' subject matter can perhaps only become part of the syllabus due to short-lived trends. For this reason, the important task of sorting out less appropriate subject matter, occasionally a matter of course for the design of modern curricula, must not be restricted to 'traditional' content.

Even just fleetingly calling to mind the syllabi in current use demonstrates the fact that for most subjects *traditional topics* are overwhelmingly predominant – at least up to the final grade levels of middle school. Compared to the radical nature of a great number of plans to revise curricula, it is astonishing to note how much subject matter survived the turbulent reforms of the 60s and 70s practically unharmed. Some of the old material simply appears in a new look. The presentation, in school textbooks for instance, has changed; in general, the terminology seems to be more 'scientific'.[19]

To conclude this section, I would like to illustrate the differentiations mentioned above with a few examples. Viennese symphonic classical music and Pythagoras's theorem were already listed above among the traditional topics of instruction. Traditional content of long-lasting relevance and topicality, even though they have already survived for many generations, would be calculations using the rule of three in mathematics and writing reports in language arts instruction. One subject that is now in the process of transition from a new one to a traditional one would be the topic of National Socialism in instruction on German history. With respect to mathematics teaching, all of the topics involving stochastic operations can be listed as new content. The slide-rule and the logarithmic table can be mentioned as traditional aids of making calculations in the school context which have been replaced by more current aids (pocket calculators, and, increasingly, computers).

2.2.3 Concretizations: An Overview of Schooling's Promotion of Cultural Competence

The discussion so far should have clearly demonstrated that cultural competence in modern society cannot be ensured by relying primarily upon traditional topics of instruction (in the sense explicated above) in schools. Even if we restrict ourselves to the diachronic aspect, that is, only aim at cultural continuity, more (sometimes, however, less) must be demanded than simply imparting a traditional canon. However, this would not yet have any effect on the synchronic cohesion of culture.

In the following two sections, I would like to concretize the discussion by considering in more detail two significant partial objectives included in the school's overarching responsibility to promote cultural competence:

- Schools should also ensure that *communication between the generations* remains possible – also in periods of accelerated development, of knowledge rapidly becoming obsolete, and of changing value systems. Here, the focus is once again on the diachronic aspect, the temporal continuity, the connection between old and new (2.2.4).
- Schools should enable adolescents to attain a *reflective cultural identity*[20]:
    - to experience themselves as part of the culture in which they grow to maturity – with its favorable and unfavorable aspects;
    - to recognize common elements within their own culture, apart from its breakdown into disparate subcultures;
    - to accept the otherness of foreign cultures as an equally justified form of human existence.

To accomplish this, in addition to the diachronic aspect of cultural competence, it is absolutely essential to take the synchronic aspect into account (2.2.5).

A third important partial objective with which the main objective of promoting cultural competence should be concretized can only be mentioned here: the *advancement of everyday culture*. In other words, this partial objective involves enabling adolescents to cope with those private and public everyday situations with which every member of society is confronted time and again – in a way that corresponds to general social practices in our society. Here, schools are called upon to complement pre-school and out-of-school (e.g., inner-family) socialization and enculturation.

Within the present context, I will not be devoting an entire section to this extremely important partial objective of schools because, to a great extent, it obviously can be subsumed under the objective of preparing for later life, which has already been discussed in detail.

2.2.4 Communication Between the Generations

The success of such communication presupposes the existence of common features that are accessible both to the younger and to the older generation more or less as a matter of course. A large portion of these common elements is produced by imparting tradition unreflectively: Norms of social action, value systems, worldviews, and elementary knowledge of the world around us are acquired in the course of common social interaction among adults, children, and adolescents, both in school and outside of school.

With reference to schooling's imparting tradition *intentionally*, here again the elementary cultural techniques are to be mentioned first. It is certainly no coincidence that the public – and, particularly, parents of schoolchildren – react to 'revolutionary' changes in the curriculum of elementary schools with fierce hostility. This was noticeable on a worldwide scale in connection with 'new math', but also, for instance, with reference to reading instruction based on the 'look-and-say' method (insofar as this was implemented in a radical way). In my opinion, this conservatism on the part of the public and parents is not to be denounced as blind traditionalism, but reflects a justified interest of the older generation in the common features they share with the younger one: They want to be able to understand what their own children are learning during their first five or six years at school; on the basis of such points shared in common, they want to be able to provide assistance to them, to answer their questions. In addition to those cultural techniques already mentioned above, this also refers to basic knowledge of a variety of topics, from our natural surroundings to songs, fairy tales, and stories that one would like to be familiar with as with part of a common heritage. Thus, a considerable portion of 'traditional' topics of instruction is unavoidable if the public school does not wish to turn a deaf ear to these demands for points in common.

If the arguments were to end here, the last section might be read as an appeal to simply let everything remain as it is – at least for the first grade levels. But this would be disastrous. For the demand that public schools promote communication between the generations on the basis of knowledge shared in common would only be met inadequately and in a one-sided way if the consequence were to be the perpetuation of the knowledge acquired at school by the respective former generation. Of course, the adults have also had more experience since their time at school, have expanded their knowledge and capacity for conscious reflection since that time, and are now living together with children and adolescents in a transformed life-world. Thus, to the extent that cultural continuity is defined in terms of communication between old and young, prudent adaptation of the curricula to transformed social conditions is indispensable. Passing on tradition in a traditionalist sense, i.e., rejecting changes regardless of the prevailing social conditions, would not guarantee the continuity desired, but, rather, would jeopardize it. As is often the case with education, here, too, it is a matter of establishing a balance.

If the task of enabling communication between the generations is allocated to schooling, this does not mean that the public school, as a representative of the parties involved, is to provide a solution to the conflict between the generations or even to harmonize it away. Youth's revolt against the authority of parents and teachers, the questioning of values and actions of older persons are essential components of younger people's search for identity and catalysts of social and cultural transformation, indeed, a part of life itself. Schools can accompany these processes; teachers can actively take part in working on and carrying out these conflicts, but they cannot bring about their resolution in advance. The liberation of the individual, which is a consequence of taking his or her maturity seriously, is inherent in the idea of education and personal development.

2.2.5 *Building a Reflective Cultural Identity*

To a great extent, the present societies in Western industrial nations are *multicultural* societies. Extremely divergent traditions exist alongside one another; lifestyles and ways of life, worldviews and religious beliefs compete with one another and, in doing so, exhibit a diversity that can hardly be grasped. Regardless of the intense national and ideological conflicts that have been carried out in the meantime, a tendency toward a multicultural world society seems to have been dominant since the beginning of the industrial age. At present, both the efforts aiming at European unity and the issues concerning integration of foreigners and policies on asylum demonstrate how difficult it is for official politics and the populations of the states involved to follow this tendency of historic import.

But in such a situation, what can cultural identity still mean to the individual? Can it be promoted by the public school in such a way that it neither simply results in a naïve identification with *one* specific line of tradition nor leads to a general relativism and skepticism? Any attempts to answer these questions must remain approximations.

In our society or in comparable ones, individual education aimed primarily at leading a responsible life can no longer be oriented to a uniform ideal of general education and personal development. Every member of society has the right to freely decide which goals he or she sets with respect to self-realization and which traditions should be upheld or discarded within the framework of one's own lifestyle. Thus, in widely diverse ways, general education in schools must be designed to make individual education and personal development possible, to provide it with a foundation.

Reflective cultural identity is only possible if the individual is capable of recognizing or decoding the phenomena he or she encounters in everyday life as culturally dependent phenomena. The diversity of cultural phenomena in our society, which tends toward multiculturalism, demands that the individual personally decide which preferences he or she will have, which phenomena will be focused on within the framework of self-realization. Originally, a child grows into the subcultures of which his or her parents are members due to their professions, social milieu, and local particularities. At school, the child must already find some response to other children who represent other subcultures. Schooling (which itself also represents a subculture that in many respects cuts across these inherited subcultures) can provide impulses to perceive such differences consciously without belittling otherness. A school is an institution that permeates (and, in the most advantageous case, interconnects) subcultures – which is one of the reasons why it is indispensable for a highly differentiated society. At the latest, subcultural differentiation intensifies further with the selection of a profession or with the decision for a more in-depth general education instead of vocational training. A reflective cultural identity cannot be confined to an identification with those subcultures one becomes a member of due to one's profession. The sort of cultural identity aimed at here would have to manifest itself in a twofold way: on the one hand, people should become aware of being – as a result of their (specialized) work, their individual interests in education and personal development, in activities and their personal preferences – part of a much broader cultural context that is, in this sense, supportive; on the other hand, people should realize that what they cannot personally identify with, namely, the interests and preferences of others, also has cultural roots.

Certainly, schools can only provide impulses for the development of a reflective cultural identity which goes along with a culturally aware faculty of perception. But such impulses are dependent upon two conditions:

First, schools must impart selected, outstanding cultural achievements of the indigenous tradition to the adolescents and prepare the way for their appropriation. This is, to this extent, the classical part of the problem, or, as it were, a part of the traditional issue of forming a canon. However, these outstanding achievements are not – and this is the difference between the claim expressed above and the postulates of traditional cultural pedagogy – to be regarded as absolutes, as unsurpassable models, but, instead, to be examined with respect to the values inherent in them and to their dependence on historical and social conditions. It would be of a certain significance to convey insights into how current trends and events (even those that seem totally new and original) are rooted in earlier developments, how we of today are also embedded in an immense flux of human feeling and perception, thought and invention.[21] This defines, as a result of the demand for a *reflective cultural identity*, a modern attendant condition for resolving the issue of canonic status.

Second, schools must put the diverse and often contradictory manifestations of contemporary culture into context with respect to one another and with respect to the outstanding cultural achievements that have been imparted in order to make the widely varied interconnections among cultural phenomena apparent. If it is true that a culture transcends its limitations with its outstanding creations (of an artistic, scientific, or social nature) and allows for a view of novel human possibilities and thus reveals new possibilities for communication between different cultures, then it is exactly this dimension that requires attention in taking outstanding cultural achievements into account. Only then is there a chance that cultural identity will not be confined to pride in the outstanding achievements of one's own culture – an attitude that readily leads to an arrogant feeling of superiority, but a chance that such achievements mutually become media by means of which one can communicate with people of other cultural backgrounds about possible common values, about what of all the peculiarities, insights, knowledge, aesthetic and ethical values is worth preserving in an emerging world culture – since the outstanding achievement is not simply some particular, but, in its particularity, points to some universal accessible to all human beings.

A number of concepts of schooling and of general education interpret and legitimize subjects taught at school and, as far as such an affiliation is postulated, the corresponding academic disciplines as avenues of access to the cultural diversity of the world around us.[22] In principle, one can certainly concur with this view. Nevertheless, in my opinion it requires some qualification if arguments of this sort are not to promote a specialist one-sidedness. The detached coexistence of various individual subjects does not produce cultural competence. This requires special didactic efforts. In every subject, main ideas are to be sought which can be used to bridge the gap between the subject and the culture external to it, which clearly demonstrate what significance the subject (or the corresponding academic discipline) has for cultural development, how it is interwoven with such development, how it is connected to everyday life in society. In the course of school instruction, such main ideas should play a much more significant role. In such main ideas of each subject – we are simply assuming that there are some to be identified with sufficient plausibility – the diachronic and synchronic aspects of cultural cohesion are combined: The main ideas have developed historically, i.e., they represent a certain history; and they also indicate the mutual influences of the subject and the culture external to it, i.e., they transcend the limitations of the individual subject.

Of course, schools cannot be exclusively directed toward making outstanding cultural achievements accessible and stimulating reflection on their universal elements. An insistent overemphasis on 'beauty, truth, and duty' strains the attention and interest of the students if their everyday interests are neglected in the process. If, for example, the everyday literary experience of a student consists in reading comic books and his or her everyday musical experience consists in the continuous consumption of various forms of popular music, then a special sensitivity is required to introduce into such a 'natural' horizon of aesthetic perception other literary and musical elements as being worthy of exploration. Much instructional material in school misses the mark in both spheres: As much as it slips past the 'entertainment' sphere that characterizes the students' leisure time, it also misses the intellectual sphere of outstanding cultural achievement. Thus, for many students it represents a third sphere, a necessary evil, a bitter pill that must be swallowed in order to attain certain professional positions. In their leisure time, they simply put this sphere behind them without any second thoughts. No longer bothered by the cultural value-judgments schools attempt to promote, they indulge in the entertainment provided by personally selected (or industrially predetermined) cultural pleasures.

This consciously pointed description already contains references to what classroom instruction and cultural competence should entail. Good teachers have always, consciously or intuitively, been masters at this trade – in the process of imparting material considered 'boring' in and of itself, by sensitively providing momentary insights into the two spheres normally neglected: by critically examining the 'cozy everyday culture' of leisure-time entertainment without wanting to denounce it or deprive the students of it; and by allowing the standard material of instruction, itself boring to a large extent, to be experienced as a component or as a prerequisite of outstanding cultural achievements, which thus become more comprehensible and, in the most favorable case, stimulate independent efforts toward their appropriation and the formation of individually creative links to them. For such active appropriation is indispensable to the establishment of a reflective cultural identity – a pedagogical insight that is by no means new.

Thus, cultural identity proves to be a dynamic element of an individual's design of his or her life within a society undergoing rapid change. As an objective within the framework of modern general education, it is of more significance as a ferment of future cultural development than as a characteristic of persistence in what has already been achieved, more as a prerequisite of the convergence of various cultures and subcultures than as a feature of differentiation, more as a way of reflectively ascertaining one's own links within a cultural context as an historical process than as an uncritical identification with a culture in which one just happens to come of age. In this sense, cultural identity should help the individual to attain a certain equilibrium in life within a multicultural society so that he or she must neither feel subjected to extremely diverse influences and trends without any orientation or secure footing nor is compelled to screen oneself off from the experience of cultural diversity by dogmatic commitments made for the sake of one's own intellectual and social self-assertion.

2.3 Developing an Understanding of the World

This objective creates a link to that pedagogical tradition which perceives supplying children and adolescents with material knowledge about the world around them as the main concern of schooling. The students are to attain an overall perspective, to be able to make sense of the phenomena they encounter and understand the relationships among them, to have a working knowledge of the world beyond the horizon of their own limited experience. Of the seven objectives attributed to public schools by the concept of general education under discussion here, no other one approaches so closely the widely held idea of what the core of general education actually should be.[23]

This idea of developing an understanding of the world has appeared under various designations in recent concepts of educational theory. These concepts implement expressions such as: "all-round orientation" (Hardörfer 1978), "understanding the world" (Helling 1963; H. Becker 1980, pp. 313-332), "systematization of conceivable human life" (Wilhelm 1969 and 1985, pp. 143ff), or the construction of an "extensive horizon of interpretation, of a body of thought that makes recognition and assessment possible" (Ballauf 1986, p. 94). The term "orientation in the world around us" is also widely used (e.g., Stegmaier 1984 passim; Geißler 1984, p. 272; Fetscher 1986, p. 21; Maier 1986, p. 28). A certain 'overabundance' of knowledge is characteristic for the idea of developing an understanding of the world. This is based on the assumption that a differentiated worldview makes a broad critical horizon accessible and facilitates both the clarification of one's own standpoint and the recognition of that standpoint's relativity. Precisely in this sense, G. Geissler (1968, p. 173) assigns schools the task of "leading to a point beyond the confines and coincidental nature of the individual situation". The reason why developing an understanding of the world, despite a certain overlap, can be clearly differentiated from the two objectives of schooling discussed above (preparation for later life and the promotion of cultural competence) is that it goes beyond a utilitarian or traditionalist evaluation of acquiring knowledge of the world. Entire subjects such as geography or extensive areas of instruction in history, the natural sciences, and literature can be more plausibly substantiated from the standpoint of this idea of developing an understanding of the world than by referring to one of the other objectives of schooling discussed above.

This task of developing an understanding of the world has also repeatedly been subject to misinterpretation and distortion. Some of these misconceptions and the corresponding 'remedies' that present themselves are to be considered here in more detail. One distortion that is bound to the concept of developing an understanding of the world almost like a congenital defect is the emphasis on encyclopedic knowledge. In the last few decades, what might be called a legitimate offshoot of the idea of developing an understanding of the world has increasingly proved to be its rival – the principle of *scientific orientation.* To the extent that one-sided and exaggerated interpretations of this principle have gained significance – which was especially the case during the German educational reform era –, important aspects of the original idea have been forced into the background: e.g., understanding the world with reference to everyday life and not primarily in a scientific sense. This can also be seen in the specific tendency to reduce music as subject matter in school to musicology.

2.3.1 Developing an Understanding of the World Through Scientific Orientation: Refinement or Reduction?

Orientation to science was a major guideword and a matter of broad consensus during the educational reform era. The structural plan of the German Educational Commission (1972, p. 33) states, "The conditions of life in modern society require that teaching and learning processes be oriented to science." A compilation in short statements of the main arguments which made this guideword seem so plausible at that time would be as follows:

Arguments in favor of scientific orientation as a guiding didactic principle

- With reference to attaining an all-embracing "systematization of conceivable human life" (Wilhelm 1969) in modern secularized society, science seemed to be the only authority to which a corresponding systematization could be oriented; for science epitomizes knowledge that can be objectified and that humankind can acquire about itself and about the world.
- Thus, the sciences seemed to provide a point of leverage which could be used to make corrections to the contingencies and arbitrary elements which had become part of the traditional educational canon throughout the course of its history.
- Scientific orientation presented itself at two levels. On the one hand, with reference to revising the canon of school subjects: if the sciences in their entirety represented something like a valid modern conception of the world, then school subjects would have to reflect this totality in a comprehensible way. On the other hand, with reference to revising curricula: it seemed necessary to adjust the content and structure of the

curricula to the latest scientific research and, in doing so, to apply the "structure of the discipline" (Bruner) as the standard.

- From scientific orientation, one hoped to gain an approach to the problem of material as well as of formal education. The elements of knowledge encoded in science promised to offer a definitive framework for the material selection of subject matter. In addition, the 'scientific method' seemed to present the all-encompassing, general type of intellectual training suitable to modern man.
- Finally, there was hope of having found in scientific orientation a means of guaranteeing adolescents a maximum of equal opportunity. For it was by no means only designed for an elite – as was the 'basic scientific training" (Wilhelm Flitner) for secondary schools discussed only a few years beforehand –, but for all students. In particular, it was supposed to help eliminate the difference between 'popularized' and 'higher' education: "Organized learning should be scientifically oriented for all students" (German Educational Commission 1972, p. 30).

Counterarguments: Pedagogical Limitations of Scientific Orientation

Generally speaking, the hopes linked to scientific orientation as a guiding didactic principle remained unfulfilled. The proposition presented here, namely, that the promotion of an understanding of the world necessary within the framework of general education cannot be effected by a reduction to scientific orientation, can be illuminated by the following reflections:[24]

- As accurate as the statement may be that there is no other authority which can compete with modern science with respect to generating objectifiable knowledge, it seems just as doubtful that the understanding of the world to be promoted by schooling – not to mention general education on the whole – should be exclusively limited to knowledge objectifiable by science: Actions and decisions in everyday life would be subject to unreasonable restrictions if one could only take recourse to 'scientifically proven' information.
- 'The sciences' are anything but a monolithic block and, for any particular system derived from them, other systems that are just as scientifically plausible can be established. The unequivocal results expected of a scientific system prove to be an illusion.
- Nor does scientific orientation solve the problem associated with the emphasis on encyclopedic knowledge. Instead, the reverse is the case: The practically exponential growth of knowledge in all disciplines, hardly to be appropriately assessed even by the experts in each particular field; the intradisciplinary competition of various paradigms noticeable in many of the sciences; furthermore, the creation of new disciplines and interdisciplinary areas of research such as computer science, cognitive science, and ecology – all of this results in a situation in which the problem of a justifiable selection of subject matter for teaching that is scientifically substantiated becomes ever more difficult

instead of easier. Apparently, if one wishes to cope with the pitfall of emphasis on the mere accumulation of encyclopedic knowledge, something like an exemplary principle is inevitable – a principle that allocates a certain significance to aspects external to the particular corresponding scientific discipline. To put this in a more concrete way: If one decides to orient instruction to fundamental or main ideas, then these ideas should not, above all, be of central importance to the subject matter itself, but, instead, the relationship between the discipline and the world around it should be reflected in them.

- An additional problem can be seen in the circumstance that with scientific orientation, the intrusion of a new variant of specialization into schools is impending. Since the effectiveness of modern science is closely connected to the principle of rigorous specialization, the teacher who considers it a duty to orient instruction in the subject he or she teaches to the latest scientific research can readily feel justified in cultivating such specialization in the classroom. In that case, the students do not acquire a fundamental 'attitude of scientific inquiry', but, rather, specific scientific methods which, under the most advantageous circumstances, can be preliminary career training for the later study of the corresponding subject matter.
- Finally, scientific orientation did not automatically help to promote fairer learning opportunities at school. On the one hand, formal standardization of curricula for various types of schools did, indeed, increase the permeability of the school system. On the other hand, new obstacles to learning were set up due to the proliferation of abstract and formal principles and to the forced retreat of lessons combined with vivid illustrations (which were disqualified as being 'popularized'). The observation that more theoretically oriented learning is easier for some students than for others was, for the most part, simply repressed throughout the course of the wholesale, standardized prescription of scientific orientation. To put it more pointedly: Whereas scientific orientation, as it was propagated and to some extent realized, did contribute to an understanding of the world for an elite of students, for other students it enlarged the gap between knowledge learned at school and the experiences of everyday life – an understanding of the world was obscured, rather than promoted.

Thus, an understanding of the world cannot be operationalized by and, most certainly, cannot be replaced by scientific orientation. On the other hand, it is obvious that the distorted forms of scientific orientation criticized here manifest themselves especially if excessive demands are made on this orientation as a guiding didactic principle. Our concern, then, is what remains of the principle of scientific orientation if it is interpreted in a more modest and cautious way. Here, I will be restricting my considerations to the *material* dimension of scientific orientation; for the *formal* dimension – essentially, the development of an attitude of scientific inquiry – can be discussed with greater plausibility in connection with the 'development of critical thinking'.

A negative principle for the exclusion of instructional content can be described as the minimum rational core of scientific orientation in teaching. This principle is often stated as follows: Nothing should be taught which, in accordance with contemporary research in the corresponding scientific disciplines, is to be considered false. More precisely, one could state: Facts that are to be taught under the stipulation that they represent objectifiable knowledge must be substantiated within the framework of the relevant sciences. What is meant can be illustrated by the historical conflict which, for a considerable period, was a heated dispute between pious Christians and natural scientists of firm convictions: May the biblical myth of creation be taught in school? According to the principle of exclusion mentioned above, the answer should be: Yes, as a *myth*. But any attempt to give the students the impression that this was a 'real event' in the sense of a scientific description and with a claim to objectification would have to be rejected.

In yet another sense, scientific orientation seems advisable as an element of developing an understanding of the world: Modern science is a part of the real world; indeed, contemporary personal and social life are inconceivable without it. Thus, adolescents may not need to be familiar with the details of the results and methods of all significant scientific disciplines, but they should be informed in each case on the main disciplinary areas of inquiry, on the competence of the discipline, its capacity to engage in problem solving, and its specific view of the world. This, in turn, cannot simply consist in knowing certain facts 'about' something, but includes an exemplary in-depth treatment of individual problems, methods of inquiry, and fundamental results. However, what in this sense can be considered exemplary for the prospective understanding of the world cannot itself be deduced from the respective scientific disciplines.

These remarks may now seem almost like a retreat to didactic positions under discussion prior to the era of scientific orientation. This indicates a circularity insofar as precisely the ceaseless pedagogical debates on the exemplary principle during the 1950s and 60s[25] and the dissatisfaction with traditional attempts to cope with the problems of overabundance of content and of the emphasis on encyclopedic knowledge prepared the way for the triumphant progress of scientific orientation. Nevertheless, without any false modesty even today we can still take a cue from some parts of the debates of that time,[26] since our experience in the meantime has shown the following: In pedagogy, the *exemplary principle* is to be understood more as a regulative principle than as a guiding principle to be put into operation for the construction of curricula. Any attempt to argue in favor of obligatory material elements of curricula designed to promote general education must aim at the exemplary character of instructional content. What this character consists of cannot be stated generally and beforehand, but, rather, must be substantiated in each concrete instance. Within the framework of such substantiation, the status of any particular instructional content in an individual discipline is only one aspect among many.

Going beyond a disencumbered scientific orientation, one released from euphoria for science, what can and must be done if the public school is to contribute to the students' understanding of the world? How can schooling create the preconditions necessary for the development of a differentiated, personal view of the world, of a "systematization of conceivable human life"[27], of a comprehensive horizon for interpretation and judgment? The following three sections focus on three aspects.

2.3.2 *Dismantling the Boundaries of Specialization*

In a certain sense, the principle of dismantling the boundaries of specialization[28] formulates a dialectical counter-principle to scientific orientation. If, on the one hand, an understanding of the world disregarding science is not reasonably conceivable, on the other hand, it is necessary to overcome narrow-minded and biased specialization, should a scientifically grounded understanding of the world become a central element of practical life experience or even of prudence in everyday life. This means that the boundaries of subject areas – and thus those of the respective scientific disciplines – are to be made *permeable* to life situations not already structured by specialization. Specialized instructors should direct some portion of their didactic efforts toward clarifying the significance of specialized concepts and structures for an understanding of material, social, and intellectual everyday phenomena; toward explaining which primarily non-specialized problems can be solved with their assistance and at what point and why the specialist perspectives encounter limitations and lead to distorted conceptions. In the case of mathematics, for example, some issues to be considered are: What can be calculated? Where are mathematical models possible or impossible? What is ignored by restricting one's view to those things that can be calculated, that can be solved mathematically?

For attempts to dismantle the boundaries of specialization, the main ideas associated with specialized subject matter are of enormous significance. The identification of such ideas must not only take place in an intradisciplinary context, but also from *outside* the specialized area, since the contribution of such main ideas to an understanding of the world outside of the discipline must be taken into consideration. Thus, the determination of what the main ideas of a certain subject are cannot be left solely to the specialists in the field. At times, the interested and informed layperson exhibits more sensitivity to the resonance of the discipline in the outside world than does the expert.

Dismantling the boundaries of specialization, then, can be understood as a *concretization* of subject-related learning, as opening the discipline to the life-world, as a bridge to experiences made outside of the classroom, as a connection between what has been learned and what can be learned in specialized instruction and with reference to issues and problems, stimuli and experiences children and adolescents encounter in their everyday lives. But the process can also be understood as a *generalization* of discipline-related perspectives: as a return of the discipline to basic activities and insights, to main ideas by means of which the exemplary character of the discipline's specific achievement within the framework of society and culture as a whole can become clear.

2.3.3 Developing an Understanding of the World by Introducing Competing Worldviews

A significant part of developing an understanding of the world consists in imparting to young people the insight that there is not only a *single* perception of the world that can make claim to some validity. Releasing the individual to a self-responsible conception of his or her own life is incompatible with passing on ideologically binding worldviews through schooling. For such indoctrination, it would be totally irrelevant whether the fixation to a certain worldview stems from a nationalistic, racist, or religious motivation – to mention only a few variants of a certain historical relevance. A distinction must be made between two issues: an individual's choice of a certain religious or political commitment is, first and foremost, a matter of personal preference. This applies even if the individual endorses extreme or narrow-minded, sectarian ideologies as long as this commitment is accompanied by a tolerant everyday practice. By contrast, putting the public school at the service of a particular religious or political movement is incompatible with the basic principles of a democratic and pluralistic society.

Perhaps it is one of the most significant advances in the history of schooling that, as a result of many painful experiences, there is, in principle, a consensus in our society and in comparable ones that schools must not be enlisted in influencing young people ideologically. Yet, this is primarily a theoretical insight; the problems involved in putting it into practice can by no means be considered to be solved.

On the one hand, influencing someone ideologically can take place very subtly, and those who exert such influence are sometimes not even aware of doing so. Dedicated teachers and educators of the past did not at all think it one-sided to primarily be concerned with the instruction of male youths and to have "naturally" allocated lesser social roles and obligations to female young people. Often enough, long-term social learning processes are necessary to even perceive this sort of historically developed blindness as such. Who can foresee which blind spots will be ascribed to us by future generations? Thus, the demand that the understanding of the world promoted by schooling should be free of ideological distortions can only be made in relationship to the particular, socially accepted level of knowledge and social insight.

On the other hand, one should be aware of the risks involved in the attempts to preserve ideological neutrality – that they can lead to indifference, a noncommittal attitude, to a loss of values and of orientation.[29] Using the metaphor of reading a map might make this clearer: If one wants to gain an orientation in an unfamiliar place, then one must have a good map and, in addition, know one's own location. Correspondingly, developing an understanding of the world means that I have an overview of possible human standpoints and, in addition, am aware of my own standpoint. Only in this case am I in a position to rationally alter my own standpoint. Developing an understanding of the world includes both components: knowing what my standpoint could be and knowing what it actually is.

2.3.4 Major Contemporary Issues and Worldwide Problems as Crystallizations of Developing an Understanding of the World

This study has repeatedly mentioned how the issue of schooling's response to the problems surrounding the future of the globe has gained central eminence within more recent educational debates. No other single issue has provided a comparable impulse for suggestions on how to design general education in schools in a fundamentally new way.

The moderate approach espoused by Klafki (1985a) has found a more welcome reception among the general public and within the educational sciences than have many of the more radical pedagogical reform concepts. Klafki expounds his idea of general education with recourse to three significant elements: general education as "a possibility and opportunity for *all of the people* of a particular society" (ibid., p. 17), as "the development of versatility or all-round skills", and as "education mediated by the general" (ibid., p. 20). In Klafki's opinion, this last element is characterized by the necessity of acquiring "a historically mediated awareness of the major contemporary and foreseeable future problems" and "the realization that *all* people share responsibility for what will happen". Klafki makes a very specific list of which "key issues" a modern general education would, in his opinion, have to be focused on, e.g. (I will cite only the first six of a total of eighteen):

- "the issue of world peace and the East-West conflict[30]
- "environmental issues
- "possibilities and risks of the advance of natural science, of technology, and of the economy
- "so-called 'developed countries' and 'developing countries'; the gap between North and South
- "social inequality and positions of economic and social power
- "democratization as a *general* principle of orientation for the design of common affairs, e.g., also of the economy, or limited to certain areas?
- "..." (ibid., p. 21)

In a fairly cautious way, Klafki then expounds his argument:

> "Thus, regarding such key issues, general education means: At various levels of the personal educational process or of the educational system, every young person and every adult should have experienced in-depth learning activity relevant to at least some of these major issues. What is indispensable in this sense is the requirement ... that a differentiated awareness of the problems involved in such issues be attained; however, the aim cannot be to promote a particular perspective or to support a specific solution proposed within the course of the debate..." (ibid., p. 21-22).

Considered in their own right, the requirements Klafki names seem reasonable and compatible with the concept of general education presented here thus far. However, Klafki's concept of general education is unclear on the systematic significance of the key issues. With a certain justification, Klafki indicates that concentration on current affairs can encompass "the risk of a certain fixation to the present, a narrowing of perspective, a lack of openness", and may even demand too much of adolescents emotionally, morally, and politically. For this reason, "polar complementation" is required – by means of learning activities "which are not solely or not directly justified by their necessary contribution to dealing with urgent current affairs" (ibid., p. 24-25). To illustrate this point, Klafki then mentions practically everything that, in the traditional interpretation, has always been part and parcel of general education in school: "Familiarity with mathematical thinking, with scientific knowledge of reality, with communication in one's native and in foreign languages", etc. (p. 25), complemented by a "considerable amount of very simple, more or less elementary practical knowledge, abilities, and skills", such as reading, writing, basic arithmetic, etc. (p. 29). Although Klafki seemed to have succeeded in providing the idea of general education with a clearer profile by developing the major significance of the key issues, he goes on to diminish this achievement in a certain sense by – to put it somewhat pointedly – placing the entirety of traditional general education alongside the new type he demands without providing a sound theoretical clarification of the relationship between the 'old' and 'new' elements. In my opinion, the substance of both argumentative strands can be affirmed. Yet, an overarching perspective does not become clear.

This difficulty does not arise if dealing with these and similar key issues is assigned to the aspect of *developing an understanding of the world*. The idea expressed somewhat vaguely by maintaining that general education should be "education mediated by the general", can then be avoided. It is obvious that an understanding of the world that deserves to be designated as such cannot be developed if global risks and dangers are ignored. Nonetheless, the systematic placement of the key issues within the framework of schooling's obligation to develop an understanding of the world precludes overestimating schooling's possible contribution to the solution of these problems and prevents the emergence of unrealistic expectations and exaggerated hopes. Schools simply provide the students with some orientation for a world in which the dangers are real and extremely critical. This world, then, is brought within the bounds of the students' judgmental horizon. But this does not by any means contradict the following: developing an understanding of the world requires additional elements, such as the systematization of phenomena illuminated by means of the sciences; and there are also other objectives of schooling, such as direct preparation for everyday life ('hands-on' skills and knowledge, in Klafki's terms), passing on cultural achievements, etc.

A socially and politically committed teacher interested in his or her students and in their future will not be able nor want to conceal personal sensitivities affected by global problems behind activities related to imparting subject matter. Discussions of current key issues between teachers and students, the expression of opinions, and reflections on possible opportunities for private and political action should not simply be emotionally charged elements that, in principle, remain external to instruction. World affairs or key current issues can become the starting-point for the kind of dismantling of the boundaries of specialization discussed in the previous section. For there is no standard subject matter that could not at least contribute certain aspects to the perception of these problems. With respect to mathematics, this will be demonstrated more directly in the following chapter.

In sum, confronting world affairs and key issues is of urgent necessity within the framework of a general education. This confrontation should be related to the development of a general horizon of ideas and opinions; it should take place within the framework of acquiring knowledge of historical, political, geographical, scientific, and ecological contexts that is not simply superficial and short-lived. And since such key issues can involve much emotional sensitivity, confronting them requires a certain calmness and the ability to view things at a distance. On their own, sensitivity and anxiety can have paralyzing effects if there are no concrete possibilities for action corresponding to such emotions.

2.4 Development of Critical Thinking

Parents who attempt to instill critical thinking in their children not only risk being called into question by them, but – at least in principle – they will also interpret this questioning of parental authority as an indication of the success of their efforts and will welcome it. A society that expects its public schools to attend to this task takes a comparable risk with the same sort of positive attitude, *mutatis mutandis*. For whoever is able to engage in critical thinking will neither be a complacent subject who simply accepts whatever teachers, politicians, and other representatives of institutionalized power in public administration, the church, the economy, and the sciences impose upon him or her, nor will he or she readily yield to presumed or actual pressure exerted by 'public opinion'. Critical thinking entails politically explosive energy – especially in the light of social relations that are easily vulnerable to critique.

This can explain why the pedagogical idea behind 'development of critical thinking' – itself an essential element of educational theory since the Enlightenment (cf. Klafki 1986) – only gained recognition as part of instructional practice at a relatively late date.[31] Only modern democracies founded on the rule of law were in a position to accept this aim for their schooling systems (if for the moment we leave putting it into practice aside). Modern democracy is the first type of society in human history that – in principle – has as its prerequisite the ability to engage in critical thinking on the part of all adult members of society. – At this point, let us give this fundamental idea more exact consideration.

2.4.1 Critical Thinking, Responsibility, Emancipation, Enlightenment

The keywords listed above already provide a rough outline of the meaning and scope of the objective to be discussed here under the heading of 'developing critical thinking'. Following the famous definition formulated by Kant in 1784 which calls enlightenment "man's emergence from his self-imposed immaturity" and calls immaturity "the inability to use one's understanding without guidance from another"[32], we can consider pedagogical enlightenment, education with the aim of furthering maturity or emancipation, and the development of critical thinking to be various expressions of the same basic idea with particular nuances. Common to all of them is that they are based upon a twofold anthropological presupposition: first, humans are creatures capable of reasoning and of reasonable self-determination; secondly, reasonable thinking does not simply develop automatically; a person can fail to achieve self-determination if education and personal development do not support the process in an appropriate way.

The concepts expressed in the keywords listed above mutually interpret one another. This will become clearer if we do not simply list them without differentiation or equate them vaguely with one another, but, rather, attempt to describe the relations among them more precisely.

Critical Thinking

Using one's own reason in a critical way means not simply accepting assertions, pre-formulated conclusions, and value judgments, but, regardless of the authority with which such assertions are coupled, to question them, to explore possible contradictions, inconsistencies, contradictory elements, and, in doing so, to have faith in one's own judgment. Critical thinking, applied reflectively to one's own thinking, makes self-criticism possible. Using one's own reason in a critical way requires the ability to think logically and make judgments, thus, presupposes a developed, functionally rational critical understanding. But it goes even further, in the sense that it is intertwined with a certain attitude and a certain insight: with an attitude of wanting to get to the bottom of things by using one's own understanding, of insisting upon a difference between appearance and reality – an old-fashioned way of putting it might be to speak of "striving for truth"[33]; with the insight that the scope of understanding is limited, that understanding is not capable of providing a solution to all problems and issues. Ideally, the habitus of an intellectual is characterized by the rationality cultivated by this attitude and by this insight.

Maturity

Critical use of reasoning powers is a prerequisite of maturity as defined, for instance, by Adorno (1971, p. 10): "That person is mature who speaks for himself because he has reflected on something and is not simply repeating what someone else has said." Someone who is *mature* is expected to take charge of his or her own life, to make responsible decisions, to be *independent*.

Emancipation

Even beyond the independence of the mature person, critical thinking facilitates emancipation if this is understood as freedom from unjustified dependencies and false constraints. This applies at an individual and at a social level. Using critical thinking, the individual person can recognize dependencies and constraints that obstruct his or her independence. At the social level, critical thinking can help entire groups of the population – women, ethnic minorities, the socially underprivileged, etc. – to demand equality and to insist on political changes in favor of it.

Enlightenment

At this point, we can refer once again to Kant and to his statement that the motto of enlightenment is: "Have the courage to use your own understanding!" (1968 [1784], p. 35). Enlightenment aims at releasing and realizing critical thinking and thus becomes a preparatory step on the way to maturity *and* emancipation. On the one hand, enlightenment can be conceived of as a communicative act between individuals. In this case, the metaphorical element 'light' means that one person casts light on something that for another person has thus far been 'in the dark'. The Platonic dialogues demonstrate this individual variant with paradigmatic forcefulness. In modern times, 'enlightenment' has taken on the more generalized form of a societal project of de-mythologizing, of modernization, of doing away with all matters resting on belief alone. However, since the Age of Enlightenment, the dangers of making the societal variant of the idea of enlightenment into an ideology have become obvious, especially if this idea is degraded to a worldview that is imposed upon others with missionary zeal.

Terminological Considerations

We can maintain that the present pedagogical issue is reflected in all of the concepts mentioned above – with certain specific connotations in each case. Why, then, should we prefer the designation 'development of critical thinking'?

In particular, the term 'emancipation' has shown signs of wear since its excessive use during the era of educational reform in West Germany. By avoiding its use as a key concept, I hope to preclude misinterpretations of the idea under discussion here and to distance myself from those theorists who see 'emancipation' as the sole and highest aim of human education and personal development.[34] For in a society as complex as ours, making a claim to emancipation with such exclusiveness can only then manage to deal with the many-faceted problem of general education at school if it is complemented by a variety of auxiliary constructions. Thus, simply for maintaining the logic of the theoretical conception, it seems more appropriate to treat the emancipatory element of general education as a central and essential, determinant characteristic of general education, but not as one of the highest priority.

The term 'enlightenment' may seem less problematical, yet it impedes ready comprehension since it is often used in divergent contexts and with widely variant meanings. The concept only attains a certain clarity by connecting 'enlightenment' to 'critical thinking'.

'Coming of age' ('maturity'), a term with a venerable pedagogical tradition, has practically disappeared from modern everyday usage. Its antiquated character obscures the continuous relevance of the pedagogical idea behind the term.

To a greater extent than these competing concepts, the prospective designation 'development of critical thinking' focuses on students' thinking. Moreover, this is significant for the concept of general education under discussion here for three additional reasons: first, in this way the aspect of material education, itself essential for an understanding of the world, is complemented by the aspect of formal education, for critical thinking can be conceived of as a universal formal ability.[35] Secondly, emphasizing independent critical thinking prevents the misconception that successful enlightenment or emancipation consists in students' adopting certain 'critical' positions already worked out for them.[36] Finally, the more specific socio-ethical and personal implications of the terms 'emancipation' and 'maturity' are left aside for the moment. This would seem appropriate since these matters will be discussed later within the context of developing a sense of responsibility and enhancing students' self-esteem.

2.4.2 *Critical Thinking and Its Self-imposed Limitations*

From the time of the Age of Enlightenment up to and including the fashionable debates on 'postmodernism' (cf., e.g., Welsch 1987), there has always been some criticism of the idea and the program connected with enlightenment, of a presumed fetishist view of (bourgeois) reason. In order to further clarify the concept of 'critical thinking', it seems appropriate to at least briefly deal with this critique. Doesn't approving of schooling's role in the development of critical thinking mean adhering to an ideology of progress that has long been disproved by the course of history?

Doubts about whether it is always 'reasonable' to rely on reason (to put it paradoxically) have been nurtured above all by the historical experience of 'dialectic of enlightenment', of self-destructive tendencies so often exhibited by developments which purported to better human conditions in the name of reason. On the other hand: Even the famous book by Horkheimer and Adorno (1969 [1947]), which was written under the direct impact of the Nazi crimes and of the horror of the Second World War and is sometimes quoted as a symbol and proof of the questionableness of the idea of enlightenment, is actually – if read carefully – a pessimistic and despairing appeal *in favor of* human reason. The decisive message contained in the book is that the world is not really enlightened and that it never will be completely. Disaster arises as a result of the semblance of enlightenment, from the hubris of rationality that considers itself to be absolute, or, expressed more modestly, from the absence of self-imposed limitations on thought.

In his remarks on the theme of enlightenment, Hartmut von Hentig (1989) put forth an argument which raises self-imposed limitations to the decisive characteristic of enlightened thought: "Thought determines the limits of thought. It leaves whatever it does not itself realize nor explain and, finally, prove, to other forms of cognition and perception, for better or worse."[37] This postulate of Hentig's (it is not to be understood as an assertion of facts) illustrates clearly: A critique of the idea of enlightenment cannot result in an abandonment of reason (for every critic, of course, will take recourse to reason for his or her own argument). Rather: 'critical thinking' proves itself to actually be critical only by also reflecting on its own limitations.

The most important aspects contained in Hentig's statement can be summarized as follows: For Hentig, one major characteristic of enlightenment is "the continuous examination of thought by thought" (ibid., p. 315). By 'thought' or 'thinking' he means "logical analysis of facts … on the basis of (a) sensory perception mediated by functioning human organs and of (b) the rules of logic", thus, "arriving at necessary conclusions" (pp. 315-316). According to Hentig, the assumption that "everything would in principle be accessible and explicable to human understanding" is itself irrational. Nevertheless, it is not only permissible, but even imperative to undertake an "attempt at explanation" – "not least in order to determine the limits of reasonable explanation" (p. 316). Hentig lists some examples of questions that present themselves for human thought, but do not allow a convincing reasonable answer:

- "Why does the world exist?
- "Why do I exist?
- "What purpose should I have in life?..." (p. 316).

Hentig points out the necessity of reformulating such questions in order to apply one's understanding in the direction of a solution, the necessity of becoming aware of limitations and of explicitly naming the uncertainties involved in speculative attempts at explanation. Such a reformulation might be as follows: "What can I know about X with respect to which preconditions and within which limits?" "Examining thought by thought" implies that these limitations are also reflected upon. At the latest, whenever reasonable action is to follow reflection, further limitations in addition to the cognitive ones become obvious: those stemming from human drives (a person acts in a way not corresponding to his or her insight); those stemming from external circumstances; and those stemming from the fact that the meaning of human action cannot be explained by thought alone. But since these limitations become visible in the course of the examination of thought by thought, since "the worst conditions are the result of non-enlightenment", of the fact that "thought has not examined thought", there is, according to Hentig, "no alternative to enlightenment" (p. 318). "A *more* reasonable world, not a reasonable one, is the workload allotted to enlightenment" (p. 319).

Regarding the conceptual clarification undertaken here, the preceding considerations demonstrate that critical use of reason ('critical thinking') is distinct from instrumental rational thinking ('simple' use of one's understanding) as a result of reflection, of thought's examination of itself. Critical thinking is characterized by self-imposed limitation. – To a large extent, this distinction corresponds to Kant's differentiation between understanding and reason: understanding as conceptual ability, reason as the ability to have ideas, to put things to question.

2.4.3 Critical Thinking and the Criteria for Judgment

Before we proceed with considerations more directly relevant to schooling, let us discuss one last (more or less) theoretical aspect of critical thinking which up to this point has only been referred to implicitly: How can we establish criteria for critical thinking?

To pose the question more precisely: If it is correct that, with the use of my ability to think, I am, in a fundamental way, not able to autonomously produce the criteria according to which I judge things, then how do I confront the usual criteria passed on to me as to any human being from tradition, religion, worldviews, and ideology? One of the possibilities open to human reason is that on a trial basis various criteria can be applied: By assuming an intellectually 'experimental' attitude, I can consider a problem that I would like to solve not only from the perspective of the 'truth' in which I happen to personally believe, but also from that of the 'truths' of other persons. By comparing consequences that would respectively result from different preconditions, I can attain new, higher-level criteria for my judgment, new insights which at the same time transcend and retain the old truths.

For example, by considering problems of leading a meaningful life from the perspective of various religions, I can come to recognize the ethical principle of religious tolerance as a value and as a criterion for judgment which does not totally neutralize my own religious commitment, but does prevent me from making this commitment the measure of all things. Critical thinking is always thinking related to some standard (the Greek *kriterion* means touchstone, distinctive feature, gauge), but in contrast to dogmatic thinking it is not rigidly attached to one certain standard for all time. And in contrast to naïve thinking, if I engage in critical thinking I am aware of this reference to standards.

The significance of scientific rationality is based not least upon this combination of being related to certain standards and at the same time being relatively independent of them. The axiomatic method provides a good example: The conclusions that I reach as a mathematician within the context of an axiomatic theory claim to be valid under the presupposition that the corresponding axioms are valid; thus, they represent a "relative truth", relative to the axioms formulated beforehand. However, I can vary the axioms, change them, and put others in their place in order to explore the consequences and thus to arrive at other relative truths.

This idea can also be demonstrated in exemplary fashion by the scientific method (in its various forms with respect to specific subjects): The systematic character of scientific work consists in – among other things – strict adherence to preliminary methodological decisions. As far as scientifically founded critique is concerned, this also provides important criteria for judging the merits of scientific work. But that flexibility of the sciences that makes the world accessible to us is only made possible by the fact that other, competing preliminary decisions of a methodological nature can be made, by the fact that reflection and research are carried out in accordance with various, mutually contradictory 'paradigms'.

2.4.4 *Critical Thinking and Scientific Orientation*

While an interest in scientifically founded knowledge corresponds to the material aspect of scientific orientation, its formal aspect is expressed in an *attitude of scientific inquiry* that precedes any specific scientific methodology. This attitude of scientific inquiry can be interpreted as a rationally molded, systematic variant of a basic attitude inherent in humans: of curiosity. Only under the historical and social conditions of the Western modern era was the enterprise of modern science, an enterprise that changed the world and humankind itself, able to arise from this anthropological precondition. In this sense, critical thinking is both a prerequisite of and a driving intellectual force behind modern science, which was only able to achieve its current prominence for everyday life by constantly questioning and skeptically examining traditional knowledge, by demythologizing traditional conceptions and images of the world – through what Max Weber calls the "disenchantment of the world".

The previous subchapter mentioned misguided developments in the scientific orientation of schooling (some of which are now history themselves) and their possible pedagogical correction. At this point, we can add: if scientific orientation does not confine itself to a misconceived sort of scientific propaedeutic aimed primarily at introducing specialist features of the respective individual sciences; if, instead, scientific orientation means focusing on the universal elements of scientific activity, the most comprehensive and far-reaching of which is precisely the critical use of reason, critical thinking, then, it does indeed have a justified place within today's schools. By exhibiting a scientific orientation in this sense, schools take into account two inevitable insights:

- The demand for rational comprehension, for an understanding of the world that is as rational as possible, is one of the most distinctive elements in the cultural self-perception of modern Western man.
- This far-reaching rationalization of the world is epitomized and demonstrated more consistently and systematically by modern science than by any other segment of society.

Furthermore: If scientific orientation in school is to serve the purpose of developing critical thinking, then one should introduce that mutual interaction characterizing critical thinking's relation to and at the same time relative independence of certain standards with examples from the scientific disciplines relevant to particular subject matter. In doing so, it is especially important to constantly establish connections to reasonable everyday thinking. The students should come to realize that in many respects scientific thought is simply a more systematic, a methodologically fortified variant of everyday thought. For where should the majority of adolescents learn this if not within the framework of their general education? Developing critical thinking in school should proceed according to the premise that this is not a repository of science and by no means some specialist activity, but an intellectual activity that is meaningful and necessary for coming to terms with everyday problems and issues. Critical thinking proves itself where it becomes practical.

Which didactical and practical consequences for instruction can now be drawn from these theoretical considerations which, up to this point, have been discussed at a rather global level? The next three sections deal with the following questions:

- What can instruction achieve in the way of generally training thinking skills?
- What value does instruction at an informal level have for promoting critical thinking?
- How can we keep the 'development of critical thinking' from being limited to promoting instrumental rationality?

2.4.5 *Learning to Think from the Perspective of Cognitive Psychology*

In an article providing an informative overview, Resnick and Klopfer (1989) list five key considerations which they use to group recent insights of psychology on promoting students' cognitive abilities. Here, I will follow their presentation, but will considerably further condense their arguments.

Learning Requires Knowledge

The fact that people learn at different rates and with varying effectiveness depends not least on their divergent levels of prior knowledge. It is considerably easier for someone who is already well-informed on a certain subject matter to supplement his or her knowledge on the subject. This person already has access to schemata, key concepts, cognitive patterns, and paradigms which facilitate the systematic placement of additional knowledge and its integration into existing cognitive structures. Attempts to teach such schemata and key concepts in isolation, to disconnect them, as it were, from their knowledge base are usually unsuccessful, for they develop their effectiveness for the integration of additional knowledge precisely *through* their diverse connections to various individual elements of knowledge, through their continuous and repeated activation. Thus, learning is to be conceived of as a constructive action that is significantly dependent on the intellectual activity of the learner. Knowing something does not mean having received and 'taken possession of' information, but, rather, interpreting it and relating it to prior knowledge. Correspondingly, mastering a skill does not consist in mechanically reproducing a fixed procedure; it is also imperative that the agent knows when to implement this skill and how to adapt it to changing conditions.

This means that the hope promulgated for such a long time by educators and psychologists that general thinking skills could be taught and trained independently of concrete subject-matter content is to be discarded. On the contrary, it seems to be necessary to explicitly refer to schemata and key concepts in connection with concrete knowledge and to thus promote an awareness of the resulting opportunities to interpret new information, to establish connections to prior knowledge, and to explain larger contexts.

Thinking Skills and Content

Making the generative power of general schemata and key concepts accessible in the way described means teaching them in combination with concrete content – and this applies to all of the traditional school subjects. Not only is intellectually demanding content to be considered in this respect, but also precisely elementary subject matter, starting with that of the earliest grades.

Hence, the still unresolved issue of whether and to what extent a *transfer* of thinking skills from one subject matter to another can be assumed (all of the empirical findings thus far have been rather disappointing), becomes less relevant. It would be much more important to establish intimate connections between a wide variety of opportunities for using intellectual faculties and the most diverse sorts of content – across the entire school curriculum. The constant training incidentally resulting from such preconditions would, at least, then cause the students to acquire a considerable number of thinking skills linked to certain content. This would be a desirable result even if there were no transfer in the classical sense, and if the contrary were true, then it would be so much the better.

Cognitive Abilities and Motivation

Simply having certain thinking skills and knowledge linked to them does not suffice; a willingness to make use of these must also be developed.

> "Substantial amounts of recent research suggest that good thinkers and problem solvers differ from poorer ones not so much in the particular skills they possess as in their tendency to use them" (ibid., p. 6).

Some researchers were surprised to find that, often enough, simply encouraging retarded students to use a certain strategy could already lead to its successful implementation (e.g., Belmont et al. 1982). This and similar findings seem to indicate that, for classroom instruction, from the outset it is particularly important to develop the habit of using or the willingness to use relevant thinking skills and strategies, and to develop the knowledge of when this use is appropriate. Moreover, whether students have a motivation to tackle difficult problems that are not yet fully assessable also seems to depend upon their view of intelligence. If they consider it to be a fixed capacity that they either can access or not, then they are quick to resign if difficulties arise; if they consider their intellectual abilities to be a potential that can be developed, then they tend to see problems as a challenge that helps them to further develop their own intelligence (Dweck/Elliott 1983, cited in: Resnick/Klopfer 1989, p. 8). Not only subject-matter content and thinking, but also motivation and thinking skills are intricately intertwined with one another.

The Role of Social Communities in Shaping Cognitive Skills

For a long time, nothing seemed more obvious than that thinking was an individual's activity, that, therefore, the analysis of thinking abilities and their subsequent cultivation had to be oriented to intra-individual activities and their optimization. Interestingly enough, again and again – e.g., in comparisons of programs for teaching higher-order cognitive abilities – results showed that opportunities for cooperative problem solving and for meaning construction in exchange with others led to greater success even for the individual learners. Here, the opportunity to observe how more experienced problem solvers deal with problems and construct arguments seems to be of particular significance. Just actively observing others already increases the chances of developing more effective thinking and problem-solving strategies. In addition, the possibility of undertaking a solution with a division of labor has a positive effect – the students can bring their individual talents into play and sometimes complement one another very well. Yet, the most important aspect seems to be that the "students know that all the elements of critical thought – interpretation, questioning, trying possibilities, demanding rational justifications – are socially valued" (ibid., p. 9). Although further research is needed on how the kinds of socialization factors mentioned specifically affect the development of intellectual dispositions, much in the existing findings indicates that long-term membership and participation in social communities in which certain expectations are placed on conduct, rational and critical arguments are valued, and in which students can experience their own ability to analyze and solve problems are decisive in the development of cognitive abilities.

Training Thinking Skills Within Classroom Instruction: 'Cognitive Apprenticeship'

There are some indications that instruction which is significantly oriented to a traditional 'apprenticeship', now usually confined to training programs in artisan professions, can conform more directly to the recent findings of cognitive psychology than does conventional classroom instruction, in which the focus is on explicit teaching. Instead of asking how instruction is to be carried out in order to produce specific abilities and knowledge on the part of the students, more attention is paid to the question of how a cognitive environment can be designed that enables the students to gain instructive experiences – in the sense of an 'apprenticeship' – with their own thinking. An approach in the direction of an 'apprenticeship' would seem desirable in the following respects:

- In the classroom, the focus should be on 'real tasks' which, even apart from instruction, have a certain significance and at the same time present a serious challenge to the students.
- The tasks to be dealt with should be presented within meaningful contexts. In particular, the emphasis should not be on isolated cognitive skills that have been removed from their normal, natural contexts. However, as with a traditional apprenticeship, novices are to be given the task of generating less complex and involved 'products' (of cognition, in this case).
- Finally, sufficient opportunity should be provided to observe how others complete tasks for which the student's own competence is to be developed. Thus, standards that allow a better assessment of one's own efforts are conveyed. With reference to cognitive (instead of manual) activities, care must be taken to make mental activities overt, to 'publicize' them.

Links to Other Theoretical Approaches

The insights emphasized here are by no means all original; but in a general sense they represent a changed view of learning that is not confined to recent cognitive psychology. On the contrary, this changed view of learning is supported by other research approaches not explicitly mentioned above. The constructive aspects of learning and thinking are especially emphasized by authors assuming the perspective of 'radical constructivism' in concurrence with von Glasersfeld and Foerster (cf. Glasersfeld 1991, Cobb et al. 1992). Studies that have an interactionist orientation have concentrated on the significant role of the social environment and of social interaction in learning and in cognitive development (e.g., Bauersfeld et al. 1988, Krummheuer/Voigt 1991, Voigt 1994). Within the framework of the theory of "domains of subjective experience" (Bauersfeld 1983, 1993), even more emphasis is placed on the specificity of learning related to particular contexts than in the approaches of cognitive psychology cited above; and this emphasis is then taken as the starting point for didactic considerations. Despite the many differences with regard to details and despite certain issues that are debated with practically ideological fervor (especially concerning the consequences of radical constructivism, which claims to be not only a psychological, but also an epistemological theory[38]), the elements common to these more recent views on human learning and thinking processes seem to predominate. This is especially the case if they are considered in contrast to the behaviorist conceptions that still dominated the science of psychology two or three decades ago or if they are compared with the popular 'naïve theories of learning', in which learning is primarily seen as the receptive absorption of information.

Interestingly enough, some of the older didactic approaches have regained topicality against the backdrop of recent research on learning how to use cognitive abilities. In particular, a renewed examination of the 'genetic' and 'Socratic' methods seems promising. Wagenschein was the last major proponent who unceasingly promulgated these approaches as methods of a form of instruction *aiming at understanding*, up until well within the 1970s. Much of what in Wagenschein's concept and in that of education specialists working in a similar vein inevitably seemed theoretically ill-founded and idealistic against the backdrop of behaviorist theories and hardly effective in the light of a contemporary utilitarian concept of schooling, now demonstrates an astonishing amount of foresight, when viewed in the context of recent research in cognitive psychology. Genetic teaching is implicitly based upon the constructive nature of human learning. The Socratic method is incompatible with a view of the learner as the passive recipient of information. Wagenschein's enthusiasm for *understanding* has its counterpart in more recent conceptions of integrated learning and in the skepticism on drilling isolated individual skills. I will return to this point presently.

However, in other respects Wagenschein was, of course, a child of his times. He makes practically no distinction between conscious didactical action and the informal level of instruction, which, as we now know, is a crucial factor in the development of intellectual dispositions such as critical thinking. The significance of this informal level for promoting critical thinking, mentioned in passing above, will now be examined in more detail.

2.4.6 Promoting Critical Thinking at the Informal Level of Classroom Activity: Cultivating 'Reasonable Behavior' toward Persons and Objects

Empirical research on teaching carried out during the last fifteen years has demonstrated considerable proof of the significance of the informal level of instruction. Widely practiced communicative structures and action-related expectations often seem to influence the process and the success of instruction more forcibly than the explicit intentions of the participants in classroom activity.[39] But from a more positive orientation, these insights seem to make new possibilities available to reflect on and improve the quality of the processes involved in instruction.

The informal level of instruction does not consist in the summation of all of the intentionally pedagogical acts; rather, in combination with other factors influencing instruction, it amounts to something like their synthesis. Within the scope of their age-appropriate powers of understanding, even elementary school children can experience critical thinking as an essential part of their way of life. Within classroom instruction 'cultivated' toward achieving this end, they can observe how reasonable argument outweighs prejudice and discover that the quality of an argument is more essential than the social status of the person supporting it, that a critical query can be of more value than the uncritical appropriation of pre-established knowledge. As an indication of the quality of social behavior toward others and with respect to the issues to be discussed, critical thinking makes high demands on all participants, especially on teachers, since this quality is by no means an achievement inherent in our society. It is often just as lacking in the conduct of adults toward one another (many discussions among academics might come to mind) as in the interaction between children and adults. If such cultivated behavior as described above can be established at least temporarily and for a certain group as a social practice, then, to a great extent, the development of critical thinking will be achieved through the socialization effected by instruction – practically as a matter of course, somewhat less involuntarily with reference to the pedagogical and social preconditions.

Yet, it would be an oversimplification to maintain a distinction between the informal level of instruction as described above and conscious didactic action along the lines of the opposition 'functional vs. intentional education'. Pedagogical intentions do certainly have some effect on efforts to create an atmosphere conducive to the development of critical thinking. However, the pedagogical focus is placed less on direct didactical intervention than on providing a certain latitude and establishing prevailing conditions in which the desired reasonable behavior toward persons and issues can – in the most favorable case – practically materialize 'on its own'.

In recent years, it has become customary – especially in mathematics education – to speak of the development of a culture of instruction in order to designate the desired changes at the informal level of instruction. This is a fitting designation, but the following clarification seems appropriate since the use of this concept is often quite vague.

The Concept of a Culture of Instruction

First of all, the concept of a culture of instruction is a very general designation for the way teachers and students interact in the classroom, for which (explicit or implicit) 'rules of the game' guide their behavior, and for how they collectively – intentionally or unintentionally – deal with the topics to be discussed. In this sense, the usage of the concept is, at the outset, *descriptive*. Any sort of instruction – whether it is arranged in an orderly way or just proceeds chaotically, whether it is interesting or boring, whether it is designed according to a certain routine or full of variety – can be classified according to characteristics pervading the situation and to specific elements of the communicative structure, to factors which in their entirety constitute the culture of instruction. – As soon as the emphasis is placed upon developing a *pedagogically beneficial* culture of instruction, a *normative* or *prescriptive* element is brought into play. In that case, the culture of instruction means a certain *cultivatedness* of instruction, of which it is assumed that it will provide better preconditions for the achievement of the pedagogical and didactic aims. This juxtaposition of the descriptive and the normative use of the concept seems justifiable as long as the context clarifies what is meant in each case. To avoid misinterpretation, I will usually employ the term 'culture of instruction contributing to general education' whenever I use the concept in a normative sense. What is meant is a culture of instruction that reflects the normative claims of that concept of general education at the base of my arguments here.

2.4.7 *Learning Thinking Skills and the Claim to Realizing Enlightening Instruction*

Thinking is not automatically critical thinking. As the course of human history has often demonstrated, rationality can be put to instrumental use in creating weapons of mass destruction, in organizing concentration camps, in sustaining undemocratic power relations by violent means. A list of such examples could be extended indefinitely. But at the moment, I am not actually interested in *ethical* obligations involved with thinking – I will discuss those within the context of 'responsibility' in the following section. At this point, attention should center on the content-related elements of thinking: Can schooling help to prevent – e.g., through the selection of topics to be reflected upon – 'common sense' and (in the case of some of the students) intellectual astuteness from being instrumentalized or wasted on trivialities? The focus here is also on the reflectivity of thought. For, following the arguments Hentig puts forth, we can conclude that the reflectivity of thought is one prerequisite of its autonomy. Thinking encompasses the possibility of making oneself an object of reflection, of critically considering the criteria of one's own judgment. Can this sort of reflectivity be promoted in classroom instruction? And if so, what is its relationship to the general development of thinking skills?

I will start with the last question. As I explained above, from a perspective of cognitive psychology, no development of thinking skills fully independent of content is feasible. The process of learning how to think is always connected to certain content; thus, the desired reflectivity cannot be practiced independently of content-related contexts. The consequence for the aim of developing critical thinking is that an intensified sense of responsibility must inform the selection of instructional topics and materials. Put more constructively (allowing for some speculation), the more challenging, the more personally astounding the instructional content is for children and adolescents or the more varied perspectives it offers, the more possibilities for learners to recognize links to their own views of the world (naturally, the challenges must not become too demanding), then the more intensely the students will become emotionally involved in the topics under discussion. And then, the chance is greater that not only the students' rational potential will be activated, but also that they will take advantage of opportunities for reflection, for a critical examination of their own thinking. Expressed in the reverse way: Instructional materials that, for the students, are isolated, meaningless and without any relevant context – rote memorization of incomprehensible formulae, vocabulary items for translations that preclude any involvement, rules of grammar whose usefulness remains obscure, or historical facts whose connection to general human issues and to the present situation remains unclear – these will usually dull thinking abilities rather than motivate them; even less will they stimulate critical thinking and reflection (unless, of course, frustration develops into protest). But even certain content that challenges the rational potential of the students (or, more cautiously, of some of the students) can be so restricted and isolated from the rest of the world that the desired reflection, pausing to consider one's own thinking, hardly has a chance to be realized. This includes such content that does, indeed, allow for astute problem solving, but at the same time promotes extremely one-sided specialization. Thus, an obsessive interest in computers or in chess can be combined with astonishing intellectual achievements.

In sum, the preliminary result of our considerations is as follows: Critical thinking cannot be promoted by arbitrary instructional content. Rather, indispensable elements include: relevance, comprehensibility, appropriateness to the students' age, stimulating emotional involvement, and perceptible links to prior knowledge.

This preliminary result should not induce us to assume once again that instructional content is the really decisive factor. Our discussion of the informal level of instruction, of a culture of instruction, and traditional pedagogical insights on the mutual influence of content and method retain their validity here: Content is always constituted within the classroom. More precisely, how much a certain content specified by the curriculum means to a particular student is, apart from his or her individual predispositions (interest in the topic, prior knowledge and abilities), dependent upon how the subject matter is treated in the classroom, which aspects are 'officially' emphasized, and how the teacher and other students evaluate student reactions to the subject. Both aspects are crucial:

- Didactical and pedagogical skills – even if combined with an encouraging classroom atmosphere beneficial to the students – do not suffice to provide long-lasting optimum conditions for the development of critical thinking if the topics to be discussed are inadequate.
- And the reverse is also true: A list of topics full of subject-related incentives and connections to the real world and aimed at developing critical thinking will not communicate to the students if the culture of instruction is not sufficiently developed.

Finally, one should call to mind the fact that critical thinking on the part of children and adolescents can neither be forcibly brought about nor conclusively prevented by pedagogical or didactic measures. Critical thinking, maturity, and intellectual independence are, in the final analysis, not products of pedagogical action, but characteristics of human self-realization which can, indeed, establish themselves *in opposition to* the intentions of educators, but which a person can also fail to attain. The polarity between reasonable and unreasonable action seems to indicate an antagonism in human beings as profound as that between 'good' and 'evil'. In both cases, education and instruction simply represent beneficial or detrimental preconditions.

2.4.8 *Tentative Summary: Interconnections Among the Objectives*

The objectives of schools which have been treated here thus far are dialectically related to one another; efforts to attain them cannot be made in isolation. Critical thinking does not develop in a vacuum; enlightening instruction always entails some subject-area content and to a certain extent requires experience, knowledge, and expertise. Without a minimum of knowledge and skills preparing for later life, without the ability to relate to one's traditional culture, without the basis of a differentiated worldview and horizon for making judgments, the individual's development to maturity is hardly conceivable. But if these prerequisites are met, then the function of critical reason, of enlightened thought as a necessary modification becomes even clearer – as a corrective to remaining confined to the constraints of leading an everyday life, to persisting in the way things have always been, to becoming ossified in worldviews that are perhaps ideologically sealed off from reality.

There is no need to further elaborate on the stipulation that the continuation of the idea of enlightenment, as important as it has proved for Western thinking, 'incidentally' also promotes the aim of developing cultural competence. This can be formulated paradoxically as follows: Enlightenment – taken seriously in its status as a tradition – requires a commitment to thought and action that potentially disrupts tradition.

2.5 Developing a Willingness to Assume Responsibility

2.5.1 The Ethical Dimension of General Education through Schooling

In all of its historical variants, the idea of education and personal development has always exhibited an ethical dimension. Today, a common stipulation is that, in light of the pluralistic fragmentation of many formerly binding moral values and in view of the difficulty, if not impossibility, of still relating to an obligatory value-system, general education in schools must now abandon this ethical dimension.[40] Instead, schools should confine themselves to neutrally imparting subject-related competencies.

In my view, this argument cannot be upheld. For the assumption that there are no common values, that there is no longer any ethical basis in modern societies such as ours, is simply false. The only observation that can be made in this respect is that a number of values that previously were able to claim universal validity (e.g., Christian doctrines, obedience to elders, patriotism) have now been relativized: Either in the sense that they are now tolerated as guidelines of certain social subgroups (e.g., religious beliefs) or that their validity is now restricted to certain situational contexts (e.g., reliability, punctuality, performing one's duty in situations which also involve others – especially in professional life). But, generally speaking, disclaiming the absolute validity of values has itself been based on higher, humanitarian, and usually more abstract values such as tolerance, human dignity, and responsibility, to put it briefly: on those values which are referred to in the constitutions of modern democracies and in the Charter of the United Nations.[41] The widespread complaint of the disappearance of value systems and of the decline of ethical principles does not so much reflect an objective fact as it expresses something else: disappointment either with the circumstance that certain values of one's own social group have lost their binding character or with the fact that moral principles which theoretically (almost) everyone would recognize and demand adherence to regarding his or her own person (justice, truthfulness, respecting human dignity) are repeatedly violated in everyday life.

Not least, it was this second type of disappointment that led to a loss of credibility for the conceptual link between education and personal development in Germany after the Second World War. Never before had the gap between ideal and reality, the fragility of supposedly irreversible humanitarian achievements of a Western society become so blatantly visible as in retrospect of the Nazi era, in which precisely the members of the 'well-educated' classes became enthused about an ideology exhibiting flagrant contempt for human beings and allowed themselves to be prodded to unscrupulous actions.

But such disappointments are inadequate starting points for conceptualization. If one appeals to them in order to do away with the ethical dimension of schooling, then one has learned the wrong lessons from history. Wherever humans have dealings with other humans, moral conduct plays a role, whether this is consciously reflected upon or not. Schools cannot avoid the problem of moral criteria for such conduct by claiming not to be competent in the matter. The stipulation that instruction is always also engaged in educating in the sense that, in addition to imparting knowledge, it also participates in the development of criteria for conduct, is a venerable pedagogical insight that has now been confirmed and highlighted once again by recent research on classroom socialization.

The issue is not *whether* a school committed to a modern form of general education should adopt this ethical dimension, but, rather, *how* this is to take place – under which guiding concept and, since moral conduct cannot be produced by 'lecturing' on the subject, by means of which practical measures. Thus, the question is: Is there an ethical norm which under the present conditions is so universal and, in addition, so widely agreed upon that a modern form of moral education could be focused on it?

2.5.2 *Responsibility as a Fundamental Ethical Category*

The category of responsibility meets these requirements to a greater extent than any other ethical principle.[42] It is no sheer coincidence that the educational theorists of the 20th century repeatedly emphasized that responsible action was a distinctive feature of well-educated people.[43] For this reason, the socio-ethical aim of general education in schools can be described briefly as *developing a willingness to assume responsibility*.

Generally speaking, we can maintain that someone who deliberates the consequences of his or her actions and is willing to be held accountable for them acts responsibly. Conscious decision and subsequent practical action mean that the individual also runs the risk of experiencing failure and guilt. This is true on a small-scale level, with reference to private and professional actions in everyday life, but also on a global scale: In the course of human history, the radius of action for humans has incessantly become larger, especially as a result of developments in the natural sciences and in technology. Responsibility is the ethical corrective that limits the immensely broad possibilities of human action to an extent defined by human reason. It acts as a counterbalance to the ambition of realizing every conceivable possibility. From the perspective of an ethics of responsibility[44], actions for which no one can *assume* the responsibility cannot be justified.

The responsibility to be demanded of the individual is related, in the first place, to fellow humans,[45] but, in addition, it also refers to our natural environment. This complement to human responsibility beyond the bounds of human interaction in a strict sense has only been explicitly explored by contemporary philosophers – in response to the ecological threats and modern risks for the survival of humankind.[46] Responsibility does not only extend to the immediate care for others – although it certainly does hold a legitimate place within this sphere whenever parents care for their children or the strong provide for the weak. Yet, in a more general sense, responsibility stands for linking knowledge and expertise to an ethically justified attitude. For schooling, the principle of responsibility links the competencies to be acquired to their actualization: only that person who is able to responsibly employ his or her competencies is to be considered well-educated.

2.5.3 Responsibility and Willingness to Assume Responsibility in School

We call the attitude that enables us to act responsibly in a certain situation a sense of responsibility or the willingness to assume responsibility. In school, there are only limited possibilities for students to act responsibly in earnest. Nevertheless, schools can cultivate the willingness to assume responsibility as a value to be appreciated, can encourage it as an attitude. Responsible action can extend to fellow students, to articles or animals entrusted to one's care,[47] but also to one's own learning development. According to the traditional view, assuming responsibility for one's own learning reaches its culmination by qualifying for higher education, but, in certain provisional forms, it is not unfamiliar to the elementary school student. In addition to responsible *action*, which is limited in the protective sphere of the school, schooling can cultivate responsible *thinking* – e.g., with regard to those global issues that schools deal with in striving for their objective of developing an understanding of the world. But there is a risk involved here that results from the relative inability to take action on the part of students and their teachers: If it is a distinctive feature of responsibility that it can only prove its worth as an ethical disposition in active conduct, then the experience of this relative inability to take action can lead to frustration and resignation. This is a problem that should be discussed openly with older students, for it can neither be suppressed nor convincingly solved in a general way.

Not the least significant requirement of a willingness to assume responsibility is an emotional component, is being emotionally rooted in a sense of responsibility. If critical thinking represents the cognitive aspect of maturing, then the willingness to assume responsibility goes a step further. Both cognition *and emotion* are prerequisites of responsible action. From the perspective of ethical responsibility, which holds me accountable for my actions, a sense of responsibility alone cannot prevent me from acting irresponsibly – if I am too badly informed. In that case, an irresponsible action is a foolish one or one based on too little information. On the other hand, knowledge and critical thinking alone cannot prevent me from acting unscrupulously if I have no sense of responsibility. In this case, irresponsible action is evil, cynical, or, at the least, careless.

2.5.4 *First Objection: Can One Learn the Willingness to Assume Responsibility?*

At this point, we should consider a significant argument against a general education aimed at developing a willingness to assume responsibility. There is some indication that an all-embracing predisposition to conduct worthy of being called 'willingness to assume responsibility' does not, in fact, exist. Just from our everyday experience, we know that many people consistently act with an enormous sense of responsibility with regard to certain segments of their active lives and yet, in other areas, are foolhardy or irresponsible, and that a number of people allow themselves to be led to acting irresponsibly in particular situations, e.g., driving home under the influence of alcohol after a party. But the realization that willingness to assume responsibility is not some general psychological disposition that one either 'has' or not (and, if one 'has' it, then, for life) or that there is no automatic transfer among situations requiring responsibility or the willingness to assume it – this realization does not imply that the willingness to assume responsibility is simply an illusion and that education aimed at promoting it would be futile (the same applies, incidentally, to critical thinking). On the contrary, the conclusion to be drawn from this realization is, first of all, that, in a number of situations, taking on responsibility should be experienced as a possibility without explicitly discussing this theme and that, secondly, one should discuss 'responsibility' as an ethical category with older students if classroom subject matter offers appropriate links, e.g., with regard to the possible contradiction between the freedom and the social responsibility of scientists, writers, and artists in the light of possible consequences of their activity.[48] However, this should not be done in a pedantic, moralizing way, but in a dialogical and reflective manner, accompanied by the common realization that responsible conduct is necessary for leading a more humane life.

Accordingly, willingness to assume responsibility as a personal attitude must be able to develop in confrontation with many diverse situations and demands. And this is precisely the chance that schools have in promoting responsibility through education: offering such situations, providing opportunities to have personal experiences involving partially self-responsible action and for reflection on these experiences. The primarily cognitive orientation that the public school is often accused of having can provide significant momentum for the generalization of a willingness to assume responsibility: It is not an indication of profound moral maturity or anything extraordinary if I act responsibly and can be depended on to be concerned about something that is a major interest of mine in any case, as long as this interest is limited and, in a certain (positive) sense, egocentric. An example would be my behavior as an enthusiastic football player if I never miss practice and make my best efforts during the games and actively support my teammates whenever I can. This kind of conduct is 'natural' and can be taken for granted if I enjoy these activities and they give me a feeling of achievement.

At least indirectly, schools can help to broaden the student's moral horizon beyond the subjective, egocentrically defined sphere of experience – by creating an appropriate climate for learning, by model adult conduct that invites identification with it, by offering a variety of opportunities for action that require a personal commitment. A characteristic of such a broadened evaluative horizon is that I now relate to things outside of my naïve, egocentric sphere of experience, that things that I had practically ignored before now challenge me or at least do not leave me feeling totally indifferent. If general education aims at achieving the objective connected to this ethical dimension, then it must make me – as a student – aware of the fact that my values not only apply to my immediate surroundings, but (at least in principle) to everything that becomes accessible to my cognitive horizon: It is up to me to decide if it appears adequate in the light of the criteria informing my value judgments. A few examples to illustrate this point: This scrutiny – examining how I judge some issue – might be occasioned by a controversial court ruling that I have read about in the local newspaper; or by a public debate on whether subsidies for cultural events should be cut in favor of public welfare or lower taxes; or by a report in the media on starvation in the Third World and my own dilemma: should I donate money, demonstrate, or what? One feature of my willingness to assume responsibility is the inner necessity of such scrutiny and the resulting impetus for action that reflects my concern. Here, for the moment it is less significant whether and how as a student I might already be able to respond to a call to political action (in the broadest sense).

At this point, it is obvious to what extent the aim of developing willingness to assume responsibility is interwoven with those of developing an understanding of the world and promoting critical thinking. Indeed, it is dependent on them: Without a differentiated view of the world, without the ability to engage in criticism, and without a judgmental horizon that goes beyond one's subjective and contingent immediate circumstances, willingness to assume responsibility can be developed only in a very restricted sense and can hardly become politically significant.

2.5.5 *Second Objection: Can One Learn the Willingness to Assume Responsibility Under Prevailing Classroom Conditions?*

Repeatedly, a further argument against the objective outlined here has been implicitly touched upon. The argument goes as follows: The demand that schools should arouse a willingness to assume responsibility is unrealistic under everyday classroom conditions. This objection deserves more careful consideration.[49] The proponents of this argument often point out the unpleasant aspects of everyday classroom instruction, especially the depressing imperfections of the agents involved – in comparison to eminent ethical ideals. What sense does a noble aim such as 'development of willingness to assume responsibility' make if day-to-day schooling is concerned with students' drug consumption and addiction, with shirking mandatory performance, if for many adolescents school is simply a necessary evil that is to be resisted with any and all available tricks, if teachers are thought of as insensitive, incompetent, and unjust? What sense does such an aim make in view of the many teachers who do not do their job responsibly, for whom leisure time is more important than instruction and pedagogical efforts focused on students, whose professional group, in its entirety, includes representatives displaying every form of human weakness and every vice – including close-minded specialization and self-righteousness, mendacity, alcoholism, even criminal conduct?

Critical inquiries of this sort are based on that fateful confusion of a projected goal with reality so often found in everyday argument. But what *should be* cannot, in principle, be deduced from what actually *is*. Realistic aims seriously examine restrictive conditions of everyday life and take them into account, but they do not allow them to dictate what is worth striving for. There is certainly often a moral consideration behind critical inquiries like those above; but the questioner usually wants to avoid disappointments resulting from the fact that moral standards are often violated in everyday life.[50] Yet, relinquishing ethical standards is a misguided strategy for dealing with the imperfections of human conduct, of one's own and of that of others. Even cynics and those who have capitulated to existing circumstances are still, for the rest of their lives, compelled to rely on (in their view, perhaps negligible) traces of a mutual human association. These would be impossible if every individual did not at least have some vague idea of what is morally right or wrong. For this reason, I would like to confront the proponents of a public education devoid of values with the reverse argument: All attempts to leave moral standards and especially an orientation to responsible action out of the scope of schooling are blind to reality.

2.5.6 Opportunities for and Limitations of Developing Responsibility in Schools

The limitations placed on the development of attitudes such as willingness to assume responsibility by the prevailing conditions in schools should not be underestimated. There is no need to emphasize that neither purely academic instruction nor simple 'moralizing' appeals are very promising. Even credible model behavior of the teacher, a disposition convincingly exhibited by him or her, cannot guarantee that a corresponding disposition will develop among the observing students. To an even greater extent than with cognitive educational goals, steps taken in school aimed at the appropriation of attitudes should be accompanied by expectations that are prepared for disappointments. An attitude is based on the entirety of often contradictory experiences students have with each other, with teachers, parents, and with their contacts outside school and the family. The life-world of adolescents that forms their character is much too complex to be molded by educational intentions aimed in a single direction and with foreseeable effects. Yet, this certainly does not mean that it would be better to limit schooling to educational aims that can be operationalized and to abstain from the difficult task of promoting personal qualities such as the willingness to assume responsibility.

Nothing would seem more detrimental to the development of a willingness to assume responsibility in the context of the school than mendacity, the breach between officially proclaimed morality and the behavior exhibited by adults.[51] As a rule, children and adolescents are very sensitive to such a discrepancy. Not the *demonstrative* model is effectively convincing. If anything, it is being honest with oneself and about the moral guidelines of one's conduct, not least, admitting one's own faults, that can be effective. Thus, schools should become places where experiences can be made in such a way that an attitude like the willingness to assume responsibility, put into practice unobtrusively, would have a good chance of being acquired by young people as a matter of course.[52]

2.6 Practice in Communication and Cooperation

The demand that schools should provide practice in communication and cooperation selects two closely related types of the many kinds of social behavior and accentuates them in a particular way with respect to pedagogy. The general remarks on the relationship between functional and intentional education made above also apply here: It is self-evident that schooling familiarizes children and adolescents with forms of social conduct. Schools are constantly engaged in this familiarization within the context of their function of socialization. But it is much less self-evident that certain forms of social conduct should be especially cultivated and that pedagogic intentions should be directed at their realization because they represent a value of social and ethical dimensions. The cultivation of 'communication and cooperation' is indispensable for a modern general education in a democracy.

2.6.1 Terminological Considerations

By 'communication' I mean interactive behavior that is aimed at both understanding others and at reconciling conflicting interests and establishing practical mutual association. Communication is directed toward understanding the viewpoints and preferences of others, their thoughts and opinions; conversely, it is combined with an effort to express oneself, to allow others access to one's own ideas. Thus, communication is a fundamental prerequisite for a rewarding social life among people with equal rights and especially for conscious collective action, whether in a private or political context.

The relationship between communication and cooperation is one characterized by interaction. Communication is the more fundamental condition. Cooperation is more of a pragmatic category. Cooperation takes place whenever a collective effort is made toward a goal to which, in principle, all participants have agreed (explicitly or implicitly). Thus, communication is a prerequisite of cooperation. Conversely, successful communication is not necessarily followed by cooperation. But cooperation can play an important part in intensifying communication.

As with other aims of global dimensions, the idea of providing practice in communication and cooperation is not unsusceptible to ideological domination.[53] By implementing the concepts 'communication' and 'cooperation', which are relatively free of negative connotations in a historical sense, I am trying to avoid misleading associations.

Hardly anyone will doubt that it is worthwhile to make special efforts to promote communication and cooperation as forms of social conduct. The ability to communicate with others, to be tolerant and maintain contact even in the case of diverging interests and opinions; the ability to negotiate a practicable compromise on controversial issues; and the ability to work together with others on problems requiring solutions, to develop collective aims and to direct one's own energies and talents to such work – these abilities are not only of great significance in private and professional circles; they are also indispensable for democratic practice in all spheres of life. They have become more important than ever before in the light of present and future threats on a global scale – they can be conceived of as preconditions for the survival of humankind.

The question is whether and how schools can promote the development of these abilities, whether they can do this in addition to or precisely in connection with specialized instruction, whether schooling (in the light of its not pedagogically desirable, but socially unavoidable obligation to selection) does not become entangled in irresolvable contradictions and totally confused goals if it commits itself seriously to this objective.

2.6.2 Competition or Cooperation?

I assume that schooling's selective function, as described by a number of educational theorists in the preceding decades (Parsons 1965 [1959], pp. 125-154; Hurrelmann 1975; Fend 1974 & 1980, pp. 29ff), is a socially *necessary* function in a modern democratic society such as ours, based as it is on the division of labor. Those who think that schooling is only then pedagogically legitimate if it abstains from making any sort of selection cling to some fictional pedagogical vision (that has been recast time and again) and they fail to recognize the dialectical nature of all pedagogical action under existing social conditions. A pedagogical critique of the school's selective function cannot aim at its abolishment without any replacement, but, instead, at the critical inspection of its implementation, i.e., at demonstrating:

- at which points the personal development of the students is hampered by rigid, degrading, and injurious forms of rating achievement;
- at which points the inequality of social opportunities is intensified rather than offset (at least in tendency) by selective mechanisms;
- and finally, which counterweights to an overemphasis of selection and competition can be employed in day-to-day instruction in school.[54]

The principle of competition and achievement is familiar to children and adolescents from a number of social spheres outside of school. Among the early socialization experiences of young people is the realization that there are situations in which it is opportune to behave in a way appropriate to this principle. For this reason, competition is also often accepted by children and adolescents as an enlivening element in the classroom, especially if it serves to confer on learning situations something of the character of sporting events. A school that preferred to totally do without such possibilities for extrinsic motivation would deprive itself of an effective means of stimulating the desired learning processes.

Students often react adversely to an overemphasis on competition. If the pressure to achieve is thought to be inappropriately high, the tendency to 'cheat' in test situations increases; excessive competition finds its own corrective in a form of communication and cooperation that is not initiated as a pedagogical measure.

By contrast, communication and cooperation that is pedagogically designed is of a different sort. A general guideline for planning instruction would be the following: As often as possible, situations should be developed in which the students are naturally obliged to communicate with one another, in which such behavior can be experienced as worthwhile and satisfying; efforts should be made to institute a culture of instruction that encompasses and requires a willingness to communicate and cooperate as natural standards for behavior. This recommendation is based on the argument that interaction aimed at understanding and cooperation defuses and alleviates competition and thus allows for a more playful attitude toward classroom activity. Moreover, we maintain that students' intellectual and psychological barriers to the required academic achievements can be overcome if students are repeatedly given opportunities to realize that such demands are not the only significant matters in the classroom.

2.6.3 Learning Social Skills and Learning Subject-Related Skills: A Contradiction?

Since schooling became a matter of conscious reflection, it has often been maintained that there is a conflict between objectives relevant to learning social skills and those related to learning subject matter. One party – for whom Herbart with his conception of *instruction promoting personal development* can serve as an early prime witness (cf. Ramseger 1991) – has constantly pointed to the necessity that, in addition to imparting knowledge, schools be concerned with the social development of the students. The opponents contend that viewing socialization as a primary goal would unrealistically demand too much of schools and would endanger schooling's real objective, namely to provide the students with sound instruction in certain subject matter. But are these really contesting goals? Or, to put the question another way: Is it even conceivable that *subject-related* skills could be learned exclusively?

Much of what was said above in reference to 'developing critical thinking' and 'developing a willingness to assume responsibility' can also be applied to 'practicing communication and cooperation'. None of these three objectives is primarily concerned with demanding the teaching of certain instructional content, but, instead, argues in favor of treating subject matter in a certain way. Moreover, all three objectives are concerned with the development of a certain *attitude*. Finally, with reference to all three we should emphasize that their realization is not made possible by *adding* something to content-related learning. There is still a widespread misconception that learning social skills related to communication and cooperation must explicitly accompany subject-related learning. Yet, learning subject-related skills and learning social skills cannot be separated from each other. Even in the most rigid form of teacher-centered instruction, in which the teacher simply presents facts, the students learn some sort of social skills. To put it somewhat pointedly: Students learn social skills primarily by means of the way in which they engage in collaborative efforts to learn subject-related skills.

2.6.4 Communication and Cooperation Between Experts and Nonprofessionals

The problem of communication between experts and nonprofessionals can be considered as one of the key issues of an enormously differentiated democratic society based on the division of labor. On the one hand, our society's ability to function and the preservation of our vital resources are increasingly dependent on highly specialized knowledge and skills available only to a few experts. On the other hand, the specialized knowledge of the experts by no means guarantees its wise implementation. This applies to everyday life, e.g., whenever illnesses compel us to rely on the medical competence of trained physicians, as well as to the context of political action on a worldwide scale. Deciding which paths the developed societies can and must take to deal with the dangers of the ecological crisis and of the worldwide gap between rich and poor cannot be left exclusively to the experts. On the contrary, these are issues that affect everyone and should be decided by all citizens – in their roles as nonprofessionals who are considerably involved. The rationality of such decisions is largely dependent upon successful communication between experts and nonprofessionals. Being affected and being informed cannot be played off against each other. Establishing a meaningful relationship between the two is, not least, an issue for general education.

With this situation in mind, Theodor Schulze (1990, pp. 37-38) suggested making the "actively involved and, in the context of this involvement, competent nonprofessional" to a new model of modern general education and thus substituting a more appropriately democratic idea for the still predominant model of the 'well-educated scholar'. Schulze assumes that we are all nonprofessionals with regard to most of the issues that concern us and that we are experts only in a few cases. Nevertheless, as nonprofessionals we are often called upon to responsibly decide when we require the assistance of experts – physicians, attorneys, school guidance counselors – and whether we wish to follow their advice. But forming our own opinions requires that we can put the right questions to the experts and can question their judgments that affect us. This not only applies to our private life, but even much more to the formation of political and public opinion if, for instance, the issue at stake is the arms race, atomic energy, environmental or social policy.

Schulze reminds us that, conversely, there are certain areas where each of us qualifies as an expert, e.g., at our workplace. At the same time, we should keep in mind that with respect to most areas which they encounter all experts are also simply nonprofessionals. In principle, every adult in our society is familiar with both roles. Much confusion could be avoided if the authority of the expert would not be emphasized so often in cases where the expert forms an opinion simply as a nonprofessional.

This clearly illustrates how closely related the necessary communication between experts and nonprofessionals is to critical thinking and to treating knowledge responsibly. A successful dialog between experts and nonprofessionals requires both participants in the dialog to be well-educated. Then, the experts can look beyond the perspective of their own discipline. The nonprofessionals, in turn, can understand the problems whose solutions necessitate experts at least to the extent that they are not condemned to trusting the experts blindly.

Much too infrequently are we actually aware of the great extent to which the teacher-student relationship is a particular variant of the 'expert-nonprofessional' relationship. A teacher who is able to appropriately communicate to his or her student-'nonprofessionals' provides a powerful model for the socially desirable conduct between experts and nonprofessionals. Students whose questions, doubts, and uncertainties are taken seriously by teacher-'experts' can learn a great deal that is significant for their future social roles as nonprofessionals *and* as experts.

2.6.5 Intercultural Education

During the last few decades, the aim of promoting practice in communication and cooperation has become even more significant for German schools due to a further development: Increasingly, students from different nations and with different cultural backgrounds converge in the classroom. This development, which can also be observed – to a lesser degree, but with the same tendency – in other European countries, makes the issue of intercultural communication a pressing educational matter and one of a certain sociopolitical explosiveness. At least a few aspects of this theme should be taken into consideration here.[55]

In certain respects, the goal of intercultural communication is simply an intensification of what we might call *intra*cultural communication in order to point out the differentiation – an indispensable part of every class, regardless of how homogeneous the group might seem. Nevertheless, the tensions that can already develop between children and adolescents of the same nationality, but of divergent social milieus and religious beliefs (including various value systems and cultural preferences dependent on their familial socialization) become much more apparent among children and adolescents of different nationalities (with different cultures, religions, and native languages). Consequently, here, there is a far greater potential for conflict.

We have already discussed at some length the idea that the target cultural identity of the individual in a society that tends to be multicultural must not be a uniforming identity that excludes everything unfamiliar to it. But such considerations do not already provide a solution to potential conflicts. If fundamentally different value systems clash, no solution can be expected from an apparently tolerant cultural relativism which declares everything to be valid as it is. A rather extreme example to illustrate the point would be the openly patriarchal family structure in the Mediterranean countries (particularly in those with Oriental influence), which blatantly contradicts Western ideas of emancipation and can also find expression in the conduct between boys and girls in school. When conflicts occur, enormous pedagogical sensitivity is required to help the socially 'progressive' perspective to prevail without denigrating the sociocultural background of Oriental children as being generally backward.

As much as the presence of foreign children and adolescents makes everyday teaching a difficult task for educators, it also offers a chance to tackle the problem of communication with people of other nationalities and cultures in a practical way instead of just theoretically making appeals for such communication. In a world that is in the process of becoming increasingly interconnected, the relevant abilities will become ever more significant. This topic is of significance for specialized instruction at school since the various subjects – due to their content-related structure – offer various possibilities for dealing with cultural differences among the students. Regarding a relatively 'universal' subject such as mathematics, we can take into account which forms of presentation independent of language might help children whose native tongue is not the official language.

2.6.6 *Practicing Communication and Cooperation in Everyday Situations*

It is no coincidence that we speak of 'practicing' here. This refers to the fact that the realization of such social values as communication and cooperation is not only dependent on insight, good will, and reflection, but at least equally on habit and experience. For this reason, the most important means of promoting these goals in schools is the development of a social practice that produces such habits and provides opportunities for corresponding experiences. The following elements can be listed as important aspects of such a practice:

- Long a familiar demand of generations of educational reformers, but still rare in day-to-day schooling[56] are methodical arrangements such as project instruction and extracurricular activities in which coordinating efforts and cooperating with others are simply necessary and require no artificial, 'pedagogical' stimulus. Regardless of whether the activity is a craft, an initiative dealing with environmental policy, the staging of a play, or making music – this joint activity, backed by a collective desire for an external effect and for success (in the broadest sense), relies on the practical realization of communication and cooperation more or less as a natural precondition.
- While such activities and projects always have the flair of the unusual and extraordinary (at best, students even forget that it is a 'school activity'), for the foreseeable future adolescents will continue to accomplish the major portion of their schooling within the context of traditional instruction: in the classroom, oriented to lessons, and with time periods divided into hours. Here, the demand for a practice promoting social conduct points to the development of a culture of specialized instruction that does not interpret subject-related skills and social skills as contradictory, but, instead, makes them mutually productive – by consciously disrupting all too monotonous rituals, by varying forms of social conduct and increasing cooperative work phases, by giving learners the opportunity to articulate their original subjective views and those they develop on the topics covered and to exchange such views with other students.

These ideas also express the hope that such school practices will have effects beyond the school situation. But a warning of excessive expectations is in order here: The history of pedagogy can be presented as the history of disappointed hopes for transferring knowledge from one area to another. Realistically, one can assume – and all of the empirical data point in this direction[57] – that students can utilize what they have learned in situations that are similar to those in which they acquired their knowledge. In a different social context, tolerance toward Turkish classmates can turn into xenophobia; cooperative conduct practiced at school can turn into dog-eat-dog behavior in professional life. Even in structurally similar situations, humans are (fortunately) free to react differently from the way they have learned to and from how their teachers would perhaps want them to. But an appropriate skepticism toward one's own pedagogic ambitions does not nullify that venerable experience of educators that credible social practice presents much better preconditions for the transfer of a willingness to communicate and an ability to cooperate to other areas of life than any sort of moral exhortation.

2.7 Enhancing Students' Self-Esteem

The demand that schooling enhance students' self-esteem brings a final specific accentuation to our concept of general education. All of the aims of schools discussed up to this point can be interpreted as demands with which society confronts adolescents – through schooling. These demands are, indeed, accompanied by the claim of representing the students' own interests, promoting their personal development, and providing this development with a foundation as broad as possible. Specifically, the demand for developing critical thinking as a prerequisite of the maturing proccss obviously benefits the students. Nevertheless, the social aspect of the issue of general education has remained predominant up to this point.

If one considers general education against the background of the dichotomy of "directing or permitting growth" (Litt 1927), then the demand for enhancing students' self-esteem leans more toward the pole of permitting growth. That pedagogical tradition interested in pursuing its goals from the perspective of the child, in developing the abilities 'inherent' in the child (and adolescent), is preserved within the context of this objective for schooling. This objective also retains the insight that, whatever their good intentions, pedagogical measures purporting to represent adolescents' interests and aspiring to realize those aims of general education discussed in detail above cannot simply be forced upon the students. Responsibility as an ethical principle, communication as a social one, critical thinking as a personal and intellectual one – all of these require a personality that comprehends oneself as a human subject, acts consciously, develops the courage to stand up for one's own convictions, and can become committed to such values.

By considering the enhancement of self-esteem as a pedagogical objective, I have no intention of encouraging egoism or narcissism on the part of adolescents, so often lamented by the elder generation, or of encouraging one of the fashionable misconceptions of 'self-realization'[58]. Enhancing self-esteem aims at developing self-awareness, self-confidence, personal identity, at the ability to clearly recognize one's own goals, inclinations, and ideas, to take action to realize them, and to realistically deal with one's own strengths and weaknesses. Just as 'critical thinking' represents the cognitive aspect of the maturity aspired to, and the 'development of a willingness to assume responsibility' represents its ethical bond, the 'enhancement of self-esteem' is directed toward the affective requirements of the maturing process. For schools, this is something to be achieved in more of an indirect way – a characteristic this pedagogical objective shares with the last few discussed above. The important thing is to create a curricular and organizational framework which provides room for personal development. Also essential are the reduction of institutional properties of schooling detrimental to the pedagogical objectives and the willingness of the teachers to seriously regard their students as independent persons – beyond the scope of any bureaucratic constraints.

Such concepts as 'ego', 'self', 'subjectivity', 'identity', 'individuality', and 'personality' are the focus of a great number of competing pedagogical and philosophical, psychological and sociological theories which illuminate the various facets of 'enhancing self-esteem' as an educational issue. Even an overview simply intent on listing and summarizing the most significant of these theories would require a book in its own right.[59] For this reason, in order to outline this goal of enhancing self-esteem, I will restrict my considerations to three aspects, each of which has been elaborated in a number of theoretical writings:

- The idea that children and adolescents are to be protected from societal domination and that their current life situations are not to be sacrificed for some future desired by the elder generation, can be considered to be a pedagogical insight of 'classical' import.
- The second aspect is more strongly oriented to the psychological component: In order to lead a fulfilling life, the individual requires a psychological equilibrium, a sufficiently sensible way of dealing with one's own feelings and drives, as well as with (perhaps already internalized) external demands.
- The third aspect refers to the social dimension of the development of the ego, to the realization that self-identity can only be acquired by obtaining a certain balance between personal and social identity. This sociological perspective has been primarily developed within the context of theories of interaction.

2.7.1 Enhancing Self-Esteem as a Safeguard Against Heteronomy

The modern age in Western society is characterized by a typical, still unresolved – and probably irresolvable – tension: On the one hand, since the end of the Middle Ages, the status of the individual has been enhanced in a way unique in the entire history of humankind. Substantial formulations of ethics, as in Kant, developed a concept of the individual's inviolable human dignity. The idea that the individual was an end in himself gained widespread acceptance. On the other hand, the breakdown of the traditional bonds of the individual, the continuous secularization of all social spheres, has been accompanied by increasing possibilities of planning the individual's life, of using him or her for purposes external to himself or herself: for economic, political, ideological purposes, etc. The dialectical relation between these two trends, their mutual relation to one another as trend and counter-trend is unmistakable: the more widespread the belief in human determination of human conditions became and the more effects this had on the world and on humans themselves, the more humans came to realize that – precisely as individuals – they were left on their own, exposed to danger and in need of protection.

Correspondingly, the more influential became a type of philosophical and ethical reflection which placed great emphasis on the fundamental right to life for every individual, on the right to personal development, and on the necessity of protecting the individual from unjustified domination and encroachment on these rights.

Since the Age of Enlightenment, this discord has also found its way into pedagogical debates. For utilitarian thought, which reached its peak during the Enlightenment, it was no great matter to demand an education of the individual which promised to form him for an ideally projected society in an optimal way, to make him 'useful' to society and the state. At the time, no one opposed such a plan more intensely and with greater historical repercussions than did Rousseau. His counter-utopia was based upon protecting the individual from ostensible societal deformation.[60] Rousseau's own demands were based upon an anthropological model that attributed everything evil in human life to the existing social conditions, everything good to not yet deformed predispositions of the individual: "Everything is good when it springs from the hands of our Creator; everything degenerates when shaped by the hands of man" (Rousseau 1978 [1762], p. 9). This fundamental idea has been taken up time and again – up to today's counter-pedagogical movement, especially apparent in the work of, e.g., Alice Miller (1980): The individual becomes evil by being tortured, constrained, oppressed, prevented from satisfying his or her real needs or from living life to the full by others. If this vicious circle could be overcome, the 'good' individuals, having achieved self-realization, would go about establishing a 'good' society.

In his major works, Rousseau was a radical educational theorist. Most of the educational theorists of the Enlightenment and classical periods held a more moderate position, as, indeed, the very concept of education and personal development manifests a goal that seeks to establish a middle ground between society and its needs on the one hand and the needs and interests of the individuals to be educated on the other. This mediating aspect inherent in the concept of education and personal development becomes apparent if we ask the following question: How can the rights and the dignity of the individual, how can his or her interest in an intact personality be preserved in the context of schooling without at the outset disqualifying society's interest in this education as illegitimate or fundamentally harmful? Among the classical educational theorists, practically no one expressed more profound insights on this topic – still seeming quite modern today – than Friedrich Schleiermacher.

In Schleiermacher, there is a special emphasis on a thought already touched upon by Rousseau, but which now becomes developed much more dialectically.

> "Life activity, being related to the future, must also be satisfied at the present time; thus, every pedagogical element which as such is related to the future must also provide satisfaction to man as he is. The more both aspects penetrate each other, the more morally perfect pedagogical activity will be. Yet, both will penetrate each other all the more, the less one aspect is sacrificed for the sake of the other." (Schleiermacher 1983 [1826], p. 48)

These statements express a program that was only to be concretized in concepts of education and schooling much later, e.g., in 'child-centered' education, in Dewey's concept of school as a place to gain experiences, or, to mention a more recent variant, in the principle of 'student-centered teaching'. Up to the present day, one major characteristic of almost every alternative form of schooling has been to attempt to distinguish itself from traditional schools in this way: by integrating future-related curricular goals into enjoyable activities geared to the students' needs, by taking students' interests into account in the selection of subject matter and in lesson planning, by allowing students to participate in pedagogical measures.

Conversely, increasing disinterest in the classroom and school-fatigue[61] do, indeed, indicate violations of Schleiermacher's principle: If students no longer succeed in integrating school experiences into the rest of their lives and in comprehending school as a meaningful part of their lives, then, the present begins to become 'sacrificed' for the sake of the future. This diagnosis is not invalidated by the fact that students' freedom in the classroom has actually increased to an extent that would have been inconceivable in Schleiermacher's time and that today's discontent with schooling is related to increased expectations on the part of children and adolescents. Schleiermacher explicitly refers to 'satisfaction' to be brought about by pedagogical elements; undoubtedly, satisfaction is a subjective category the content of which is subject to historical transformation.

In sum, one aim of schooling is to provide children and adolescents with a place where they can live in conformity with their present needs and interests, where they are respected as independent individuals, as full humans with a corresponding personal dignity.

2.7.2 *Enhancing Self-Esteem as Support for Realistic Behavior*

The terms 'self-esteem' and 'weak self-concept' originally come from the psychoanalytic tradition, even though they have long diffused into ordinary language and have lost something of their conceptual clarity, as often happens in such cases. Since enhancement of self-esteem is a major element in the concept of general education under discussion here, the widespread associations of the vague everyday understanding of the term are to be taken into account. To clarify the concept, one can start with the main idea from psychoanalysis – regardless of one's view of psychoanalysis as a whole. This main idea also contains stabilizing elements for a modern pedagogical perspective.

Freud contrasts a human's 'ego' with the 'id' and the 'superego'. Accordingly, the 'id' represents the entirety of the (mostly unconscious) drives. The 'superego' represents internalized (also often unconscious) ideal norms to which behavior is oriented. Thus, the superego encompasses that normalizing element which we call the conscience in ordinary language. The 'ego' is that element which we perceive as the center of our personality in conscious action. The 'ego' operates on the reality principle; it relates our wishes and impulses to act to the demands of the real situation. If it is functioning 'normally', it helps us to avoid a self-defeating excessive fulfillment of our drives by giving reasonable consideration to the consequences of our actions, but it also helps us to avoid frustrations which would arise from completely refraining from drive-fulfillment. To put it briefly, then, the psychologically healthy human who is not 'neurotic' has achieved an equilibrium between 'ego', 'id' and 'superego'.

Psychoanalytic theory speaks of a 'weak self-concept' if the ego is not able to operate on the reality principle. The goal of psychoanalytic therapy is to reengage this function by enhancing self-esteem. A personality with a good sense of self-esteem is characterized, among other things, by being able to resist external captivation and temptation in such cases where succumbing would violate one's own principles. In "The Authoritarian Personality", Adorno (1973 [1950], p. 52-53) took recourse to this personality model and interpreted the willingness of so many Germans to betray moral principles under the pressure of external circumstances during the Nazi era ('acting under orders') as a 'weak self-concept' in the sense described above.

There has been no lack of attempts to make the aim of enhancing self-esteem contained in the psychoanalytic model of personality productive for the analysis and solution of educational problems. From his experiences in dealing with extremely difficult children, Austrian-American Fritz Redl (1951, German trans.: 1979; 1971) developed a concept that still holds a certain fascination today. Some of his insights can also help us to better understand problems with 'normal' children.

Redl attributes most of the difficulties that educators experience in dealing with children and adolescents to various forms of 'ego disturbances' (1971, pp. 27ff; 1979, pp. 75ff). Such disturbances can manifest themselves in the "inability to cope with aggressions arising from frustration" (1971, p. 28) or in the "loss of self-control effected by group exhilaration" (ibid.), in a "seduction through infection". Redl's interpretations are focused on a hypothesis supported by a great number of clinical observations and stating that a disturbed ego cannot fulfill its natural functions, that it is too weak to exert reasonable self-control over the excessively powerful domination of instinctual drives. For this reason, the pedagogical measures Redl took in dealing with extremely aggressive children who were filled with hate and apparently prone to acting totally amorally, aimed at strengthening their egos in a specific sense. Their goal was the step-by-step (re-)learning of a minimum of self-control which was supported by a 'therapeutic milieu', by the attempt to establish positive social bonds. Worth considering and also quite plausible in view of the theoretical reflections involved is Redl's experience that a total release from control, allowing the children to do as they pleased, did not at all lead to regaining self-control and personal autonomy. Apparently, certain external rules and social obligations are necessary for the development and maintenance of the essential quality of self-esteem.

Certainly, such experiences stemming from extreme situations cannot be applied to everyday schooling without any sort of modification. Nevertheless, the inner structure of 'normal' school problems such as lack of discipline, verbal and physical aggression, and taking pleasure in destruction might reveal itself much more clearly if we consider the situation of children with severe ego disturbances. Without wanting to play down the social causes of such phenomena, and, in particular, without wanting to make a premature decision on whether the causes behind usual school problems are internal or external to the school, we can maintain the following: On the basis of Redl's observations, there is a high degree of plausibility in the stipulation that a 'weak self-concept' in the sense of a lack of psychological equilibrium does, indeed, play a significant role, and that, for this reason, enhancing self-esteem in the sense of efforts supporting (re-)gaining such a balance is an indispensable pedagogical objective.

These considerations allow us to refute the following argument that appears to derive its soundness from ordinary usage of the word 'self-esteem': An aggravated teacher might remark, "There may certainly be some shy and less self-confident students whose self-esteem should be enhanced, but why should I enhance the self-esteem of a student who is impudent, obstinate, and offensive? On the contrary, every possible means must be used to put such a student in his or her place!" No further comment is necessary. But this misinterpretation should demonstrate clearly how far removed the concept of individual enhancement of self-esteem under discussion here is from that of a primitive pedagogy aimed at breaking the 'evil' self-will of the child. This idea is still perceptible in the fictitious objection mentioned above.

2.7.3 *Enhancing Self-Esteem in Support of Determining One's Identity*

The concept of identity has found its way into pedagogical debate, most especially in connection with the writings of George Herbert Mead and his theory of symbolic interactionism. It has proved to be an integrative concept, allowing for a useful description of the individual's development within a social context as well as for an adequate outline of the tension between individuals' interests and societal demands. Although the literature does include important attempts to relate the concept of identity to psychoanalytic theory (e.g., Erikson 1966), I will confine myself to an integrative discussion of new aspects of research, gaining insights from several distinct theories.

The human search for unmistakable, personal identity is a dilemma that can be expressed in terms of the opposition between 'being like everyone else' and 'being like no one else'. The belief that 'I'm unique' – especially widespread among adolescents – can make me just as unhappy as the opposite assumption, that in a fundamental way I am replaceable and that thus my personal life makes no sense.

The variations of the concept of identity inspired by interactionism interpret this basic dilemma as an opposition between social and personal identity. I acquire social identity in my various social roles, with which I respond to the societal demands of my environment. In the final analysis, my social identity is a combination of all of my partial social identities, which are attributed to me from my various roles. My social identity, which is connected to certain external expectations, contrasts with my personal identity, which I become aware of through inner impulses, desires, and perhaps interests diverging from the established norms. I can only achieve a stable identity that involves self-confidence and a positive self-concept if I succeed in establishing a balance between my social and personal identities. This sort of balanced identity is never a fixed property, but remains somewhat precarious, something that must constantly be attained and secured anew. In this context, there is even occasional talk of the "identity work" to be done.[62]

There is a (perhaps only fairly approximate) correspondence between the 'social identity' of the interactionist personality model and the 'superego' of the psychoanalytic one. In the latter case, the focus is on the external demands and social norms that have been internalized by the individual. In the former case, these demands and norms are practically left outside: They remain within the social reality that medially surrounds the individual. The pedagogical consequences implied by these two perspectives diverge. Enhancing self-esteem in the sense of bringing about a psychological equilibrium primarily describes the solution to a private problem – with external social or pedagogical support. Enhancing self-esteem in the sense of bolstering the process of establishing one's identity demands an appropriate formation of a social sphere of experience that helps the individual to attain a balanced identity. It seems pointless to play these two perspectives off against each other and to consider the theories behind them to be totally divergent in their ability to adequately describe reality. I see both perspectives as generalized interpretations of distinct types of problems and situations. These interpretations can complement each other in teaching practice since teaching itself does indeed involve both types of problems, yet often fails to include either strategy of enhancing self-esteem.

2.7.4 Promoting Self-Esteem and a Balanced Identity in the Classroom

'Being like no one else' – adolescents can realize this aim in school in various ways: by outstanding academic performance, but also by a refusal to perform as required; by model cooperative behavior, but also by sabotage and aggressive actions directed at persons and objects; put briefly, by conforming to the official institutional norms, but also by opposing them. Daily encounters in class with a group of children and adolescents of (almost) the same age demand, in some way or another, strategies of self-assertion, the confirmation of one's own identity. Underlying such strategies is the conscious or unconscious desire or urge to make a special contribution to the group – something specific and unique.

The development of our system of general education has often seen a very restricted and one-sided interpretation of the assertion of the general validity of subject matter, i.e., the stipulation that first and foremost students should learn things that are *relevant to all* of them, thus something that is the same for all of them. This aspect seemed to legitimate the questionable ideal of group learning at the same pace and, in extreme cases, has the result that students can only secure their own identity at school in categories of 'better' or 'worse', at the most, differentiated according to subjects ("I'm better at math, but then worse at English" etc.).

'Open' Forms of Inner Differentiation

General educational measures directed against uniform learning in class and designed to pay more attention to the differences among the students are usually described with the use of such concepts as inner differentiation and individualization. Of course, we cannot simply presume that such measures will in themselves already benefit the students' establishment of their identity and enhance their self-esteem. Individualization can also lead to isolation – if, for instance, individually working out a pre-determined learning program, with or without a computer, is considered its ultimate form. If inner differentiation is exclusively oriented to academic achievement in a certain subject, it can even intensify the one-sidedness of the opportunities for establishing one's identity in school that was criticized above.

Everyday educational practice rarely exhibits convincing realizations of inner differentiation. The term itself causes an emotionally charged reaction among many practitioners. Reasons for this rejection can be readily identified: There is a lack of adequate instructional materials; too much time is required for preparation; realization in the classroom proves too nerve-racking.

Yet, this sort of evaluation is often based on a concept of inner differentiation that could be termed a 'closed' differentiation (cf. Heymann 1991, p. 64-65). This is the idea that, taking a student's abilities and learning level into account, a curriculum tailor-made for this student should be assigned to him or her by an external source. For this type of inner differentiation, the responsibility for the individual student's progress is primarily in the hands of the teacher (or of the person who designed the learning program). It is not difficult to understand why the required corresponding instructional organization for classes of thirty or more students is considered to be an unreasonable demand.

The opposite ideal type, 'open inner differentiation', demonstrates other characteristics (ibid.): Open differentiation requires that the students themselves learn to assess their own strengths and weaknesses, that they learn to determine which type of learner they are. Within learning settings that provide adequate opportunities for learning, that appeal to various types of learners, and allow various possibilities for becoming interested in the subject matter, the students themselves can determine their own emphases and make selections according to their own preferences and abilities. The crucial point is: Where instruction is oriented to a model of open differentiation, the students themselves assume the responsibility for their own learning process. In such a case, the form of interaction is important, but not decisive. To mention an extreme example: A didactically well-planned lecture by the teacher that contains various examples, provides impulse for further thought, offers broad possibilities for identifying oneself with what is presented, and activates various modes of perception can adequately realize open differentiation.

At the outset, forms of open differentiation relieve the strain on the teacher – not necessarily with regard to the required didactic creativity, but at least in view of the preparatory work and the pressure involved in carrying out the actual lessons. A further compelling argument is as follows: As a measure of formal education designed to help students establish their own identity and enhance their self-esteem, open differentiation offers considerably more advantages than closed differentiation. In actual teaching practice, elements of both types of differentiation can be combined (cf. Krippner 1992).

Specialization as an Element of General Education

The possibility of developing expertise with respect to certain subject matter or practical abilities and to contribute to lessons with this special ability provides students with a much greater potential for establishing their own identity than under the standard conditions in school, which are centered around making one's mark by good academic performance. If I can become helpful to others with my own expertise or particular abilities, the status of my personality is enhanced in a way that is not primarily based on some standardized, competitive form of 'being better', but on a socially acceptable form of 'being otherwise'. This experience is also the basis for the influence exerted on the self-concept of adults by competent and enthusiastic performance in a socially acceptable profession, promoting their sense of identity and enhancing their self-esteem.

While older students can be given opportunities for such interest-oriented specialization by presenting papers, participating in projects, or even to the point of becoming absorbed in a rudimentary form of independent scientific or artistic work, for younger students – those in elementary school or at the lower grade levels of secondary school – the focus is on other forms of participation. A great deal depends on how successful the teacher is in relating instruction to the experiences and interests of the students apart from school, in involving what is important to them, their subjective insights and intentions, and interconnecting these to the 'official' subject matter. The more the school becomes a place for experience that is accessible to the students' experiences outside of school, the more it offers them opportunities to make individual contributions and to establish their own identity.

Thus, general education, as we understand it, by no means excludes specialization. However, 'specialist instruction' at schools would not be compatible with the idea of general education itself. 'Specialist' instruction would demand that the students preoccupy themselves with matters that only seem significant from the perspective of a singular discipline or of a specific profession. But prescribing a uniform specialization for *all* the students is to be clearly distinguished from allowing for the varied specialization of *individuals*. This latter sense of specialization can become a significant element of general education. If students are given the opportunity to intensely pursue their inclinations, to nurture their talents and abilities, then the process of establishing their identity will be supported meaningfully by schools.

Fostering Creativity and the Use of One's Imagination

Enhancing the students' self-esteem requires that they are provided with sufficient opportunities to develop their imagination and creativity. The entire history of the critique of schooling is permeated by the rebuke that compulsory education has a stupefying effect and stifles the intellect. In principle, the necessity of counteracting such effects, of allowing for students to develop their imagination and to experiment with their creativity, is no longer a controversial issue. Despite the limitations placed on the efforts of schools in this respect, this objective has become increasingly significant in the awareness of the general public – not least as a counterweight to the wide variety of opportunities for passive consumption in leisure-time activities.

What does this mean in a concrete sense? During childhood, pupils enjoy developing their imagination in a variety of ways at play. Playful activities, especially role-playing, offer students opportunities to imaginatively relate self-presentation and self-discovery to the discovery of the world around them and to social interaction. Small-scale research projects can provide older students with a foretaste of scientific creativity. Artistic activities ranging from drawing, painting, photography, designing, making music and enacting plays, up to making movies and writing ambitious texts provide students of all ages with opportunities to link their creative self-expression to the appropriation of the widely diverse aesthetic dimensions of their cultural tradition. The more actively the students participate in such activities, the smaller the risk is that traditional 'high culture' will simply be reduced to an object of familiarity and enjoyment for an educated elite. Thus, by promoting such activities, schools can equally take steps to realize their objective of developing cultural competence as well as that of enhancing self-esteem.

Corporeality – Sensualness – "Head, Heart, and Body" (Pestalozzi)

In the last few decades, these and similar catchwords have often been employed as critical concepts of internal educational reform.[63] Within the context of the concept of general education under discussion here, and, more specifically, with regard to the objective of enhancing self-esteem, the significance, legitimacy, and scope of such phrases can be determined more precisely.

In my opinion, these terms are not substantial enough to designate the core of a new conceptualization of schooling: They cannot replace a definitive scheme of general education objectives for schools. Nevertheless, they do correctly indicate actual deficiencies and suggest necessary corrections to the traditional understanding of instruction.

The motive common to educational critics who argue along these lines can be found in a rejection of the too intellectually-oriented school, devoted (almost) exclusively to cognitive learning. The critics maintain that this one-sided orientation to cognitive achievement diminishes the students' ability to perceive themselves as complete entities. Often, the contention is that this risk is all the greater

- because the quality of experience in surroundings apart from school is also constantly decreasing,
- because adolescents now perceive the world around them only through technological media – especially through television,
- because active orientation in the context of the interplay of the senses and demands made on one's entire person (consisting of body and mind) are becoming less and less important for coping with everyday life in our highly industrialized society.

If one concedes that general education should be oriented to the 'entire person', then such arguments cannot simply be dismissed. On the other hand, here, too, it would be demanding too much of schooling and overestimating its actual potential if one expected that it would, above all, compensate for the loss of sensualness in our so highly civilized everyday life.

A further aspect should also be taken into consideration. One of the historical achievements of schooling is that it enables children and adolescents to learn in a systematic and abstract way that is unique in comparison to all other societal spheres. In a sort of time-lapse, schooling permits an acquisition of knowledge in fundamental areas that, outside the context of such systematic courses and then limited to learning by participating in activity, would only be possible for very few people. There is no reason to assume that the loss of sensualness in life apart from school would automatically be accompanied by the type of systematic and abstract learning so typical of traditional schooling. On the contrary: The flood of information that in our society overwhelms children, adolescents, and adults through printed media, radio, 'canned music', television, videos, and computer games is to a great extent fragmentary, creates sensory illusions (illusory worlds produced technologically), numbs the senses (listening to music at a deafening volume), and dulls our judgment (advertising).

Thus, in the worst case, if one drew radical conclusions from these critiques for restructuring the public school, one would deprive it of its specific achievement, of systematic, symbolically mediated learning processes aimed at abstraction – without any assurance that some new, more spontaneous form of experiential learning involving the body and all of its senses would be able to effectively counteract the fragmenting tendencies of our civilization.

There must, then, be other consequences to be taken from the legitimate critique of the tendency to place too great an emphasis on the intellect. In my opinion, the two most important ones are as follows:

- Systematic learning and learning related to attaining higher levels of abstraction is not to be equated with learning that is too exclusively oriented to the intellect. Systematic courses can also involve (e.g.) the senses; instead of boring handouts and information written on the blackboard, they can employ evocative and aesthetically interesting pictures and demonstrate creative (not simply dryly informative) linguistic practice. Systematic courses can also involve more of the student's physical capacities than simply the ability to remain seated and to use one's writing hand. Integrated and systematic learning are not mutually exclusive.
- In addition to systematic learning, schools must offer opportunities for spontaneous, playful, bodily oriented activities – not so much in order to compensate for general trends of our civilization, but more to establish a counterweight within the school setting to the (still) unavoidable practice of sitting at attention. The generally pronounced popularity of sports among the students indicates that, within its traditional spectrum of subject matter, schooling has, indeed, managed to find outlets for the excessive pressure from orientation to the intellect.

Summary

The educational objective of enhancing self-esteem implies that students are to be taken seriously as persons, that they are to be given opportunities to involve their particular interests and develop their specific abilities, and, finally, that an integrated type of learning that also takes their various senses and their physical nature into account should be provided. All of these demands are not to be understood as absolutes, but as orientation parameters for an internal and external educational reform in small steps which can also be realized particularly in specialized classroom instruction.

2.8 Concluding Remarks on the Scope and Status of this Concept of General Education

With this detailed discussion of seven objectives for schooling, I have been outlining how general education in school can manage to mediate between the necessities of life in present society and the individual's right to self-realization and to responsibly taking charge of his or her own life. The dominant theme was the idea that general education should achieve a social universalization of culture, that it is to be thought of as the prerequisite of individual personal development. The double aspect already emphasized by Schleiermacher (i.e., 'becoming capable' of life in society, but also contributing to society's 'improvement', only possible by means of the self-realization of the individual) is preserved in this concept of general education and reinterpreted in the light of present conditions.

In conclusion, I would now like to illuminate once more the inner cohesiveness of these seven objectives – their mutual interconnections – by relating them to three dimensions. One demand that is indispensable to any human society, namely, the *ability to participate* in already existing states of affairs (here, above all: preparation for later life, the promotion of cultural competence) is linked to a concern that has only become a significant issue of general interest in modern societies, namely, a *cognitive ability* that transcends one's immediate affairs (combining material knowledge of the world around us with judgmental abilities: developing an understanding of the world and developing critical thinking). The final aspect of this concept of general education, in certain respects also a counterweight, is a dimension that could be called the *development of humaneness* (promotion of a willingness to assume responsibility, practice in communication and cooperation, enhancing self-esteem) and that once again accentuates particular socio-ethical and personal elements of our culture.

The concept that has emerged from this discussion of the seven objectives is a framework for the orientation of educational theory involving suggestions of what a school should offer to all students and how it should do this. It states more precisely what must be taken into account if the public school is to provide preconditions of the individual's personal development and his or her reflective participation in social, economic, and cultural life.

Thus, the seven objectives do not describe some educational ideal. They allow for various, even competing ideals of individual life-styles. General education in school cannot guarantee an active, self-fulfilling adult, much less 'produce' one. The logical status of these seven individual aspects, their designation as *objectives* of the school and not, for instance, as characteristics of a 'well-educated personality', corresponds to my interpretation: General education is simply to be understood as the socially universalized premise of individual personal development.

This means that despite the coercive means with which the state can at least ensure compulsory education, a general education cannot be forced upon the individual. Rather, general education is to be understood as an invitation that may or may not be accepted by the individual person. If through schooling the adult world makes demands that are too far removed from what children and adolescents desire or demand for themselves, then they will refuse this invitation.

The starting point for many concepts of general education is a categorization of the cognitively accessible world, of spheres of human life, or of the sciences (e.g., Derbolav 1975, 1976; Nicklis 1980; Westphalen 1982; Wilhelm 1982, 1985). Usually, such a categorization can be transformed more or less directly into a theoretical justification of an existing or alternative canon of subject matter. For good reasons, I have abstained from such a categorization for the concept of general education under discussion here. It would only detract from the intention of using the idea of general education as a critical measure of specialized instruction. For if I presume *a priori* that a certain segment of reality is to be given special consideration within the context of general education, then I am at a loss to formulate a fundamental critique of a discipline that claims to represent this segment.[64] Formulated pointedly with a view to the next chapter: A fundamental critique of the conventional type of mathematics instruction can only be developed if at the outset there is no presumption that mathematics (or mathematical thought or the accessibility of the real world to mathematical interpretation) is a necessary component of general education. Perhaps it is even possible to become 'fully human' in our society without learning mathematics[65] – this should, at any rate, be a permissible hypothesis.

The theoretical frame of reference constituted by the seven objectives described above can be used at various levels as a critique of actual conditions in our schools and classrooms and for the discussion of alternatives. Proceeding from the more comprehensive to the more specific and concrete, we can list the following: the organization of the school system; the structure of the canon of subject matter for general education (with reference to various types and levels of schooling); the selection of content and methods within existing subjects (curricula and syllabi); the design of lessons in view of specified syllabi. Intersecting these levels is a further one: that of (specialized) teacher education programs. Even if a related more comprehensive level is simply assumed as given, significant statements with reference to the seven objectives described can be made for every subordinate or auxiliary level. Since at every one of these levels the most fundamental issue of schooling is of some relevance (i.e., decisions are also made on the content and method of public school instruction), this frame of reference meets, in principle, the requirements that were placed on it. It explicates the idea of general education as a criterion (or, more precisely, as a set of criteria dialectically interrelated to one another) that is applicable to instruction and to the curricular specifications for instruction – a criterion that, in particular, also permits a critique of specialized instruction. In other words: The seven objectives of schooling, as they have been discussed here, concretize general education as a qualitative standard by which specialized instruction and the requirements of such instruction can be judged.

CHAPTER 3

MATHEMATICS INSTRUCTION FROM THE PERSPECTIVE OF GENERAL EDUCATION

Whether mathematics counts pennies or guineas, whether rhetoric defends truth or falsehood – both disciplines are totally indifferent towards this question. ... Mathematics cannot do away with a single prejudice, cannot lessen idiosyncrasies or placate partisanship; it cannot resolve a single moral issue.[66]

Johann Wolfgang v. Goethe, around 1820

Computation is a way of dealing with numbers that mathematically alert children, with a small amount of guidance, can understand fairly well by exploring it themselves. But in the face of the prevailing lethargy among adults as well as among children, computation was usually a subject for rote memorization in school. The paradoxical situation arose that the only school subject that could truly be thoroughly understood (empirical laws of nature, vocabulary, grammar, social structures must usually first simply be assumed as factually given) became the subject matter for noncomprehending rote memorization par excellence.[67]

Carl Friedrich v. Weizsäcker, 1974

If, some years ago, one out of ten drivers still drove too fast, now 'only' one out of five does this. But even five per cent is too much, and thus, the controls persist, and the fast drivers will continue to pay fines.[68]

Norderneyer Badezeitung, 1991

What should be taught in school can be deduced neither solely from the overall idea of general education nor from any one particular concept of general education alone. For mathematics instruction, this means that the concept of general education developed here does not itself produce content for instruction, but, instead, presumes mathematics to be a cultural achievement, a fact of social life, an academic discipline, and a domain of knowledge that can be imparted to students.

Nevertheless, neither is it possible to establish the entire complex of mathematics as a subject for classroom instruction nor does it make sense to teach arbitrarily selected areas. A simple continuation of traditional school mathematics is equally dissatisfactory. What is needed is a selection based on sound arguments. The criteria for such a selection cannot be derived from mathematics since this is not a mathematical issue. The pedagogical criteria that I have grouped under the guiding principle of general education and have related to one another represent in their entirety a viewpoint *external* to the discipline. They can further critical examination and evaluation within the complex of mathematics; they provide orientation for the selection of that sort of mathematics and those teaching methods specific to the subject which are most appropriate for realizing the aim of general education in public schooling.

A socially established practice such as mathematics instruction – with a venerable tradition and a habitual status in everyday life – demonstrates a certain tendency toward persistence in the light of attempts at change. The predominant social conditions under which classroom instruction in general and mathematics teaching in particular take place cannot readily be changed. To vary the argument: The same teachers who support the present practice must essentially support every reform. For this reason, the path to instruction oriented more strongly toward general education cannot be enforced from external sources (school supervision, university education departments), but can only consist of small steps involving many participants for whom these steps make good sense. Even if certain characteristics of the type of mathematics instruction aimed at here may seem utopian at first glance, one should keep in mind that almost everything that I will be discussing in accordance with this guideline in what follows is already present – at least in rudimentary form – in current educational practice. The arguments stemming from educational theory and pedagogy being applied here can help to identify such 'seeds' of general education in teaching practice, to cultivate them, and to systematically develop them into points of crystallization for a practice that is more meaningful and more satisfying to all persons involved. For a number of details, I will be relying on approaches that have already been extensively elaborated in the literature on teaching. The theoretical frame of reference developed here cannot replace didactical theories. Yet, it can make a pedagogical perspective available that permits a sound assessment of competing didactical approaches.

One advantage of the concept of general education used as the basis of our discussion here is that it prevents us from seeing universal remedies for the widespread discontent with mathematics instruction in educational innovations that, in the final analysis, prove to be too one-sided. To list a few examples: From the perspective of general education, it would be a one-sided innovation to restrict teaching to mathematical content relevant to everyday life, but, conversely, we could find no justification for neglecting such content (preparation for later life). It would be one-sided to confine mathematics instruction to applications (or solely to computer applications) – but it would be foolish to ignore such applications and to reinforce current anxieties about them (developing an understanding of the world). It would be one-sided to want to train cognitive abilities only in a general way, e.g., by teaching heuristics for problem solving and thus ignoring the characteristic 'material' component of all of mathematics, but it would be a fatal mistake to consider formal cognitive training to be worthless (development of critical thinking).

The dialectical structure of the concept of general education at the basis of our considerations can help to avoid such one-sided tendencies.

Naturally, the objectives of schools discussed above are not all relevant to mathematics instruction to the same extent. There are subjects that allow for teaching responsibility in more explicit and direct ways (e.g., social studies, ethics, religion, literature). On the other hand, the first four objectives, from 'preparation for later life' to 'development of critical thinking', outline an interconnected complex of human knowledge and ability to which mathematics instruction makes a specific and unique contribution.

Focusing on the Lower Secondary School

The concrete observations in what follows refer primarily to mathematics instruction in the lower secondary school. There are objective and personal reasons for this focus.

Whereas a moderate specialization in the upper grades of the secondary school, particularly in advanced courses, is compatible with the idea of general education and the general significance of most mathematical themes treated in elementary school is readily acknowledged, the lower secondary school, especially the grades 8 to 10, constantly presents teachers and curriculum designers with exceptional problems in achieving a balance between the specifics of the discipline and the claim to providing general education. The intersecting line between a mathematics that every citizen undoubtedly requires simply to cope with everyday life and one whose relevance for general education is not so readily perceptible runs directly through the lower secondary school. – The personal reason for this focus is provided by certain experiences I gained as a scientific adviser for the development of new mathematics syllabi for the lower grade levels of college-preparatory secondary schools (*Gymnasien)* and for comprehensive schools (*Gesamtschulen*) in North Rhine-Westphalia.

Similarities and distinctions between mathematics curricula for comprehensive schools and for college-preparatory secondary schools pertain to the following difficult questions: Must there be a fundamental difference between a mathematics that is designed for all students and one that is only directed to those who aspire to an academic profession or, at least, to the qualification for university studies (*Abitur*)? Or, should we make a distinction between a mathematics for students who will later be engaged in work not closely related to mathematics and for those who as adults want to work in mathematics-intensive fields? The concept of general education at the basis of our discussion demonstrates that such dividing lines cannot be of a fundamental nature – what is good 'for everyone' must also be good for those with marked cognitive abilities. But at the same time, it is obvious that not all students can be satisfied by the same mathematical requirements. Thoroughgoing uniformity ignores the needs of more gifted students just as much as it exceeds the possibilities of less gifted ones. A core mathematics curriculum must allow for differentiations and enrichment within the class, within one type of school, and among various types of schools. Thus, a conceivable distinction between a mathematics specifically designed for college-preparatory secondary schools and that geared to comprehensive schools is not a distinction that excludes a portion of the students at the outset, but one that facilitates determining the latitude available for designing mathematics instruction that genuinely aims at general education and personal development.

3.1 Mathematics Instruction and Preparation for Later Life

In light of the discussion in the preceding chapter, 'preparation for later life' in the present context always means 'preparation for later life in a strict sense'. The focus is on the practical usefulness of what is taught at school for everyday life. The matter at hand is one of teaching and appropriating qualifications that are applicable away from school, in professional and private everyday life, and are necessary insofar as leading a 'normal' life could be constrained in their absence. I have explained in detail why it seems reasonable to analytically distinguish this pragmatic, utilitarian component of general education from other aims of schools: In this way, both an underestimation (viewing education as an esoteric practice) and an overestimation (utilitarian reductionism) can equally be counteracted.

Determining the contribution of mathematics instruction to preparation for later life, then, means judging the mathematics to be learned according to its practical usefulness for the next generation. Thus, mathematics and mathematical methods are not examined as ends in themselves or as cultural achievements, but simply as potential aids in coping with everyday life.

Even very casual reflection on this topic leads to the discovery of a peculiar contradiction: On the one hand, the stipulation that the continued existence of our technological civilization depends on a sound mathematical education of the next generation is constantly stressed, and every few years alarming reports on the decline of basic mathematical knowledge and skills among children, adolescents, and adults are circulated.[69] On the other hand, we are often confronted with the sobering fact of how little the majority of adults actually use, in both private and professional life, what has been part of hard-core school mathematics for decades. Regardless of school type and state boundaries (within or without Germany), the following seems to hold true: Almost everything that goes beyond the standard subject matter of the first seven years of schooling can be forgotten without the persons involved suffering from any noticeable disadvantages. Accordingly, a large part of the rest of school mathematics, indeed, particularly of the mathematically more demanding topics, would not be vital in the sense described above.

This contradiction cannot be readily resolved. For better clarification, I will develop a more comprehensive assessment of this complex issue step by step. To begin with, I will provide an overview of those basic qualifications in mathematics and the related curricular elements that are actually important for the majority of today's adults in private and professional situations (3.1.1). Then, a few interjectory remarks serve to develop three distinct categories of mathematics with significant functions for everyday life: mathematics as an 'inventory of life circumstances', as a 'tool', and as a 'means of communication' (3.1.2). These distinctions allow a more precise characterization of the deficits of traditional mathematics instruction in light of preparation for later life: Which type of mathematics with practical significance for everyday life or which treatment of precisely this form of mathematics is usually neglected (3.1.3)?

Furthermore, I will be concerned with the requirements of vocational training: If general preparation for a profession is part of the preparation for later life (in a strict sense), then what status does the preparation for that type of mathematics that is integrated into vocational training programs have? Since mathematics instruction in the latter part of lower secondary schooling or at the upper level of college-preparatory secondary schools is, in some instances, one element of or even the basis of professional training, then perhaps this instruction would appear in a different light (3.1.4). Finally, I will discuss the following issue: Does recruiting a new generation for mathematical or mathematics-related professions, important as this is for society, require a type of mathematics instruction for everyone that significantly exceeds the knowledge and abilities needed by the average school graduate? If so, then the aspect of *usefulness for the individual's life* would have to be downgraded with reference to a *collective usefulness* (3.1.5).

On the whole, an insight that from the perspective of educational theory already proved to be of major importance in Chapter 2 can be further examined at a disciplinary level:

- On the one hand, the aspect of preparation for later life cannot be thought of as an absolute. As the primary or even unrivalled starting point for the development of curriculum, it would inevitably lead to insurmountable difficulties. Formulated pointedly, *mathematics instruction that attempts to confine itself to immediate preparation for life will only inadequately prepare for later life.*
- On the other hand, the aspect of preparation for later life is an indispensable corrective to every disciplinary development of curriculum. A neglect for practical application in everyday life, so often to be observed in usual classroom mathematics instruction, is incompatible with a modern interpretation of the idea of general education and personal development. Again, formulated pointedly: *Excluding the usefulness aspect deprives mathematics instruction of its potential effectiveness with regard to the personal development of the majority of the students.*

3.1.1 Mathematics as an Aid in Private and Professional Everyday Life

Which sort of mathematics or which qualifications encompassing mathematical elements do adults in our society use as aids in their private and professional everyday lives? The professional everyday life of that minority engaged in mathematics-intensive occupations can be neglected here.

Although to my knowledge there is no empirical research that has examined the relevant issues in a *representative* way, the results of limited surveys dealing with more specific issues have so much in common that the list I will be presenting below can be taken as a fairly useful frame of reference.[70] In addition, any adult in our society who attempts to form his or her own opinion on the basis of personal observation will be able to confirm these results to a large extent. The relevant facts, as elements of a common social experience, are, so to speak, readily available to every member of society through reflection on everyday knowledge. The list included below is based on the following studies:

- Raatz (1974) interviewed trainers and personnel managers on the subject of minimal mathematical knowledge and abilities of employees; he also directly questioned employees, most of them skilled workers, "at selected jobs requiring a high level of mathematical qualification" (ibid., pp. 43ff). This survey pertained to the fields of data processing, administration, engineering, and production.
- In England, the Sheffield Region *Centre for Science and Technology* surveyed a representative sample of regional industries on the knowledge of mathematics required by young people during their first year of work (Knox 1977).
- Conditions in England were also examined by the large-scale project "Mathematics in Employment (16 – 18)", which was based on observations at the workplace and on interviews (Fitzgerald/Rich 1981). Some of the results of this study were incorporated into the 'Cockroft Report', which attempted to provide a new basis for the reform of mathematics instruction in Great Britain (Cockroft et al. 1982).
- An Austrian study examined the use of mathematics at the workplace by employees with high school diplomas (*Abitur*), i.e., a group quite different from those in the studies mentioned thus far (Borovcnik et al. 1981; cf. Peschek 1981).
- In a series of case studies in the summer of 1989, I employed semi-structured, 90-minute interviews to question a non-representative group of ten adult employees of both sexes, university and college graduates, – but no mathematicians, natural scientists, or teachers – about their use of mathematics in professional and private everyday life.[71]

Despite the major differences among the groups with reference to their level of formal education, present job, age, and nationality, the results concur to an astonishing degree. The following list presents an extract of the findings:

Mathematical content and content-related qualifications occasionally utilized in everyday practice by non-mathematicians after completion of their formal education:[72]

- *Arithmetic*: counting; mastery of basic arithmetical operations ('in one's head' or with paper and pencil, depending on the complexity); calculating with quantities, knowledge of the most important units of measurement, making simple measurements (primarily of time and distance); calculating fractions with simple denominators in unambiguous contexts; calculating decimal fractions; computing averages (arithmetic mean); computing percentages; computing interest rates; using the rule of three; completing arithmetical operations with a pocket calculator; basic skills in estimating and making rough calculations.

- *Geometry*: familiarity with elementary regular figures (circle, rectangle, square, etc.) and objects, as well as with elementary geometrical relationships and properties (perpendicularity, parallelism); ability to interpret and draw simple graphic representations of quantities and their relationships (charts, diagrams, maps) and the relationships between given points using Cartesian coordinate systems.

Naturally, this list does exhibit certain borderlines that are not quite distinct. There are many adults who are inhibited about mathematics and for whom even such operations as calculating percentages or interest rates or employing the rule of three (i.e., arguably the 'highest level' of the mathematical activities listed above) present considerable difficulties and who, for this reason, prefer to rely on others in situations in which personal decisions would be based on corresponding calculations: e.g., on an investment adviser of their bank or on a salesperson who can explain in non-mathematical terms which investment or which product would be best for them. Conversely, simple applications of elementary algebra (e.g., transposing terms) do have a certain, if subordinate, significance for some artisan and technical vocations (cf. Raatz 1974, p. 60). Generally speaking, we can maintain that in the case of proportional relationships, non-mathematicians prefer calculating with the rule of three to using the (mathematically more elegant and more abstract) procedure of formulating and solving a linear equation. This was also demonstrated without exception in the case-study interviews I conducted even with test persons who were indeed capable of working out equations.

As a first result of our discussion, this certainly confirms the stipulation that there is very little use of mathematics in everyday life – judged by the extent of mathematics instruction in the average secondary school curriculum (regardless of type of school). And although during the last few decades, more and more segments of society have been subject to an intense process of 'mathematization' (from industrial production and company planning to the marketing of products; from statistical surveys of all areas of life to predicting the outcome of elections; from scientific research in fields traditionally closely related to mathematics, such as physics, to linguistics and history), there are practically no indications that an increasing need of mathematical qualifications *in everyday life* goes along with this process. Much to the contrary, relegating demanding mathematics to the computer or to extravagantly designed software provides the user with seemingly problem-free 'tools' whose exterior does not disclose the extent of mathematics necessary for their production. At any rate, effective use of such tools by no means requires complex mathematical qualifications.

A careful examination of the list above reveals a further phenomenon: Even though the subject matter of mathematics instruction usually far exceeds what is practically needed for everyday life, some of the basic qualifications mentioned are fostered only marginally, casually, or even half-heartedly. Specifically, this applies to the following: abilities and skills involving quantitative estimations, approximating and recognizing orders of magnitude, as well as interpreting and implementing tables and graphs. Common to both areas is the fact that the relevant qualifications cannot simply be reduced to working out algorithms (i.e., unambiguous step-by-step procedures), as is the case for most school mathematics instruction. Accordingly, one cannot simply design exercises with predetermined recipes for working out unambiguous solutions. The next section should explain why a modern mathematics instruction aiming at general education must not neglect these areas in this way.

3.1.2 Interjectory Remarks: Mathematics in Everyday Life as an Inventory of Life Circumstances, as a Tool, and as a Means of Communication

Children growing up in our culture learn to distinguish and designate humans and animals, trees and clouds, houses and cars, books and television sets as elements of their life circumstances. Similarly, prior to and during their formal education, they eventually become familiar with words and the 'normal usage' of words for integers, the corresponding numerals and other mathematical symbols, for circles and rectangles, for spheres and pyramids, etc. For mathematics in the form of mathematical symbols for numerals, numbers, and arithmetical operations is also a component of these life circumstances. Such mathematical elements, which provide a preliminary basis for the conscious appropriation of mathematics, can be characterized as an 'inventory of life circumstances'. Without some familiarity with them, our everyday surroundings would remain a great deal more alien and incomprehensible to us. Conversely, from the knowledge of this inventory comes a basic orientation that can already be understood as a rudimentary level of that understanding of the world to be developed through education.

Some of the qualifications listed in section 3.1.1 are simply related to a mathematical 'inventory of life circumstances' in the sense just explained. Knowledge of this inventory is an important prerequisite (often not reflected upon) for any intensive work in mathematics, which, in our society, already commences during elementary school. The learning problems of children of preindustrial peoples who are suddenly, from one generation to the next, subject to schooling clearly demonstrate the importance of this sort of basic sensitivity, for these children are confronted with a mathematics curriculum exhibiting significant Western influence without having everyday surroundings that offer a corresponding mathematical inventory (cf. Gay/Cole 1967, Mitchelmore 1980, for a summary see also Bishop 1988, pp. 20ff).

The practical usefulness of mathematics as an inventory remains preliminary to the use of mathematics as an aid. This latter use, in the context of everyday utilization, is our primary concern here. A classification of aids as 'tools' or as 'media' can be transferred from the realm of technological devices to the domain of mathematics and would then allow us to identify two distinct uses of mathematics in everyday life. Let us first consider a few general characteristics of technological devices which serve as aids for coping with our daily lives and for making life more comfortable.

Whenever I employ some technological product as an aid (tool, prosthesis, vehicle, etc.), I am usually not interested in this product in and of itself. Only so long as I am not really adept at appropriately using this aid am I attentive to the product itself and to the handling of it. As soon as this stage of learning is completed, I tend to 'forget' the aid itself and concentrate on what I want to achieve with it. In a very similar way, this also applies to our use of bodily organs: While walking, I do not concentrate on my legs; if I hear something I do not concentrate on my ears. In learning how to play the piano, I do concentrate on my fingers (and fingering), but once I am familiar with playing the instrument, I 'forget' them. Gehlen's interpretation of technology as a means of relieving, replacing, and surpassing bodily organs (1961, pp. 93ff) offers a plausible suggestion for transferring this lack of concentration from the original context of bodily organs to that of artificial aids.

Furthermore, effective use of technological products (or of bodily organs) usually does not require that their functions are understood in any profound sense. I need not have comprehended the biological and physical functions of my glasses or my television set (as technological equipment) or of my eye (as an organ) in order to make effective use of them.

Apparently, these characteristics apply to mathematical aids used for practical purposes in everyday life only in a restricted sense. Even a very routine application of mathematics as an aid – such as typing numbers into a computer – requires a certain attentiveness, a certain intellectual control of the mathematical aspect of this activity. While adding columns of numbers with paper and pencil, for example, – not very long ago still a routine activity of salespersons – one turns away from the particular situation (in this case, the sale of merchandise) and focuses on the mathematical operation of addition. Only if I assign the operation to some instrument such as a pocket calculator can I, as a user of mathematics, allow my attention to be withdrawn from the procedure. Put more generally: In comparison to utilizing technological aids, the use of cognitive aids requires reserving at least some remainder of concentration for the aid itself. However, we can assume that also for the use of cognitive aids, a profound understanding of the 'functioning' of this aid is not necessary if its implementation can be represented as an algorithm.

How does everyday usage distinguish between a tool and media? The 'tool' metaphor has been adopted from spheres relating to craftsmanship. A 'tool' in the original sense of the word is literally 'handled', i.e., put into use manually to achieve a purpose that would be very difficult or even impossible for the bare hand. In a figurative sense, the term 'tool' designates an aid for active operation: A tool allows an agent to intervene in his or her surroundings or facilitates such interventions, which then partially and purposefully alter these surroundings. In addition to the well-known tools of artisans, examples of tools in this sense are: forks and spoons, writing and drawing utensils, pocket calculators, etc.

By contrast, media facilitate access to information or an exchange of information: They help us to perceive, to express ourselves, and to communicate. Technological products that are better understood as media than as tools include: glasses, hearing aids, telephones, television sets, blackboards, notebooks, books, etc.

Whereas this classification of technological aids and devices as tools and media remains somewhat incomplete (e.g., means of transportation have not been taken into account here) and, in addition, does not always provide clear distinctions (e.g.: computers), this differentiation can characterize two major uses of mathematics reasonably well.

If we consider the list in section 3.1.1 once more, everything there that pertains to techniques of calculation can be subsumed under the category 'tool': Using calculations, I produce from given numbers and dimensions those that I require to achieve a particular purpose – e.g., to determine the total price when purchasing several items (addition of decimals), to check the interest on my savings account (calculating the interest rate), or to determine the gas mileage of my car (rule of three, division, multiplication).

By contrast, quantities (amounts, results of measurements) and their arrangement in tables, graphs, and illustrations of quantitative and topological contexts through pictures, drawings, and three-dimensional models all serve the purpose of communication, of providing information. At times, the emphasis is on making a qualitative linguistic statement more precise (saying, "There are seven cars in the parking lot" instead of just: "There are cars ..."); conversely, at other times the purpose might be to replace very specific quantitative information with perhaps less detailed information better suited for an integrated understanding (e.g., if at an election instead of the number of votes counted, the relative portions of the vote gained by the political parties are represented by a pie chart). In such instances, mathematics serves primarily as a means of communication. Deciphering such mathematics-related messages itself requires certain basic mathematical qualifications and sufficient practice in using them. One prerequisite of the active use of mathematics as a means of communication is the use of mathematics as a tool.

In sum: using mathematics as a tool means implementing its operative possibilities; using it medially denotes employing its presentational possibilities. In both cases, mathematics in and of itself is not the focus of attention, as it is only used as a means to achieve a goal in some context external to mathematics. Yet, by contrast to technological tools, the use of mathematics as a 'cognitive tool' does require a certain attention to the mathematical activity. Passive medial use of mathematics does not rely on this attentiveness if the relevant use is sufficiently internalized and routine. Nevertheless, by contrast to the use of many technological products as media (e.g., glasses, television set), a sensible use of mathematics as a means of communication requires intensive and conscious learning processes: A graph only seems to speak for itself; deciphering it appropriately, understanding its 'message', is greatly dependent on previous learning processes.

3.1.3 Mathematics Related to Practical Everyday Life: Desiderata of Traditional Curricula

Clearly, the focus of traditional mathematics teaching is on the operative aspect, thus, on the use of mathematics as a tool. The representational aspect, the use of mathematics as a means of communication, is learned in a more casual way and is only rarely an explicit topic of instruction or looked upon as part of 'genuine mathematics'. Here, a problem arises since in applications related to practical everyday situations there is a recognizable shift away from the use of mathematics as a tool to its use as a means of communication. If our immediate starting point was the stagnation in adults' everyday application of mathematics both in private and professional spheres, more careful consideration shows this assessment to be too indiscriminate. Certainly, on the whole, everyday use of mathematics has not increased substantially during the last three decades. But the application of mathematics as a tool has clearly decreased, while its use as a means of communication has gained significance. As a *tool*, since the rise of handheld calculators and of computers, everyday mathematics has increasingly been relegated to these technological devices; on the other hand, precisely these products have made an abundance of diverse and novel possibilities available to use mathematics *medially*.

The content of representational mathematics in everyday social life is related to elementary geometry and to functions applied descriptively on the one hand, and to statistics and statements of probability on the other. The quantified information heaped upon us daily by the press and television usually relates to statistical data or predictions. Although they have been firmly established in the syllabi of all types of schools for some time now, statistics and probability theory are still gravely neglected in comparison to the 'classical' topics of school mathematics (arithmetic, algebra, geometry).

In sum, then, a neglect for the representational and communicative aspects of mathematics in everyday life can be observed – an insufficient practice of and reflection on mathematics as a means of communication. This indicates a first desideratum of school mathematics with a view to its practicality for everyday life.

The transfer of 'operative' everyday mathematics to technological equipment makes a further desideratum apparent: The need to control the processes now relegated to machines is increasing. The emphasis is no longer on precise completion of computations in one's head or with paper and pencil, but, instead, on controlling inputs, estimating expected results or, at least, their magnitude. This shift is paralleled by processes, which, with increasing industrialization and automatization, have led to changes in vocational qualifications in the production sector. There is a constant decrease in the need to actually manipulate tools – in the literal or figurative sense. Much more important are controlling technological instruments that do this work and identifying erroneous processes or defective products, i.e., exercising 'quality control'. For a reasonable treatment of mathematics in everyday life, such changes are not simply to be judged as negative. When teachers and parents lament the continual decrease of a certain facility in elementary computation among today's children and young people, they often implicitly deny or at least downgrade the possibility that something else has taken the place of these heretofore vital qualifications. Indeed, this development offers a chance for more detailed classroom discussion on contexts that can make such computations meaningful in the first place.

In sum, we can characterize this *second desideratum* as follows: In connection with the operative use of mathematics, more emphasis should be placed on control and on categorizing activities such as estimating, approximating, and correctly assessing magnitude.

Reflections on those mathematical qualifications indispensable in everyday life often overlook the fundamental significance of a third desideratum of conventional school mathematics. I will digress here a bit.

In everyday life, mathematics always has some applications context. Most of the time, the applications contexts providing the frame of reference for usual everyday uses are simply learned implicitly in elementary mathematics teaching. Thus, for all of the basic mathematical qualifications listed in section 3.1.1, there are common situations – I call them 'standard situations' –, in which the particular use of mathematics, whether as a tool or as a means of communication, is reasonable and appropriate. In other words, for such standard situations, there are mathematical standard models whose appropriateness in everyday life is taken for granted: e.g., if the total price for a certain amount of articles is determined by adding the individual prices of the items (if a discount is given, this model is no longer appropriate); or if, by using the rule of three, from a given price of an amount of products the price of a different amount of the same products is deduced (the proportionality of product and price is simply assumed); or if the area of the floor of a room (of rectangular shape) is determined by taking the 'length' times the 'width'. Not least because conventional mathematics classes only rarely reflect on the issue of whether the prerequisites of models are actually met in standard applications, many students encounter difficulties in applying the mathematics they actually already know in situations that differ from the standard situations they are accustomed to. This problem arises even with the usual tasks couched in everyday terms, which often only pretend to refer to practical applications. Many students cheat their way through, using superficial strategies that seem to lead to success: e.g., they try the sort of computation last covered in class on the numbers mentioned in the problem, or, they simply apply the computation from a model problem recently discussed in class, etc. Recourse to such strategies has been taken even for such nonsense problems as: "There are 10 sheep and 25 goats aboard a ship. How old is the captain?", and this has often given rise to a critique of traditional mathematics teaching (cf., e.g., Winter 1985, pp. 7ff; Baireuther 1990, pp. 215-216; Baruk 1989, pp. 29ff).

In what follows I will be distinguishing between the 'ideal types' of 'standard applications' (applications in very common and habitual 'standard situations') and 'nonstandard applications' (related to situations in which it is not at all immediately obvious if and how mathematics can be applied). Mathematics teaching should deal with these applications in distinct ways.

Children, adolescents, and adults generally engage in everyday use of mathematics in standard situations: making counts, buying something, measuring something, determining portions, etc. In such cases, the connection between the situation and the mathematics required seems more or less natural. The mathematical operation to be carried out in the particular situation is not experienced in isolation or limited to school learning situations, but has repercussions on children's everyday surroundings, e.g., counting dishes to set the table for guests, shopping, measuring the height of a room to wallpaper it, computing one and a half times the amount of a recipe's ingredients, etc. In other words: In the case of such standard applications, mathematics is 'automatically' linked to instances of standard modeling that require no further reflection since they repeatedly prove themselves to be highly practicable and are vividly experienced through concrete everyday actions. This is not due to any peculiar characteristic of the particular mathematics, but, instead, is a societal phenomenon. In their entirety, these standard applications define our *everyday mathematical culture*. They are validated by social conventions and are normally interpreted the same way by everyone. Socially established expectations determine which particular mathematical activities are to be assigned to everyday mathematical culture. While displaying a lack of understanding for mathematics as a whole is considered a trifling offence, mathematical failure in standard situations, within the course of standard applications, is embarrassing. The person this happens to is considered to be somewhat incapacitated, like someone who is illiterate.

By contrast, in the case of 'nonstandard applications' appropriate instances of mathematical modeling must first be discerned, looked for, or, perhaps, even invented. A given situation must be interpreted. The objective structure of the situation or of the problem must be determined or reconstructed; it must be reformulated using mathematical concepts, illustrated by mathematical structures; and mathematically attained results must, in turn, be interpreted regarding their significance for the original problem. Apparently, this is quite difficult without supportive everyday experience. For this reason, many students, especially those having difficulties, prefer 'pure' mathematics to the applications they experience as being too demanding. In the sections on developing an understanding of the world and on critical thinking, I will be discussing the problems of a school mathematics that is oriented to applications and, at the same time, seriously concerned with nonstandard applications. For the moment, we can maintain the following: A mathematically reflective approach to nonstandard applications, as desirable as this may seem, is certainly not among the qualifications that are vital to the majority of school graduates.

Standard applications cause fewer difficulties for most students and young people, but are normally approached in an unreflective way. We have already drawn attention to the fact that the (fluctuating) border between standard and nonstandard applications depends on the state of everyday mathematical culture. This state can only be affected indirectly by school mathematics teaching. My argument is that the decisive factor for enhancing everyday mathematical culture is not at all the subject matter imparted in the classroom: For generations, all of the various types of schools have offered significantly more subject matter than has actually been applied in everyday life. If the subject matter imparted and assessed by schools were decisive, then we would already have a totally different everyday mathematical culture. It seems to be much more decisive how we treat the subject matter, which experiences with mathematics and its applications are made possible by mathematics instruction. The following idea may be speculative, but it is certainly not implausible: If we could succeed in fostering a more reflective approach to the mathematics relevant to everyday life, in developing a stronger awareness of the relationship between objective problems and corresponding mathematical models, this might have some impact on dealing with nonstandard applications. One result might be that school graduates who do not opt for mathematics-intensive careers would make better use of the rational potential of mathematics. Or, in nonstandard situations, factors inhibiting the consideration of whether using mathematics might lead to better solutions can be reduced. Conversely, one might more readily question whether a seemingly obvious use of mathematics is really appropriate (on inappropriate applications of mathematics cf., e.g., Dewdney 1994). Of course, a precondition would be that, in general, mathematics applications would play a more significant role in classroom instruction, that school mathematics would relinquish its self-complacency and its rather extensive isolation from the 'real world'.

At this point, the *third desideratum* mentioned above can now be formulated as follows: More attention is to be paid to elementary applications of the mathematics to be learned. Standard applications should become a matter for classroom reflection; they should be examined critically and provide occasion for in-depth consideration of the relationship between an objective situation and a corresponding mathematical model – in the sense of a propaedeutic of mathematical modeling.

Obviously, the ideas just explored go beyond the perspective of preparation for later life in a strict sense. Given the current state of everyday mathematical culture in our society, anyone who has learned to cope with the usual mathematical standard applications through more traditionally oriented mathematics teaching and practical experiences away from school is still sufficiently prepared for the requirements of everyday life, even if he or she is not capable of reflecting on the relationship between practical problems and mathematical models and thus naively accepts such models as given. Instruction that fosters more extensive reflection already imparts more than a mathematics that prepares for later life in a strict sense. On the other hand, we will be demonstrating that this enhancement is indispensable for the concept of general education under discussion here when we consider the contribution of mathematics teaching to developing an understanding of the world and to fostering critical thinking.

To sum up what we have been discussing thus far: With respect to preparation for later life, it is mandatory that mathematics teaching equip students with those basic mathematical qualifications that are currently in actual use in our society, in private and professional everyday life. But if we take the developments connected with technological advances in microelectronics into account, then certain new emphases in imparting these basic qualifications seem inevitable: For operative mathematics (mathematics used as a tool), greater emphasis should be placed upon activities related to control, such as approximating, estimating, and recognizing magnitudes. Students should be able to reliably make use of technological aids for purely algorithmic processes. Furthermore, in comparison to operative mathematics, greater significance is to be allotted to representational mathematics (mathematics as a means of communication). In particular, the basics of statistics and of probability calculus along with their applications in social, political, and ecological realms are topics to be treated much more conclusively. In general, more attention should be paid to elementary applications of the mathematics to be learned. Working out standard applications relevant to everyday life can be enhanced along the lines of a propaedeutic of mathematical modeling.

Such emphases not only imply a (relatively mild) modification of traditional instructional topics, but, even more so, a different treatment of the topics to be covered in the mathematics classroom. Compared with traditional mathematics teaching, many of the required mathematical activities are more flexible; they are processes to a much greater extent; and they are more intensely oriented to communication than is working on problems emphasizing algorithms and results. Thus, the main area where innovation is needed is the design of mathematics lessons.

3.1.4 Mathematics in Vocational Training

Up to now, I have simply been questioning how schools can prepare students for the demands on mathematical abilities generally common in professional and private *everyday* situations. A further distinct issue is the question of whether and to what extent the mathematical requirements of vocational training are to be taken into consideration. For the time being, mathematics-intensive careers (of mathematicians, physicists, computer scientists, engineers, etc.) are to remain excluded from our discussion.

Mathematics is a major component in the educational programs designed as training for a number of practical and academic professions. In addition, mathematical skills and knowledge are often among the subjects tested in job interviews. One problem resulting from this is that the extent to which mathematics is required, taught, and tested in vocational education has no plausible relationship to the mathematical knowledge and ability required on the job in the corresponding professions (cf. Sträßer 1984, pp. 57ff).

Yet, the question does arise whether, in determining the really vital mathematical qualifications, we have not made all too many restrictions. For, naturally, we can demand of schools that they make as broad a spectrum of career options as possible available to their graduates. Any school graduate who fails a job interview test due to the math problems or who cannot cope with the mathematics he or she encounters on the job (whether this be the computations of an optometrist or the descriptive statistics used by a psychologist) will be justified in accusing the school of not adequately preparing him or her for later life. This applies even if the mathematical part of their vocational education should prove to be grossly dysfunctional compared to the actual requirements of the workplace. Thus, we should distinguish between a *first-degree necessity* (which mathematical qualifications are actually relevant to practical life circumstances, whether private or professional?) and a *second-degree necessity*, which, as it were, is artificially produced by the educational system.

To deal with this issue adequately without going into too many details of vocational education, I will proceed as follows: I will describe three fundamental characteristics which tend to distinguish mathematics teaching within the framework of vocational training (i.e., mathematics at part-time vocational schools, at full-time vocational schools, and in the course of university education) from mathematics teaching at schools aiming at a general education. Then, I will briefly deal with the question of mathematics problems in aptitude tests. Finally, I will pose the question: To what extent should schools that provide general education explicitly (and perhaps more fundamentally than is currently usual) deal with the mathematical requirements of vocational education within the context of their obligation to prepare for later life?

Three Common Characteristics of Vocational Mathematics Teaching

In the following characterizations, I will be neglecting the enormous differences specified by the level of mathematical qualifications required by particular forms of vocational education or through the mathematical qualifications imparted to the average graduate of distinct types of schools aiming at general education.

(1) There is a distinct discrepancy between the claim to imparting vocation-related mathematical qualifications and the actual realization of this claim. For most on-the-job activities, only very superficial information is available pertaining to whether and which mathematical qualifications are required in everyday professional practice. Often, a relation to a particular profession is only established by connecting standard school mathematics topics artificially (by formulating problems in related terms) to the relevant fields. A number of factors indicate that specific vocational mathematical qualifications tend to be learned more implicitly on the job and that thus the persons involved often remain unaware of them.

(2) In vocational education, the tool aspect of mathematics is often the main focus.[73] For classroom practice, this means that mathematical procedures are often imparted as recipes and practiced by the implementation of largely standardized applications. Neither reasoning internal to mathematics nor reflections on the appropriateness of the applications play any significant role. In comparison to average vocational instruction, the much-criticized standard instruction at schools demonstrates for comparable subject matter considerably more reflection and argumentation and considerably less schematic and recipe-like procedures. Thus, the transition from mathematics teaching at schools aiming at general education to that at vocational schools tends to be accompanied by a decrease in the level of reflection. (One can find a similar problem within the context of academic education in courses where mathematics is offered as an ancillary science – e.g., statistics for physicians, social scientists, psychologists.)

(3) In view of the lack of some correspondence between the mathematics imparted and that required at the workplace, a vocationally unspecific, generalized selection comes into practical effect, whether intended or not. Those who can cope with the mathematical requirements of their vocational education are not necessarily better prepared for their specific profession, but, instead, more generally adaptable, more flexible, more intelligent, or more mathematically talented, depending on the level and orientation of the particular mathematics courses. In this way, in mathematics courses within the framework of vocational education, a selection that has already taken place at the schools aiming

at general education is duplicated, confirmed, and in some cases perhaps corrected. In other words: The mathematical components of vocational education (not exclusively, but especially these) serve the purpose of ascertaining the general aptitude of those aspiring to particular professions. One might also comprehend the debate on key qualifications as a clear indication that if it is indeed not possible to impart those qualifications immediately needed in everyday professional life with some precision, general aptitude might be better determined by the use of multiple criteria.

Aptitude Tests

The aim of aptitude tests is to select capable applicants for the educational opportunities available. Such tests usually also include mathematical problems. To a great extent, these consist in traditional topics of first-level secondary curriculum, e.g., written computations with decimal fractions, computing dimensions, using the rule of three, computing percentages, doing geometric calculations (cf. Lörcher 1980, p. 130). Thus, the mathematical qualifications tested correspond for the most part to those listed in section 3.1.1 as relevant to the everyday life of non-mathematicians. Accordingly, the selection relies on a set of qualifications that might be labeled 'fitness for everyday mathematical culture' – but with the drawback that this is primarily a bygone everyday mathematical culture. Most of the changes noted in the previous section have remained practically neglected by such aptitude tests. School graduates often fail such tests because the routine procedures (algorithms) required for solving the problems are not readily available and, due to a lack of practice, there is some uncertainty concerning the details.

In sum, there is indeed a gap between mathematics teaching at schools aiming at a general education and the mathematical requirements of aptitude tests and of vocational education. This gap would become considerably smaller if the desiderata of general education examined in section 3.1.3 would be taken more seriously, thus, if 'first degree' vital mathematics would be more present and become more intensively integrated into the topics beyond its own scope. This would also do away with the necessity of offering special 'crash courses' as preparation for vocational aptitude tests shortly before the end of lower secondary schooling. The gap that would nevertheless still remain could not reasonably be closed by standard public schools. Here, the only sensible solution would be a reform of mathematics teaching at vocational schools (and, similarly, a new concept of introductory mathematics courses for non-mathematicians at universities). Since imparting mathematics for specific vocational purposes in the way of distributing recipes seems to be increasingly obsolete, the only sensible demand must be that mathematics in the context of vocational education should also be oriented to aspects of general education. However, the mathematical applications should be very closely related to the particular vocational field.[74] The argument behind this is that mathematical competence for a particular vocation cannot really be optimized by amassing isolated, individual mathematical qualifications – these are acquired on the job –, but, instead, by becoming fit for everyday mathematical culture while using the specific vocational field as a frame of reference; by developing a feeling for numbers and geometric (especially spatial) ideas; by actively using tables, diagrams, and drawings; by attaining insight into the function of mathematics in various objective contexts (specific to certain vocational fields).

3.1.5 Recruiting the Next Generation for Mathematics-Intensive Careers

Up to this point, we have been examining the usefulness or vital necessity of mathematics exclusively with regard to those students who will later *not* be involved in a mathematics-intensive career. Although these students account for the large majority, the minority of students who, having graduated from school, will major or minor in mathematics at a postsecondary institution should not be neglected. Indisputably, an in-depth mathematical education is also of considerable *practical* benefit to the members of this group, who represent the next generation for mathematical, scientific, and high level technology-related careers and are especially important for the reproduction of our technological and scientific civilization.

Here, a problem arises that is often either underrated or even ignored. With regard to preparation for later life, mathematics instruction of a generally compulsory nature for that majority of students not becoming involved in mathematics-intensive careers should be optimized differently than that designed primarily for the recruitment of the next generation of those opting for mathematics-related professions. The main issue here is not the idea that the sort of mathematics teaching desirable for the majority would be detrimental to the future mathematician. But we cannot rule out the possibility that he or she will be offered too little fundamental mathematical knowledge and specific disciplinary training. It is also possible that the point in time for being able to firmly maintain this fundamental knowledge and disciplinary training will already be too late if it only emerges after having finished general education schooling. On the basis of such considerations, it becomes evident that the current practice of mathematics teaching is an unsatisfactory compromise. To put this pointedly:

- In the traditional mathematics classroom, the majority of children and adolescents must acquire a considerable amount of very specialized skills and knowledge which they will never again put to use, from which no recognizable transfer occurs, and which to a great extent will very quickly be forgotten. Due to these circumstances, the majority of the next generation see traditional mathematics teaching as a largely superfluous activity.
- Conversely, traditional mathematics teaching offers select mathematically sensitive students the opportunity to concentrate on training for a mathematics-intensive career with no additional effort after they have completed their general education in school.

Thus, formulated pointedly again, the dilemma of traditional mathematics teaching can be seen in the fact that those who do not at some later date become mathematicians must study a great deal of specialized mathematics so that future mathematicians do at least become acquainted with the minimum requirements of their profession. To get some idea of how to avoid this dilemma, I will proceed as follows: For an exemplary contrast, I will examine the musical profession to demonstrate how the following generation is recruited for a demanding profession which, as the evidence demonstrates, can only then be expected to be successful if specialized training is begun at an early age. Considering the similarities and differences of the preparation for musical and mathematical careers at school can be quite revealing. Subsequently, I will outline a scenario in which some of the elements of the musician model are transferred to the recruitment of the next generation for mathematics-intensive careers.

Recruiting the Next Generation of Musicians

Apart from very few exceptions, a career as a musician is an impossibility for anyone who does not begin to play an instrument at an early age. In this case, 'early' means: at best, before one reaches the age of ten, at the latest before the age of thirteen or fourteen. In our society, preparation for musical careers is not organized by the state (by means of compulsory instruction in the mastering of instruments at schools), but on a private basis. Parents arrange for music lessons for their children at schools of music or from private music teachers. For this reason, music instruction within compulsory schooling can concentrate on the future non-musicians and, in doing so, can also involve those instrumentalists and potential music 'experts' trained outside of school and productively incorporate their abilities into the planning and realization of lessons.[75] Thus, the problem of recruiting the next generation for musical careers can be considered to be solved. However, the circumstances that make this solution possible deserve careful attention:

- For parents, allowing their children to have music lessons is linked to ideas of social prestige and is not considered to be a bad investment even if the children do not decide to become musicians. Much to the contrary: Being able to competently play an instrument is rated in much the same way as being able to dance – as an enhancement of normal civilian occupations or as the key to self-realization outside of one's profession. By contrast, the musical profession itself is sometimes still seen as a vocation with no money in it.
- In view of the high costs for private music lessons, one can hardly speak of 'equal opportunity' for children from less wealthy families.
- The societal need for professional musicians is considerably smaller than that for mathematically qualified professionals.
- The productivity of a modern national economy is much more dependent on the qualifications of a mathematical, scientific, technological elite than on the qualifications of a musical elite.

A Scenario for Future Mathematics Instruction

The reasons listed above demonstrate clearly that the recruitment of future musicians cannot serve as a model for recruiting future mathematicians. Nevertheless, it seems worth considering whether a more consistent external differentiation taken from the musician-recruiting model might not alleviate the dilemma described above, even if both the general and more specialized mathematical preparation for later life were to remain within the responsibility of schools aiming at a general education. A three-level scenario for such a purpose might be as follows:

First Level. At elementary school and at lower secondary schooling up to the 8th grade (regardless of type of school), mathematics teaching oriented to general education should be offered to all students as a group. This instruction is compulsory and consistently avoids topics that are only relevant for subject-area specialists. (What I mean here are topics that no one dares to remove from the standard mathematical curriculum primarily because, within the framework of the standard curriculum, recourse is made to them at a later date. Examples are mentioned below.) With this common instruction, much emphasis is placed on *fitness for everyday mathematical culture* (i.e., fluency in the pragmatic, preparatory content and techniques listed in section 3.1.3, including those which prepare for careers in a general way); on exemplary consolidation in accordance with the ideas I will be expounding below under the headings of *cultural competence*, *developing an understanding of the world*, and *developing critical thinking*; and on *a culture of instruction* as discussed in the following sections of this chapter.

Second Level. Then, from the 9th grade on, an external differentiation begins.

- Mathematics teaching for those students who wish to maintain an option for a mathematics-intensive career, who display mathematical inclinations and are judged by their teachers (?) as being sufficiently mathematically competent, aims specifically at providing in-depth coverage of subject-area aspects. Among other things, a mathematician's basic routines and skills are to be trained (from transposing terms to establishing proofs), and systematical topics that are no longer compulsory for 'non-mathematicians', but are considered to be prerequisites for more intensive study of mathematics, are to be dealt with, e.g., quadratic equations, trigonometry, computation with exponents, and logarithms.
- For the rest of the students, instruction oriented to general education is continued according to the general goals already described for grades 1 through 8 – of course, with modifications regarding the increased cognitive abilities and the changed horizons of the now 14 to 17-year-old adolescents. Descriptive statistics (e.g., in the form of explorative data analysis) could now play a more significant role than in traditional instruction. Work with new computer tools such as spreadsheets and dynamic geometry software would also be conceivable. Within the course of this instruction for the majority, it would naturally also be possible (assuming the corresponding capabilities and interests of the learners) to undertake in-depth investigations (à la Wagenschein) of mathematically significant topics and of themes related to the history of mathematics, to engage in studies of Pythagoras's theorem or of number theory phenomena, and to examine non-linear functions in connection with intriguing applications.

The differences between the various levels could be defined specifically, depending on the type of school.

Third Level. Finally, in the upper grades of secondary schools, those students aspiring to postsecondary studies should be clearly and consistently separated from the others, approximately along the lines of today's basic and advanced courses. But as far as content is concerned, there should be significant differences from current instructional practice:

- The basic courses are no longer to be conducted as 'diluted' advanced courses, but are to be largely detached from the traditional upper level curriculum. Analysis and linear algebra are no longer compulsory. Instead, the focus is on a more profound understanding of applied mathematics in everyday life, primarily in connection with probability topics and with the use of the computer as a mathematical tool.[76]
- A modification of priorities to the benefit of probability topics could also be considered for the advanced courses, which, as far as content on the whole is concerned, need not require such extensive alteration. Yet, the primary goal here would now be to provide the appropriate preconditions for the study of mathematics as a major or minor subject at the postsecondary level.

To forestall any false impressions: The commitment to general education is by no means relinquished with regard to the more subject-oriented, intensive mathematics instruction in grades 9 and 10 and to the advanced courses in the upper grades of secondary schooling. In particular, the differences between the various instructional levels must not affect the requirements of a culture of instruction contributing to general education as these will be discussed below. Thus, the desired stronger orientation to disciplinary aspects is not at all a justification of specialist complacency in the context of classroom instruction. Conversely, establishing courses 'for the majority' does not mean that the distinctive features of mathematical thought in comparison to mathematically naïve everyday thought should be banished from such classrooms. – In much the same way, it is also an objective of music lessons within general education to impart some conception of the spirit of music to the musical novice. One important aim of mathematics teaching for the majority would be to prevent the separation of everyday thinking from mathematical thinking, so often perceived in those who have enjoyed traditional mathematics instruction, by orienting lessons to the potential horizon of future non-mathematicians.

This scenario is to be interpreted as an incentive for discussion. It is not fully developed, and it leaves a number of questions unanswered: What justifies the hope that students, teachers, and parents will make a sensible choice at the beginning of the 9th grade? Aren't students still too young for such far-reaching decisions affecting their future vocational options? What possibilities are there to subsequently correct a choice already made? – On the other hand, we should consider the fact that future vocational options are influenced to a much greater degree by the early choice of the type of school to attend (at least in the German school system). In my opinion, the scenario above does actually uphold the principle of equal opportunity.

It is important to weigh the problems just mentioned against the fundamental problem that the scenario presented above attempts to deal with: the fact that mathematics teaching for all students that takes requirements for practical everyday life seriously is inadequate mathematics instruction for future specialists; and that, conversely, mathematics teaching that prepares future specialists adequately is of very little practical value to the life circumstances of all other students. The assumption that school mathematics teaching is indispensable for general education per se is the most widespread justification of traditional mathematics instruction. This assumption and the conclusion drawn from it, namely, that for the greatest part of their time at school all students should receive the same instruction, only obscure the fundamental problem. For this reason, the widespread discontent with traditional mathematics teaching seems more justified than the attempts to placate this discontent on the part of that teaching's apologists.

3.1.6 Concluding Remarks

In their private and professional everyday lives, adults who are not involved in mathematics-intensive careers make use of relatively little mathematics. Everything beyond the content of what is normally taught up to the 7th grade (computing percentages, computing interest rates, rule of three) is practically insignificant in later life. However, modifications of our everyday mathematical culture, especially those resulting from the entrance of computers into all spheres of life, are hardly reflected in conventional mathematics curricula. School mathematics is still centered on the use of mathematics as a tool; by contrast, in our social surroundings, mathematics has become increasingly important as a means of communication. School mathematics is still focused on working out algorithms; in our society, less exacting activities that are often not even considered part of mathematics are becoming ever more important: estimating, approximating magnitudes, interpreting graphs and tables, simple forms of mathematical modeling. Mathematics teaching should be more explicitly concerned with these developments.

A reform of mathematics instruction in schools must not neglect the mathematical content required in vocational education. But since here the general aspect of selection is usually more decisive than actual needs in vocational practice, there is little to be said against trimming down the curricula in the area of classical school mathematics. On the contrary, it would be much more appropriate to extend the reforms suggested for general education instruction to mathematics within vocational education – a step that would also correspond to the demand for promoting key qualifications.

One fundamental problem that perhaps cannot be tackled by a uniform design of mathematics instruction for all students – even if various intellectual requirements are taken into consideration – pertains to the fact that an appropriate mathematical preparation for later life for the majority of future non-mathematicians is not compatible with that which would be ideal for the future mathematicians (in the broadest sense). A possible solution to consider would be an earlier external differentiation to be carried out at about the 9th grade level. To be sure, this would entail taking leave of the still predominant ideology which alleges that traditional school mathematics is, in and of itself, already a vital component of a general education. General education elements justifiably attributable to mathematics would, then, have to be made productive for both groups of students despite the external differentiation.

3.2 Mathematics Instruction and Cultural Competence

What can mathematics instruction contribute to the promotion of cultural competence? – On the basis of the distinctions made in section 2.2, I will first address the *diachronic* aspect, i.e., to what extent mathematics teaching – in transmitting significant cultural achievements – helps to promote cultural *continuity*. Subsequently, I will be broadening the horizon of these reflections to include the *synchronic* aspect and will discuss possible contributions of mathematics instruction to cultural *competence* in transmitting *main ideas*. The greater part of this section will deal with this latter topic.

3.2.1 Imparting Mathematical Tradition Via Mathematics Teaching

Unquestionably, school mathematics instruction does impart the mathematical tradition – regardless of how well or how badly actual instruction is carried out and of whether students and teachers are aware of this function of imparting tradition. A much more complex issue is whether and to what extent this imparting of tradition through mathematics teaching helps students to acquire a reflective cultural identity as defined by the concept of general education under discussion here.

Three levels of imparting the mathematical tradition can be distinguished: that of everyday mathematical culture, that of school mathematics, and that of mathematics as a scientific discipline. These levels can be examined individually with regard to the mathematics classroom's function of imparting tradition.

Schooling's Contribution to the Continuity of Everyday Mathematical Culture

As already noted in chapter 2, there are certain elementary topics of mathematics instruction that are not imparted to the next generation simply for the sake of perpetuating tradition, but, instead, to prepare students for their future lives. The content of this part of the curriculum corresponds to the minimum canon outlined in section 3.1.1.

The mathematical topics and the ways of treating them listed in this minimum canon cover a large part of what is required for participation in the *everyday mathematical culture* of our society. By dealing with these topics, mathematics teaching provides for the reproduction of a certain status that our civilization has achieved, a status for which a core repertoire of mathematical abilities is characteristic. Alterations in this minimal canon reflect (usually with a considerable delay) *change* in our everyday mathematical culture – as was illustrated in the previous section with reference to the changed role of statistics, of dealing with data, and, more generally, to the increased use of mathematics as a means of communication.

Thus, imparting that standard of mathematical ability that is widespread in private and professional everyday life in our society through schooling – including cautious adaptations to more recent social and technological developments – represents the *promotion of cultural continuity* in a very elementary sense. This contribution of schools to the continuity of everyday mathematical culture will also remain indispensable in the foreseeable future. In addition, it serves a purpose independent of the subject matter, the pedagogical significance of which has already been noted (section 2.2.4): communication between the generations – by cultivating a body of knowledge and abilities transcending a specific generation.

Can school mathematics teaching also endanger the continuity of our everyday mathematical culture? A notable example of this is a now historical reform, the over-hasty introduction of 'New Math' at the end of the 1960s. Certainly, a number of various reasons can be given for the failure of this reform (cf. Damerow 1984, pp. 36ff). But one probable reason that is often overlooked was the following: There was a flagrant violation of the principle of promoting cultural competence; the continuity of everyday mathematical culture was called into question by this reform. Many parents no longer understood what their elementary school children were required to learn in mathematics; the communication between the older and younger generations in a presumably basic field of knowledge took on unexpectedly difficult forms. Moreover, any applicability of the new content in everyday situations was simply out of the question.[77] In this case, the conservatism on the part of the parents (and, to a large extent, of the teachers) was, at its core, rational. The partial withdrawal of the reform reestablished elementary mathematical instruction as a mediator of the continuity of everyday mathematical culture, of a continuity that had seemed to be in danger of disintegrating in the meantime.

In sum, we can maintain the following: With regard to imparting the tradition of elementary mathematical content through schooling – apparently necessary for sustaining our everyday mathematical culture –, interaction takes place between schools and the society outside of school. Unquestionably, elementary mathematics instruction makes a fundamental contribution to promoting cultural competence.

Schooling's Contribution to the Continuity of School Mathematics

It is more difficult to assess the following phenomenon: Within school mathematics there is a way of imparting tradition that demonstrates itself primarily in a marked tendency of content toward persistence as soon as this content has been established as part of the curriculum. The most obvious result of imparting tradition in this way is that one might speak of a 'standard curriculum' of school mathematics. How should this form of continuity of the standard curriculum be evaluated from a standpoint of educational theory? Is it – in some way that is perhaps not immediately perceptible – an expression of the desired promotion of cultural competence? Or is it simply the result of institutional constraints and thus more of a hindrance to the sort of general education aimed at here and to the corresponding promotion of cultural competence? Let us recall the main reasons for this tendency toward persistence:

- In the classroom, teachers prefer mathematical topics they have been well trained to treat or, if this precondition is not fulfilled, that are still familiar to them from their own schooling. (One example would be the limited acceptance of probability topics.) This tendency is reinforced by the structure of academic teacher education, which for the most part separates mathematical education from that of practical instructional and didactic/methodical education, and by the structure of practical teacher education programs, which are primarily oriented to adapting to existing instructional techniques.

- The persistence of content that has already been introduced is additionally stabilized by the usual procedure in developing new lesson plans and guidelines, for the main say in this matter is allocated to practicing teachers.
- Finally, the necessity of selection privileges content that is readily adaptable to use as tasks in an examination. This has a double effect: Assessment is particularly important in mathematics because what is learned seems to be more objectively assessable than in many other subjects and because mathematical achievement and general intelligence – as it is normally measured – display a relatively high correlation. And for these reasons, within the discipline the perpetuation of such content in the curriculum that has proved itself in examination procedures is fostered – content that allows the construction of problems with unambiguous solutions (easily corrected) and for which collections of problems are already available (tests are easier to develop).

Thus, the continuity of school mathematics is understandable from the structure and social function of the system of schools and teacher education. Consequently, this way of imparting tradition hardly supports the promotion of cultural competence as it is aspired to here. On the contrary, mathematical activities that are more interesting from the view of educational theory – exploratory, heuristic, and problem-solving activities, and those with uncertain results or involving social cooperation – have only a slight chance of asserting themselves against the standard topics of school mathematics. In contrast to the contribution that elementary mathematics instruction makes to the continuity of our everyday mathematical culture, the continuity of content so characteristic of mathematics instruction as a whole contributes little to promoting cultural competence apart from the scope of the elementary sector.

Schooling's Contribution to the Continuity of Mathematics as a Scientific Discipline

Undoubtedly, mathematics as a scientific discipline is a cultural achievement of the highest order. Mathematical and similar activities can be traced to the earliest stage of written records, to about 4,500 years ago. People have been systematically engaged in mathematics in a strict sense for about 2,500 years, since in the Greek city-states philosophers such as Thales and Pythagoras developed and taught mathematics as an 'abstract' science (i.e., independent of practical considerations). The epoch-making development is to be seen in the fact that the principle of deduction became established as a method for acquiring and ensuring mathematical insight.

Since that time, the development of mathematical theory, having once been launched,– despite intermittent stagnation and despite the support of mathematical culture by various, dissimilar groups (considering the substantial participation of Arabs, by no means a purely Western achievement) – has never completely come to a halt. Innumerable anonymous accomplishments and those attributable to certain individuals have contributed to mathematics, have changed it and made it a dynamic development. The scope of the body of mathematical knowledge accumulated has long superseded the capacity of any one individual. Developments in the sense of an increase in the extent of this body of knowledge, of an extension of its internal connections, or of making it available to ever new applications are practically exclusively the domain of highly specialized experts at universities, research institutes, and industrial laboratories. There are mathematical researchers in virtually every country of the world. Worldwide, international exchange among the members of this community is taken for granted and is much less impeded by cultural, ideological, or language barriers than in most other scientific disciplines.

In comparison to other distinguished cultural achievements in fields such as literature, the fine arts, or music, fewer accomplishments that can be accurately dated and attributed to certain individuals (in the same way as novels, paintings, or symphonies) are incorporated into the tradition in mathematics. Instead, individual achievements (proofs, theories, or computational procedures) are integrated into the entire corpus; prior and newfound insights are related to one another and interpreted anew. Knowledge in this field is, as it were, constantly in the process of being updated. For this reason, a great deal of cultural continuity – despite occasional fundamental crises – is inevitable for the further scientific development of mathematics simply due to the given preconditions.

Even this brief recollection clearly demonstrates that we cannot interpret schooling's objective of promoting cultural competence in such a way that schools should be immediately concerned with the continuity of mathematics as a scientific discipline. This internal continuity develops according to its own dynamics. On the one hand, it depends on the creative potential and activities of a highly specialized group of experts, on the practices that this group as a scientific community develops; on the other hand, it depends on societal necessities and the social acceptance of what this group produces. If mathematics instruction is able to contribute to the continuity of mathematics as a *scientific discipline*, then its contribution must be that a sufficient number of motivated and competent future scientists will emerge from this instruction.

3.2.2 Promoting Cultural Competence by Orienting to Main Ideas

Yet, in which way can mathematics teaching actually contribute at all to promoting cultural competence? My reply to this question is to be substantiated and explicated in this and the following sections:

The decisive contribution of mathematics instruction to the promotion of cultural competence is *to allow for the specific universal features of mathematics and their significance for culture as a whole to be vividly experienced in an exemplary fashion on the basis of main ideas.*

With this argument, the restriction to the diachronic aspect of cultural competence has been abandoned. This aspect can be attended to more appropriately by integrating the *history of mathematics* into mathematics instruction more intricately, especially in connection with a genetic method.[78] Here, I will be dealing primarily with the synchronic aspect of cultural competence.

The proposal to orient mathematics teaching to a limited number of *main ideas* is hardly original (others have sometimes spoken of 'fundamental', 'basic', 'universal', or 'guiding' ideas). For many decades and in quite a few variations, this proposal has been brought forth by mathematicians, mathematics educationalists, and teachers again and again without having had any lasting effect on the curriculum and on classroom practice (cf. Schweiger 1992)[79].

The discussion thus far demonstrates that, above all, the culturally significant universal features of mathematics must find adequate expression in such main ideas, and students must be able to understand this on the basis of their experiences. This is not at all self-evident. For which particular ideas are appropriate to according mathematics instruction a didactical guiding function can be solely deduced neither from mathematical nor from didactical considerations. Reasonable arguments in favor of the selection of such ideas must be supported by an (explicit or implicit) interpretation of mathematics as a segment of our entire culture. Main ideas should allow the relationship between mathematics and nonmathematical culture to become perceptible – especially connections to what are familiar experiences of students in their everyday lives. To put it in computer jargon, main ideas must designate the interfaces between mathematics and culture as a whole. If the selection of such ideas is motivated by the structure of the discipline or if the 'main ideas' are even equated to 'basic concepts of the discipline', then the risk of entrenching the so often lamented cultural and "social isolation" of mathematics (Lenné 1969, p. 165) will be high. Then, for the future, too, mathematics would remain a rather preposterous specialist activity in the opinion of the majority of students and future adults. From this, we can conclude that:

- with the main ideas for instruction oriented to general education, the universal features of mathematics should be expressed in a way comprehensible to students;
- they should be meaningful for various individual mathematical topics;
- they should be something other than simply basic mathematical concepts, i.e., they should not exclusively have a significance internal to mathematics;
- above all, they should demonstrate how mathematics is interrelated to other aspects of the culture of our society.[80]

Before I proceed to examine a number of possible main ideas, I will digress here a bit and recapitulate some thoughts put forth by Bishop (1998). On the basis of cross-cultural studies, Bishop attempts to identify and classify transcultural basic mathematical activities. In contrast to approaches to be discussed subsequently that proclaim to be *normatively* oriented – they attempt to determine which main ideas are to be dealt with in mathematics teaching –, Bishop's line of inquiry is a *descriptive* one: His concern is to determine which forms of mathematical or similar activity can actually be observed in different cultures. In my opinion, Bishop's inquiry offers an interesting framework for the subsequent discussion since many of the potential 'main ideas' exhibit a pronounced affinity to the basic cultural activities he describes.

Transcultural Basic Mathematical Activities – A. J. Bishop

Is mathematics or at least some rudimentary form of it present in every culture? Does cultural development inevitably lead to that sort of mathematics that has developed in Western culture? Is Western mathematics 'universal'? Are its concepts and conceptions, in principle, comprehensible to anyone, regardless of individual cultural background? According to Bishop (1988, pp. 20ff), recent cross-cultural research can provide a number of tentative answers to these questions. Studies on mathematical activities in non-European cultures[81] disclosed notable intercultural contrasts and concurred in finding that mathematical thought in the Western sense does not exist at all or only in very rudimentary form in the cultures investigated.

Bishop argues that instead of emphasizing the differences between Western mathematics and similar activities of other cultures – which would often encourage a devaluation of non-Western mathematics –, it would be more productive to look for points in common. His working hypothesis states that there are mathematical activities in all cultures. The related ideas and concepts within a certain culture could then be interpreted as culturally or environmentally dependent responses to activities or processes, which, in principle, would be transcultural (or: universal). According to Bishop, then, a classification of such mathematical key activities should attempt to avoid specifically Western mathematical concepts and terms.

In this sense, Bishop distinguishes six key activities that can be found in some form in all of the cultures investigated and, for this reason, in his opinion can be thought of as 'universals': counting, locating, measuring, designing, playing, and explaining.

The activities within this classification that seem most familiar are counting and measuring. Bishop justifies their distinction with the argument that in all cultures there are social contexts that are associated with the idea of counting discrete objects, just as there are contexts that suggest comparing continuous phenomena by measuring. The fact that Western arithmetic (or the Western concept of number) provides a response to both problems is something specific to our own culture and occasionally obscures the principle distinction between the two key activities mentioned above. For similar reasons, Bishop distinguishes between 'locating' and 'designing' – although in Western mathematics both activities are considered geometrical ones. It might be surprising to find 'playing' as a separate category within this classification. In all cultures, Bishop finds approaches to guiding social behavior by explicit rules, thus making 'hypothetical' ('as if') behavior possible. Finally, the activity of 'explaining' is an expression of the cognitive aspects of exploring and conceptualizing.

For a more exact description of the six key activities and their particular cultural configurations, one should consult Bishop's book. Let us return to the question that stimulated our recapitulation of Bishop's thoughts: Can the activities that he distinguishes help to identify main ideas to which mathematics instruction aiming at general education could be oriented, as instruction that should contribute to promoting cultural competence? One thing is certain at this point: The key activities mentioned do not already represent the actual ideas that we are looking for. As Bishop rightly emphasizes, various cultures have produced various 'responses' to the necessities expressed in these key activities in the form of culturally specific 'symbolic technologies'. Of course, mathematics instruction in our society deals primarily with those 'responses' that Western mathematics has produced.[82] All of our suggestions for 'main ideas', which are now to be examined more closely, are related to Western mathematics.

In the following sections, I will explore in more detail the approaches of Whitehead (1967 [1913]), Bruner (1970 [1960]), and Wittenberg (1963).

Main ideas for Mathematics Instruction According to Alfred North Whitehead

One suggestion to center mathematics instruction around only a few fundamental ideas was already made at the beginning of the 20th century and comes from A. N. Whitehead (1967 [1913]).[83] More clearly than many of the more recent authors, Whitehead indicates the danger for mathematics, as practiced by the experts, of becoming isolated from other aspects of culture. He maintains that for this reason mathematics instruction aiming at general education should acquire new emphases. Despite the expanse of time separating us from Whitehead, in principle, his thoughts on this subject still seem quite up-to-date.

Whitehead maintains that the science of mathematics, "as it exists in the minds and in the books of students of mathematics", is esoteric. He explicitly differentiates between "mathematics as a subject for profound study" and "its use as an educational instrument".[84] His argument is that many of the properties that are of inestimable value for the experts – e.g., the "boundless wealth of deductions from the interplay of general theorems, ... the variety of methods, and their purely abstract character" – are educationally disastrous. "The pupils are bewildered by a multiplicity of detail, without apparent relevance either to great ideas or to ordinary thoughts." For this reason, Whitehead demands that mathematics instruction "deal directly and simply with a few general ideas of far-reaching importance" (all quotes: ibid., pp. 78-79). He explains this demand as follows:

> "The main ideas which lie at the base of mathematics are not at all recondite. They are abstract. But one of the main objects of the inclusion of mathematics in a liberal education is to train the pupils to handle abstract ideas. ... For the purposes of education, mathematics consists of the relations of number, the relations of quantity, and the relations of space. ... These three groups of relations, concerning number, quantity, and space, are interconnected." (ibid., p. 80) "But what at all costs we ought to avoid, is the pointless accumulation of details. As many examples as you like ... But these examples should be direct illustrations of the main ideas. In this way, and this only, can the fatal reconditeness be avoided." (ibid., p. 81)

Whitehead is convinced that the general treatment of mathematics he suggests, dealing with the fundamental mathematical ideas mentioned, is worthwhile for *all* students – thus, just as much for future non-mathematicians as for future users of mathematics:

> "This general use of mathematics should be the simple study of a few general truths, well illustrated by practical examples. This study should ... be completely separated ... from professional study..., for which it would make a most excellent preparation. Its final stage should be the recognition of the general truths which the work done has illustrated" (ibid., p. 81).

At this point, the skeptical reader may have formed the impression that, for the sake of the perspicuity and of the fundamental nature postulated for the basic ideas he propagates, Whitehead wishes to banish any profound mathematical insights whatsoever from classroom instruction. But the further course of Whitehead's essay – when he explains, with reference to the "idea of quantity" how the illustrations he calls for can be filled with content – shows that this is a misconception (ibid., pp. 82-83). On the one hand, he recommends a treatment of Euclid's fifth book (which deals with, among other things, the problem of incommensurability) – at least for 'advanced' students; on the other hand, he recommends treating the idea of 'functionality' up to the concept of 'rate of change' and the role of mathematics in the exact formulation of natural laws, for "the physical ideas and the mathematical ideas illustrate each other" (ibid., p. 82).[85]

What makes Whitehead's reflections still worth reading today is their stimulating potential, their application of forward-looking standards (from the perspective of his time) for mathematics instruction aiming at general education. In the last eighty years, mathematics has undergone profound transformations, both in its fundamentals and in its applications[86]. This fact alone requires that we give new consideration to the question of *which* individual basic ideas could allow mathematics instruction to contribute to promoting cultural competence.

Fundamental Intellectual Experience According to Alexander I. Wittenberg

No other German-language author of the postwar period has so vehemently furthered the cause of laying a foundation for mathematics instruction based on educational theory than Wittenberg in his most important work, "*Bildung und Mathematik*" ("Education and Mathematics") (1963). With a background as a mathematician with extensive philosophical interests – similar to Whitehead, to whose position he felt a pronounced affinity –, Wittenberg fought passionately for an orientation of mathematics instruction that was *not primarily concerned with subject matter.*[87] We can only correctly determine which of Wittenberg's ideas are still relevant for establishing 'main ideas' for mathematics instruction today if we briefly summarize his position on educational theory and, from today's perspective, rectify certain biases connected with his position that were characteristic of the time. One general point to keep in mind for such a review is the fact that, to a large extent, Wittenberg's book is not a scientific text in the strict sense, but a polemical treatise on educational policy, which early on opposed certain guiding principles of the era of educational reform, e.g., the one-sided interpretation of *orientation to science.*

Wittenberg had far-sighted premonitions of many of the negative consequences for mathematics instruction that resulted from the temporary excessive preoccupation with 'New Math' – consequences which were only beginning to suggest themselves when Wittenberg's book went to press.

Wittenberg's pedagogical reflections, especially those regarding the function of 'higher education' for sustaining a stage of cultural development already attained, display a certain proximity to the approach of Wilhelm Flitner. For today's readers, this can help explain one distinct feature of Wittenberg's approach that might seem quite peculiar at first – if 'general education' is understood as 'education for all': the primary role that Wittenberg assigns to the college-preparatory secondary school as an idea and as an institution. At first glance, he seems to be solely concerned with the education of an intellectual elite. Wittenberg actually equates the idea of general education to the idea of the *Gymnasium* (ibid., p. 15). Later on, however, he does relativize this fixation to an elite somewhat;[88] and the associations and concretizations with which he explains his idea of general education are by no means only worth considering for instruction at this type of school. For example, Wittenberg emphasizes a "commitment to truth" (p. 16) that cannot be separated from the "acquisition of the ability to have experiences, an ability to comprehend emotionally", and personal development into a "homo democraticus" (p. 29). Here are a few abbreviated statements of points that Wittenberg emphasizes and that explicate his idea of general education:

- The original reasons for a science must not be confused with the simple, but essential insights of a science (p. 10);
- An abundance of content makes specialized instruction both more complicated and more trivial (p. 11);
- Striving to teach as much as possible ('encyclopedic ideal of education') discredits education's commitment to truth (p. 17);
- Instruction aiming at general education will always convey an experience of the complexity of truth to the students and educate them toward independent, critical, original, and responsible thinking (p. 25);
- Every disciplinary field should commit itself to these higher goals of general education (p. 43).

Only after Wittenberg has already elaborated his views to this point does he begin to speak of mathematics. For him, the major consideration of what mathematics can contribute to general education is contained in the following question: "Which experiences that are so fundamental and so significant that they are essential to a general education does this science make available to human beings?" (ibid., p. 45) Wittenberg concludes that mathematics contributes "two fundamental intellectual experiences to the entirety of human experience of one's own existence". Expressed again in abbreviated form, these are as follows:

(1) Mathematics is a 'reality' constructed by human thought (a "fauna of mathematical entities") which, nevertheless, is not of an arbitrary nature, but is formed by necessities and allows for "discoveries".

(2) There is a correspondence (adequacy) between mathematical thought and the human ability to experience the 'outside world' (nature).

Then, the purpose of mathematics instruction is to make both of these fundamental experiences available to students, to let them become vivid experiences for the students through their own researching, rediscovering activity, through actively dealing with given mathematical phenomena (e.g., geometric figures), through discovering and trying out elementary mathematical methods. In this sense, these 'fundamental experiences' are actually *main ideas* which Wittenberg would like to have guiding mathematics instruction.

Oriented to the 'Structure of the Discipline': Jerome S. Bruner's Contribution

Some of the proponents of a concept of 'main ideas' for mathematics instruction (Wittmann 1974, Schreiber 1979 and 1983, Fischer 1976, Picker 1985a/b) explicitly refer to Jerome Bruner, whose books have been accessible to the German-language discussion since the first translation in 1970 (1970 [1960], 1973, 1974).

Bruner had a rather ambivalent influence on this discussion – whether because, contrary to his intentions, certain of his controversial arguments were misunderstood or because his writings themselves were predisposed to such misinterpretations. Let us recapitulate Bruner's most important statements on curriculum theory. The core of his arguments can be summarized as follows:

- Students should not learn individual facts pertaining to a certain subject matter, but, instead, the fundamental structure of the subject, through which they can then classify, assess, and deduce individual facts.
- The fundamental structures of the subjects to be taught are contained in a limited number of basic concepts or fundamental ideas that should be used to structure the curriculum.
- Curriculum should be structured in a spiral manner. The fundamental ideas of a subject can be imparted to students of any age in an intellectually honest way. For this reason, students should encounter them again and again, each time in a way corresponding to their increased cognitive abilities – with appropriate depth and involving 'higher' levels of representation.

Bruner's curricular approach involves a rather immediate connection between the psychology of the human subject and the content to be learned, while largely abstaining from further modifications due to pedagogical or didactic considerations or to those of general educational theory. He has elaborated such considerations in other contexts. Depending on whether Bruner's curriculum theory is examined in isolation and its main points are taken literally or if they are seen in relationship to ideas Bruner has expressed while exploring other educational topics, conclusions are drawn and interpretations formed that are practically contradictory to one another.

In one extreme interpretation, Bruner's curriculum theory is said to maintain the following: The knowledge to be acquired by young people is structured in an optimum way in the modern sciences. The primary task of schooling is to transform the 'structure of the disciplines' into the cognitive structure of the learner. Two things are necessary in order to optimize this process: In accordance with insights stemming from the psychology of learning, concepts to be learned are to be acquired step-by-step, mediated by different levels of representation; to avoid overwhelming the students with an abundance of unimportant details, one must concentrate on the basic concepts of the particular discipline. The most competent experts in each field are to develop such basic concepts.

It is no surprise that precisely the proponents of a radical orientation to science often appeal to such an interpretation of Bruner's theory and that, then, the construction of school curricula – especially in the area of mathematics and the natural sciences – often neglects didactic and educational theory issues of a broader scope and is restricted to a list of technical terms based on academic systematizations, perhaps supplemented by suggestions for a categorization according to Bruner's levels of representation (cf. Frey/Isenegger 1975, pp. 159-160). Incidentally, at a later date, Bruner himself spoke out against such limited interpretations of his approach (Bruner 1971).

The proponents of a 'main ideas' concept of mathematics instruction mentioned above appeal to an interpretation of Bruner's approach that is in many ways opposed to the one just examined. In this view, Bruner's advocacy of learning involving discovery instead of simply being receptive assigns him a position similar to that of the proponents of the genetic principle. Thus, the 'structure' of a subject does not simply result automatically from the adoption of basic technical terms from the corresponding academic discipline. According to this interpretation, Bruner's "basic concepts" and "fundamental ideas" become crystallization points for realizing what in a particular subject is of general significance, for recognizing 'ideas' precisely in the sense proposed by both Whitehead and Wittenberg. Then, Bruner's curriculum theory attains exactly what seemed to be totally lacking in the first interpretation – a dimension related to educational theory. The decisive question, the answer to which leads either to the first or to the second interpretation, is the following: Are Bruner's "basic concepts" and "fundamental ideas" thought of as being internal to one subject or do they transcend it; is Bruner aiming at conventional scientific terminology or at concepts that connect specialized knowledge and methods with the 'world outside'?

With reference to Bruner's remarks on mathematics instruction, I would like to demonstrate that on this point he provides contradictory answers himself, that thus there are, indeed, good reasons for both interpretations, but that, in the final analysis, the second interpretation seems more appropriate within the context of the entirety of Bruner's work.

With the very first explication of what he means by the "structure of a discipline", Bruner provides a mathematical example. He refers to algebra as a method of

> "arranging knowns and unknowns in equations so that the unknowns are made knowable. The three fundamentals involved in working with these equations are commutation, distribution, and association. Once a student grasps the ideas embodied by these fundamentals, he is in a position to recognize wherein 'new' equations to be solved are not new at all, but variants on a familiar theme. Whether the student knows the names of these operations is less important than whether he is able to use them." (Bruner 1965, pp. 7-8)

Several things related to this example can be criticized. First, one can object on grounds inherent in the discipline: Within the context of solving equations, the "commutative law" (of addition and multiplication), the "associative law" (ditto), and the "distributive law" facilitate transposing the terms of a particular equation. By contrast, the two decisive 'inventions' involved in solving equations – the fact that one can, in principle, compute with 'unknowns' just as with numbers, and that the validity of the equation remains unchanged as long as the same operations are carried out on both sides of the equation – are on a totally different level of symbolic operation. This makes it doubtful that a student develops insight into the essence of solving equations if he or she has only understood the computational laws mentioned and can implement them.

A second critique: When Bruner notes that it is not necessary for the students to know the names of the operations carried out, we can agree with him up to a certain point: He is justified in opposing the didactic tendency to want to meet the requirements of scientific orientation by excessive use of scientific terminology – a tendency which took on rather peculiar forms not least within the context of 'New Math'.[89] On the other hand, we can ask: To what extent do the students really "comprehend" the "ideas embodied by these three fundamental operations" if they only need to use them (in a technically correct way)? Doesn't grasping the 'idea' of, for instance, commutation require that I explicitly make what I am doing a subject for observation, for reflection? Furthermore, should not what has been grasped also be given an explicit designation? That conceptualizing is dependent on the interplay between synthesizing verbalization and content-related concretization has been a familiar theme since Kant's time: "Thoughts without content are empty; intuitions without concepts are blind" (1956 [1781], p. 95 [A51/B75]).

A third critique: The interesting thing about solving equations – also beyond any applications inherent in mathematics – is something that Bruner appears to totally overlook in this context. Since I will be dealing with this topic systematically later, I will only mention it briefly here: An enormous number of widely varied situations that can be described quantitatively in some form or other can be mathematically modeled using equations. Formulated in symbolic mathematical language, the equation is, as it were, a model of the non-mathematical problem under consideration, which is reduced to only a few aspects in the equation. At least in the simpler cases, which are the usual content of classroom instruction (e.g., linear and quadratic equations), the mathematical methods of solving equations are practically a technology that allows the determination of unknowns in a standardized way. This technology functions totally independently of the content of the original problem. The relevant procedures for solving equations can be completely transformed into algorithms and, for this reason, can, in principle, be worked out just as well by machines. Nevertheless, the usefulness of the solutions discovered are, of course, dependent on the appropriateness of the model selected. Very often, this is what causes students so much difficulty in conventional mathematics instruction. Thus, if Bruner wants to attribute the idea of solving equations to the ideas of commutation, of distribution, and of association, then he cheats the students of two things (probably without even vaguely being aware of it[90]): of the realization that in this case the mathematical technology is only a subordinate aid which, in principle, even a 'stupid' machine can master; and of insight into the connection between mathematics and the non-mathematical world, which can be experienced by modeling with equations. In the final analysis, this insight is nothing other than the second fundamental experience as designated by Wittenberg (see above).

Thus, considered by itself, the algebra example would seem to indicate that, in the case of mathematics, Bruner's argument in favor of teaching fundamental structures is solely concerned with a focus for mathematics instruction on the major technical terms from the perspective of structural mathematics or on the corresponding mathematical operations. His faith in the competence of the best specialists of a discipline to determine such basic concepts for the purpose of general education (ibid., p. 19) seems to support the assumption that he sees "fundamental ideas" exclusively as ideas inherent in the particular discipline. Yet, there are also indications that Bruner would like to connect the "structure of the discipline" and the "fundamental ideas" representing it to the non-mathematical world. Thus, referring to mathematics and attempting to clarify his "spiral curriculum", Bruner writes,

> "If the understanding of number, measure, and probability is judged crucial in the pursuit of an exact science, then instruction in these subjects should begin as intellectually honestly and as early as possible in a manner consistent with the child's forms of thought." (ibid., pp. 53-54).

This statement can be interpreted as being closely related to Whitehead's line of thought and, at the same time, far removed from a reductionist educational standpoint based on structural mathematics. Moreover, when Bruner emphasizes the importance of intuitive and heuristic thinking and of understanding for the exact sciences in comparison to analytic thinking along exclusively formal lines (ibid., pp. 55-68), this is also more in congruence with a broad interpretation of his curriculum theory as opposed to a narrow, specialist one.

If the ambiguous statements on the "structure of the disciplines" and on the nature of the "fundamental ideas" are seen in the context of the entirety of Bruner's work (e.g., in the context of his remarks on "discovery learning"), then the broader interpretation of his theory is to be preferred. Those examples that seem to indicate that a curriculum developed from a structural mathematics perspective in a strict sense would realize Bruner's intentions actually only demonstrate that at certain points Bruner uncritically adapted ideas and exaggerated hopes of the 'New Math' movement which were encouraged by the *Zeitgeist* of the 1960s.

3.2.3 Synopsis and Evaluation of Catalogs of Main Ideas

The approaches briefly examined below interpret the concept of 'main ideas' in very different ways and, on the whole, present a rather disjointed spectrum of such ideas. Apart from three exceptions, the suggestions to be examined were publicized within the last two and a half decades. The exceptions are, on the one hand, the approaches of Whitehead and Bruner, which I will be including in the synopsis as points of reference, and, on the other hand, a syllabus (for the *Gymnasium* in North Rhine-Westphalia) from 1963, which incorporated the concept of main ideas – indeed, without any effects on the subsequent syllabus of this German state. Setting different priorities, the proponents of the concept of 'main ideas' pursue the following three objectives:

(1) The often lamented phenomenon that from the viewpoint of the students, mathematics instruction dissolves into a simple collection of separate, individual activities and topics, is to be avoided. Orientation to main ideas should help students to assign a more generalized meaning to their activities in the mathematics classroom, a meaning that does not first become perceptible at the end of a long learning process, but that accompanies this process and can motivate the students' efforts.

(2) With the main ideas, the students should acquire an appropriate image of mathematics, should be able to recognize the "structure of the discipline". As with objective (1), the main ideas are to help interconnect the often isolated individual topics; yet, the emphasis here is not on helping students understand, but on clarifying the specific characteristics of the discipline and illuminating the distinctive features of mathematical thought and of the formulation of mathematical concepts.

(3) The main ideas should allow the learners to perceive how the mathematics taught in school is connected to their other experiences of the world and to their own thinking. In comparison to objective (2), the emphasis is not so much on the 'distinctive features' of mathematics, but on its meaning and function for recognizing and forming the world around us, on enabling students to experience how mathematics is also relevant outside the scope of mathematical specialization.

Obviously, for the issue of how school mathematics teaching can promote cultural competence, objective (3) is especially significant.

Synopsis: Catalogs of Main Ideas

Source	*Catalog of Main Ideas*	*References, Theoretical Basis*	*Priority of Objectives in Consecutive Order*
Whitehead (1913/1967)	• number • quantity (measurement, functional graph) • space	none mentioned	(3)–(1)–(2)
Bruner (1960/1965) (incomplete list for clarifying the concept 'structure of the discipline' and spiral curriculum)	• number • measurement • probability	cognitive psychology, general curriculum theory	(2)–(1)–(3)
Guidelines for Instruction at Secon-	• concept of number • extremes	none mentioned	(2)–((1))–((3))

Source	*Catalog of Main Ideas*	*References, Theoretical Basis*	*Priority of Objectives in Consecutive Order*
dary Schools – Mathematics (Minister of Culture, North Rhine-Westphalia 1963)	• function • geometrical representation • vector • set, structure		
Jung (1978)	• algorithm • infinity • measurement • … (suggestion: a maximum of seven)	Whitehead, Rumelhart (schema, argumentation), Vollrath (1978) (opposing view)	(3)–(1)–(2)
Vollrath (1978) (The ideas listed are only examples pertinent to analysis – from an incomplete list.)	• extremes • convergence • speed of convergence • development of a function • …	Rademacher/Toeplitz (1968[1933]), Wagenschein (1965) [V. refers primarily to mathematicians]	(2)–(1)–((3))
Schreiber (1979)	• algorithm • modeling • invariance • optimization • function (graph) • characterization	Whitehead (1962[1913]), Bruner (1970)	(3)–(2)–(1)
Tietze et al. (1982)	• algorithm • approximation • forming models • function • geometric transformation • linear transformation	Bruner (1970), Heitele (1975), Jung (1978), Vollrath (1979), Schreiber (1979)	(2)–(3)–(1)
Baireuther (1990) (examples from an incomplete list)	• geometry as a representation of reality • geometry as an instrument for planning • numbers: meaningful, since	none mentioned	(1)–(3)–((2))

Source	*Catalog of Main Ideas*	*References, Theoretical Basis*	*Priority of Objectives in Consecutive Order*
	oriented to practical needs • numbers: written in decimal form, put onto number lines		
Steen (1990) and Coauthors	• dimension • quantity • uncertainty • shape • change	no single list of references: various authors	(3)–(2)–(1)

Commentary

Even at first glance, it is obvious that the catalogs of main ideas listed here cannot simply be integrated into a single conception. Nevertheless, I wish to use the catalogs as a sort of 'excavation site' for a preliminary draft of a further catalog that is explicitly committed to mathematics teaching's goal of promoting cultural competence. To this end, I will review the authors mentioned once more in relation to the discussion thus far in order to select individual aspects, to emphasize them, or to discard them.

Even today, we can still agree with Whitehead's major concern (1967[1913]), which can be characterized as having a strong orientation to objective (3). In any case, the ideas he mentions – number, quantity, space – should constantly be taken into consideration in an up-to-date mathematics curriculum. They are, as it were, archetypes[91] of mathematical thought to which there are corresponding transcultural basic mathematical activities, as can be surmised from the previous section. However, Whitehead's catalog needs to be expanded since it inadequately reflects more recent mathematical developments that have gained significance in our society.

With his mention of 'probability', Bruner (1965[1960]) – to whom I refer here explicitly in the sense of a non-specialist interpretation of his curriculum theory – already indicates a possible meaningful supplement to Whitehead's catalog. The mathematical treatment of stochastic phenomena has acquired such significance that it deserves considerable attention within mathematics instruction designed for general education.

The guidelines from 1963 are to be understood as an historical attempt to explicitly incorporate elements of the 'main ideas' concept into syllabus design. The limited selection of basic concepts should be criticized: Essentially, these are purely mathematical concepts instead of integrative ideas going beyond disciplinary systematization. At the time, most teachers and textbook authors interpreted the proposals outlined here in exactly this restricted sense: In the mathematics classroom, certain concepts were to be more strongly focused on than others. Thus, this selection appears open to question.

To the ideas already mentioned, Jung (1978), whose main intention is to further develop Whitehead's approach, adds those of 'calculation' or 'algorithm' and of 'infinity'. Calculation and algorithm represent an aspect of mathematics that has increasingly acquired practical significance due to the possibilities of carrying out computations and symbolic manipulations with the use of computers. In an algorithm that can be worked out by humans and machines without any comprehension, the mathematical intelligence indispensable to the development of the algorithm is stored and made available in 'crystallized' form. While standard mathematical curriculum often focuses on the mastery of (more or less complicated) calculations and algorithms, reflecting on the idea of algorithm offers the student an opportunity to distinguish between the (mathematically demanding) development of algorithms and simply working out algorithms and to gain insight into one essential reason behind the universal role of mathematics in our present culture. – Among the authors mentioned, only Jung explicitly suggests paying considerable attention to the 'idea of infinity'. He makes this suggestion primarily because of the manifold possibilities he sees for intensifying the philosophical dimension of learning mathematics. According to Jung, the question of how the idea of infinity can be comprehended at all in a finite world presents itself: In which sense and with what justification can we speak of the existence of the infinite? This is one of the controversies between supporters of Platonic theories and constructivists. Jung does not even mention the connections to the history of mathematics (starting from the first efforts of ancient Greek thinkers and continuing to the 'invention' of infinitesimal calculus and further up to the antinomies of set theory), but they are obvious. Although profound discussions on the subject can certainly be carried out with philosophically interested young people, the critical question to be asked here is whether such a prominent position should be granted to the infinite within the framework of 'mathematics for all', whether it is not much more the case that this would be an obvious supplementary topic for students especially interested in mathematics.

Vollrath (1978) – who only speaks of 'ideas' (without characterizing them as 'main' or in some similar way) – pursues the concept of working out the ideas behind mathematical terms in the course of instruction. He is, indeed, also interested in allowing central themes to thus pervade the individual topics of mathematics instruction; yet, since with this approach practically any mathematical concept of a certain importance can be related to an 'idea' (e.g., the idea of extremes, the idea of convergence), the structure of the entire curriculum aimed at can hardly be distinguished from the usual ones based on systematic arguments inherent in the subject matter. However, Vollrath is less interested in such a generalized structural plan than in a more intellectually stimulating treatment of the usual topics and concepts in the classroom. Following these suggestions and orienting mathematics instruction to such mathematical 'ideas' would more likely promote coherence within the subject (which is lacking often enough in standard mathematics instruction!) than further cultural competence in the way aimed at here. – Similar comments could be made on the suggestions put forth by Fischer (1979), Schupp (1984, 1992), Picker (1985b), Hering (1985), and Kütting (1985), which, for this reason, have not even been included in the synopsis. All of them focus on selected topics or individual ideas (e.g., Schupp emphasizes optimization), not on a set of main ideas for mathematics teaching as a whole. It would certainly be worthwhile for further studies to investigate the possible development of plausible hierarchies or of networks of guideline ideas within which ideas specific to particular topics would be variedly connected to those found to be of major importance.

By contrast, Schreiber's remarks (1979, 1983) are explicitly oriented to an all-encompassing perspective for the school mathematics curriculum. To a great extent, the ideas he mentions disregard individual mathematical topics. Only the idea of function refers directly to a concept of the standard curriculum. Although Schreiber, too, cites Whitehead among his sources, such mathematic activities as counting, measuring, and structuring space, as significant as they are for competence in everyday life, have no explicit part in his catalog. They are included within more abstract considerations, if, indeed, they are not obscured by them. Although such topics may be enlightening in certain contexts, the "method of exhaustion", "invariance", "optimization", and "characteristic function" seem hardly imaginable as elements of a conception of mathematics that would be accessible to young people. The comments on Jung have already clarified the significance of algorithms. The idea of function or, more precisely, of functional dependence would seem to be continually relevant and can be imparted to students in various ways. One aspect to which Schreiber makes casual reference in order to clarify the "method of exhaustion" should be reassessed and given higher priority: Isn't it possible to synthesize connections between mathematics and the 'real' world of experience that are perceptible in counting, measuring, etc. in a more all-embracing way that is also well-suited to the students' situation by using the idea of 'modeling'?

Within the context of their textbook on mathematics instruction for the upper grade levels of secondary schooling, Tietze et al. (1982) only mention their catalog in passing. They attempt to "summarize on a more general level" (ibid., p. 41) ideas specific to certain mathematical topics and treated in different chapters of their book. I have included this catalog in the synopsis primarily because it takes more 'modern' developments in school mathematics into account and because explicit reference is made to the idea of forming models.

The main ideas are of an entirely different nature for Baireuther (1990) than for the authors commented on so far. He is concerned with allowing students to experience meaningfulness in their concrete actions with respect to as many subjects as possible. For this reason, the ideas around which Baireuther would like to center instruction are explicitly directed toward the students: For every unit, students can write down the ideas as generalized mnemonic formulas in their notebooks. Despite what thus might be called an accent on teaching methods in this conception of ideas and despite the somewhat arbitrary selection and varied significance of the ideas formulated by Baireuther, his approach is worth considering in reference to promoting cultural competence. Baireuther is concerned with constantly linking the mathematical activities of the students, the applied mathematical concepts and techniques, to the 'real' world and to the intellectual horizon of the students, that is, with cultivating that connection between mathematical and non-mathematical culture that is often neglected in standard instruction. Baireuther suggests illustrating the mathematical abstractions as often, as creatively, and as diversely as possible with the use of concrete ideas and actions (i.e., those familiar to the students or subject to their sensory experience). This way of treating main ideas can certainly also apply to catalogs that, in other respects, are more convincing than Baireuther's selection in demonstrating the cultural significance of mathematics.

Steen's publication (1990)[92] represents the current attempts at a reform of mathematics instruction in the USA. In five individual articles that can each be read separately in their own right, each author focuses on a particular mathematical idea, its integrative power, intellectual depth, and manifold applicability, and demonstrates these aspects with a considerable number of examples. Even this structure of the book already indicates that the emphasis is not on a systematic examination of the resulting catalog of main ideas, but, instead, on demonstrating for each idea selected the abundance of possibilities for establishing connections to other mathematical and non-mathematical phenomena. Since the ideas examined in the individual articles cut across the usual systematic classifications of the discipline, certain topics and areas of school mathematics (numbers, functions, three-dimensional shapes) are explored and illuminated from various viewpoints. Again and again, applications of mathematics that are socially relevant and important for everyday life are considered, and thus, in refreshing contrast to the suggestions commented on so far, this American publication proves to be an interesting contribution on the issue of how mathematics instruction can promote cultural competence – even though this issue is not an explicit theme of the book.

3.2.4 Main Ideas for Mathematics Instruction: Some Suggestions

It would be unreasonable to assume that we could now, as a final step, submit a catalog of main ideas for up-to-date mathematics instruction that would avoid the weaknesses of the suggestions already present in the debate, would be theoretically consistent and, at the same time, would be an aid for the concrete development of curricula. Nevertheless, I would like to make some suggestions for further discussion.

My primary concern with respect to the desired contribution of mathematics teaching to promoting cultural competence is the following question: Which ideas illuminate for students the universal characteristics specific to mathematics that are based on abstraction and on symbolic techniques attained through abstraction? In considering this issue, I attempt to satisfy the following conditions: None of the ideas should exclusively represent a particular mathematical topic. Each of the ideas should be explicable at various cognitive levels and should pervade the mathematical curriculum as a central theme from basic instruction up to higher-level mathematics (Bruner's "spiral curriculum"). At least some inkling of the 'transcultural basic mathematical activities' discussed above, as the universal basics of doing mathematics, should be perceptible in the ideas. Finally, the ideas should be susceptible to in-depth exploration at any point of the learning process: Students should also be able to profit from dealing with an idea even if they do not reach the final goal previously established for their mathematics instruction.

First, here is a list of the entire catalog of ideas:

- the idea of number
- the idea of measuring
- the idea of structuring space
- the idea of functional relationship
- the idea of an algorithm
- the idea of mathematical modeling.

Each of these ideas appears in at least one of the catalogs discussed above. I will comment on them in consecutive order to demonstrate that each one in and of itself takes the primary concern mentioned above into account and that, in addition, they are all reasonably compatible with one another.

The Idea of Number

The counting of concrete, distinct objects can be found in all known cultures and can be considered to be *the* fundamental mathematical activity. In our culture, children learn the basics of counting just as naturally as their native language, and the boundary between colloquial speech and mathematics is flexible. The abstraction expressed in my considering the 'number' of a quantity of concrete objects requires just as little reflection on my part as the fact that with the words of informal speech I abstract from the material nature of the objects designated. But it is characteristic for the advance in mathematical thought that at the next stage of abstraction, at the level of computation, the new objects of thought, the numbers, become independent and agents of their own operations – of adding and subtracting, multiplying and dividing, comparing and classifying. Even for the levels of abstraction considered thus far, the idea of number determines and pervades our everyday culture in manifold ways. For example, we might call to mind dealing with numbers in connection with time and money – an activity that is already of major significance for preschool children. If promoting cultural competence is part of a general education, then mathematics instruction should stimulate reflection on this far-reaching impact of numbers on our everyday culture.

To briefly list other important forms and stages of abstraction connected to the 'idea of number': working with numbers that do not represent any concretely available quantities of objects; counting 'abstract objects' such as (possible) results of a chance experiment; computing with variables, with which we abstract from 'concrete' numbers; describing rules stipulating the relationships among numbers and the operations possible with them; supplements to the realm of numbers, etc. Very complex progressions lead from the naïve application of numbers as the fundamental mathematical abstraction to the axiomatic reasoning concerned with natural or real numbers and, further, to profound hypotheses on number theory and algebraic structures. At another level, possible applications come into view since numbers are symbolically represented in a specific way: codes for purposes of classification or recognition (such as telephone numbers, house numbers, zip codes, article or item numbers) have become indispensable aids for everyday life in our society.

The Idea of Measuring

Activities that in the broadest sense can be subsumed under the concept of measuring are also fundamental, present in all cultures, and in this sense universal. If counting immediately proceeds from distinguishable, discrete objects, measuring concerns continuous phenomena that must be quantified. Then, the idea behind measuring, comparing what is to be measured with a defined unit, allows us to also express the results in numbers – an invention of enormous consequence. On the other hand, the idea of measuring has repercussions on the idea of number – this idea now demands numbers other than natural numbers. As the problem of incommensurability (treated as early as in the work of Euclid) demonstrates, the assumption of being able to measure any amount of distance inevitably leads to postulating the existence of 'more refined' numbers than the rational numbers (or, to avoid the implicit Platonism in this statement: postulating that it makes good sense to construct 'more refined' numbers than the rational numbers).

In measuring activities, mathematics is connected in manifold ways to the 'real' world of sensory perception. This is immediately obvious where something can be measured 'directly': e.g., measuring length by repeatedly implementing some unit of measurement (pacing out an area), measuring an area by partitioning it, measuring time by counting periodically repetitive processes, or weighing by balancing against standardized weights. Even at this early stage, the physical nature of the world around us has an apparent impact: When do periodic processes have the 'same' duration; how do scales work? This applies especially to the next level of abstraction, where the measurement is indirect or where we work with derived units of measurement. For example, what does the 'speed' that we can read on the speedometer of a car signify? Only the possibility of measuring makes the formulation of quantitative natural laws possible in the first place – laws which then enable us to make exact predictions. Thus, the mathematics attained by abstracting from the 'real' world has concrete repercussions on this world.

As with the idea of number, the idea of measuring pervades and influences our entire everyday culture – even as far as the experiences of preschool and elementary school children are concerned. But whereas numbers are obvious everywhere, the idea of measuring obscures itself in the world that has been measured. The kilogram cartons in the supermarket have been weighed and packaged; the distances between cities are indicated by signs, etc. All of the usual instruments used for measuring – digital scales for fruit and vegetables, digital clocks, thermometers, odometers and speedometers in family cars – immediately display complete numbers. Children can no longer vividly reconstruct how these numbers have come about. For this reason, to promote cultural competence, mathematics instruction should provide opportunities for vivid elementary measuring experiences. But equally important is the explication of and the reflection on theories of measurement at the upper grade levels. Here, an interdisciplinary emphasis, taking the natural sciences or the empirical social sciences into account, is of the greatest significance.

The Idea of Structuring Space

Cross-cultural studies among preindustrial peoples have made us more sensitive to the significance of an abstraction that we often take for granted. This abstraction is concealed in the transformation from the space of our sensory perception to 'Euclidean space', thus, from the space in which I move around and which I visually perceive in an unreflective way to an imagined space in which the relationships between idealized objects only constructed in the imagination (such as points, lines, and planes, circles and triangles, spheres and tetrahedrons) can be theoretically described according to a strictly logical classification. The mathematical thought of those who are capable of this level of abstraction has access, as it were, to a 'secondary visualization': The enormous abundance of relationships and correspondences between arithmethical and algebraic regularities on the one hand and geometrical ones on the other allow spatial visualization to become both a corrective and a source of inspiration for mathematical thought that is not primarily concerned with geometrical aspects. Incidentally, the geometrical thought of creative mathematicians goes far beyond Euclidean space and can even arrive at a 'tertiary visualization'. For students, Euclidean space would normally be the appropriate framework for the acquisition of the 'idea of structuring space'.

No other mathematical idea has literally shaped the known advanced civilizations (not only Western culture) in such a conspicuous way as has the idea of structuring space. Basic geometrical forms and design principles such as symmetry can be found all over the world – from simple, everyday objects to religious architecture; aesthetic principles of the fine arts (even of such an 'abstract' art as music) can often be traced to geometric principles – even up to the point of the conscious constructive use of geometric knowledge in perspective painting (this last example does, indeed, represent a development restricted to Western culture and its advanced Euclidean geometry).

The idea of structuring space finds expression in the basic cultural activities distinguished by Bishop as 'locating' and 'designing'. This distinction seems less relevant to mathematics instruction. It would seem more important to pay sufficient attention to both constructive activities (drawing, forming, drafting, geometrically constructing) and to the cultivation of geometric perception, to the interpretation of our surroundings in geometrical concepts and relationships. The usual restriction of geometry to plane geometry from lower secondary schooling onward should be avoided as often as possible. The idea of structuring space can only be developed if three-dimensional Euclidean space is also repeatedly brought into play as a frame of reference in both a theoretical and a practical sense.

In conjunction with the ideas of number and of measuring, an abundance of interconnections can be discovered by means of structuring space – which itself also includes geometrical operations on a two-dimensional level. Thus, the visualization of data, for example, by different forms of graphical representation (in statistics or as results of measurements) has become a decisive factor in our everyday mathematical culture, the understanding and critical examination of which should be furthered by mathematics teaching that promotes mathematical cultural maturity.

The Idea of Functional Relationship

The fact that the world around us does not simply appear chaotic to us, but, to a great extent, ordered and predictable in its development, is a result of our being able to postulate relationships between phenomena originally perceived as distinct. We can describe regularities and their possible causes because we often notice that similar events or similar human actions have similar consequences: thunder follows lightning, night follows day, eating a meal satiates us, sowing a field leads to the possibility of harvesting crops. In all known cultures, the interplay between experience and reflection has brought about an extensive accumulation of knowledge concerned with how to take advantage of these regularities and thus secure life and survival.

It was reserved to Western modernity to significantly expand the scope of this knowledge of relationships. The discovery that certain aspects of the 'real' world can be observed and quantified in isolation and that the relationships between these quantified aspects can be described by mathematical means marked the actual beginning of the exact sciences in the modern sense. Not only the natural sciences, but also many social sciences and disciplines in the humanities increasingly make use of the possibility of describing empirical relationships quantitatively. With the idea of the functional relationship between directly or indirectly measurable dimensions, an additional main idea beyond that of measuring is brought into play, an idea that is indispensable for understanding the cultural role of mathematics. Without the mathematical formulation of functional relationships, natural laws in the modern sense would simply be inconceivable, and, for this reason, without comprehending the idea of functional relationship there can be no profound insight into scientific and social progress.

The study of functions as new and considerably more abstract subjects than numbers or geometrical objects has provided theoretical development within the mathematical discipline with decisive impulses. The more detailed study of functional relationships characteristic to modern mechanics led Newton and Leibniz to the 'discovery' of differential and integral calculus. Far beyond immediate applications in physics, further theoretical exploration of and research on functions has also been productive for other mathematical fields, such as algebra and geometry.

More so than is the case with the three ideas discussed above, the idea of functional relationship goes beyond the everyday mathematical culture of our society. Functions cannot be 'seen' in the same way in which everyone in our society is constantly confronted with numbers, measurements, and geometrical forms and representations. If young people become familiar with the idea of functional relationship, then they have access to a theoretical concept that can help them detect and clearly recognize 'latent' mathematics in their everyday lives.

Being familiar with the idea of functional relationship is something different from the mathematically correct handling of the types of functions usual in classroom instruction. The idea of functional relationship connects everyday knowledge to a powerful mathematical method. In the course of instruction, students must be given the opportunity to discover this interconnection anew or to make it a matter of conscious reflection and comprehension. For example, they can then discover that an observation formulated at first rather vaguely (as 'the more there is of one thing, then even more there is of another') can, under certain conditions, be described much more precisely as a proportional or linear function; or they might find that if the increase of a certain dimension – as with many rates of change – is proportional to the value of this dimension itself, exponential functions describe this sort of relationship in a reasonable way. The cultural significance of the mathematical abstraction behind the concept of function becomes accessible in such experiences. Thus, the mathematical formulation of functional relationships proves to be a universal means of theoretically interconnecting and symbolically working with measurable changes in the world around us. – Certainly, for a small portion of young people, the following will also be of significance for their general education: The increasing level of abstraction so characteristic of advanced mathematics can be comprehended on the basis of functions. In differential equations, variables no longer represent unknown numbers, as in usual equations, but unknown functions; and functions can also be interconnected with one another and comprehended as elements of algebraic structures.

Some of the manifold interconnections to the ideas of number, of measuring, and of structuring space have already been indicated. Functions operate on numbers and produce new numbers; they describe and interpret series of measurements; they can be represented by geometric curves and as such geometrically studied, and the converse is also true. Graphs used in elementary geometry (to illustrate rates of change, reflection, or rotation) can help to prepare a general understanding of functional relationships that emanates from the idea of structuring space.

The Idea of an Algorithm

Expressed in terms of ordinary language, an algorithm is a sequence of unambiguous instructions that describe the route to the solution of a problem exactly and completely. There are algorithms not only in mathematics and computer science (such everyday actions as looking up a telephone number in the telephone book or baking a cake can also be presented as algorithms), yet, they do attain special significance within these two fields. Even elementary school children already learn how to handle algorithms, for instance, in acquiring the written, basic types of computation.

Although developing and working with algorithms was already familiar to the Babylonians (Heron's method) and to the ancient Greeks (Euclidean algorithm, sieve of Eratosthenes) and has been one of the standard tasks of mathematicians at the latest since the beginning of the modern era, certainly even forty years ago no mathematician, mathematics teacher, or educational theorist would have thought of suggesting the 'idea of an algorithm' as a main idea of a mathematics curriculum designed for general education. Algorithms have gained considerably more attention since programmable computers became accessible as mathematical tools. For with the use of the computer as a "symbol-processing universal machine" (Bussmann/Heymann 1987, pp. 15ff) any algorithmic problem can, in principle, be solved. Conversely, this fact also indicates the limitations of the computer, as mathematicians such as Gödel and Turing proved early on in their brilliant theoretical writings.

In traditional mathematics instruction, working out algorithms correctly is a decisive factor up to and within the upper grade levels of secondary schooling. One example of this would be debating the topic of functions or curves, still a very popular problem for tests and final exams. The status of such problems within the framework of a mathematics instruction designed to further general education should be reconsidered – especially since fully developed computer software has become available that provides answers to all of the usual questions concerning a given term of a function (zero of the function, extremes, points of inflection, the progression of the curve for given intervals).

Programmed machines can work out given algorithms more quickly and more reliably than humans; on the other hand, even in traditional elementary school instruction there are already certain tasks and problems that cannot be put into the form of an algorithm. Thus, we should make a fundamental distinction between working out specified algorithms and the creative activity of constructing algorithms for solving a certain problem. Not that the former is always simpler than the latter – applying a complicated algorithm without any errors can be very difficult and can require enormous concentration. But an understanding of the idea of an algorithm will only come about if students can amass experience in both activities and can reflect on their experience.

In what respect does an algorithm represent a universal idea illuminating the significance of culture? In exemplary fashion, algorithms can be used for comprehending the separation of applied knowledge from the larger context of insight on some subject. While working out an algorithm, I need not be concerned with (and need not even understand) why the algorithm works. The intelligence and creativity put into its development are 'stored' in the algorithm; the user need not have access to the knowledge invested in constructing the algorithm. This applies to carrying out written division and to determining the root of a number; it also applies to solving a quadratic equation as well as to differentiating a given function. But it equally applies to the user-friendly interface of a personal computer (which, below the surface, is algorithmically structured) and to commercial software for solving standard problems of all sorts. The principle of an algorithm can be generalized even further: It reflects the principle of industrial production. Industrial production methods permit the manufacture of standardized, high-quality products relatively independently of the skills and knowledge of the workers involved. The development of an industrial production method, which combines mathematical, scientific, economic, and technical expertise, is essentially nothing other than the construction of an algorithm. Conversely, a mathematical algorithm is nothing other than a (more or less industrially standardized) solution to a class of structurally related problems which, with intellectual effort and sufficient expertise, could all be solved individually, i.e., without applying the algorithm. To illustrate this with a simple example: A division problem that I have standardized with the written algorithm and can conveniently solve can also be solved 'directly' if I understand the principle of division – by making trials and checking with multiplication. With the idea of a mathematical algorithm, the students are confronted, as it were, with one of the major ideas of modern civilization – including the limitations of its applicability. If this can be made clear in the course of mathematics instruction, then, this would indeed be an essential contribution to a general education and, more specifically, to promoting cultural competence.

The Idea of Mathematical Modeling

The idea of mathematical modeling as a main idea of mathematics instruction is also a fairly recent arrival to the debate. It is situated at a different level than the ideas considered beforehand, but precisely for this reason, it allows a synthesis of our considerations up to this point. A sound idea of mathematical modeling permits us to put the more 'classical' ideas (discussed first above) and the idea of an algorithm into a common framework.

The concept of modeling can describe the applicability of mathematics and its relation to the 'real' world in a very general and, at the same time, quite elementary way.[93] Whenever mathematics is applied to describe and clarify objective situations and to solve real problems, a mathematical model is constructed (or, recourse is taken to an already existing model). Assertions about the relevant situation or solutions to the problem under examination resulting from the use of the model are not valid in isolation from the model. They are in need of interpretation and must be checked for their appropriateness. – To illustrate these thoughts, I will provide two simple examples and relate them to an everyday situation.

Natural numbers can be conceived of and studied as independent mathematical subjects; in conformance to the mathematical laws recognized as being valid, one can use these numbers to count, compute, and operate at will. But as soon as we count real objects (e.g., potatoes), we make use of the numbers as a mathematical model. If I am concerned with buying the potatoes counted, the count probably will be of no use to me. Since the potatoes do not all weigh the same and the seller originally purchased them by weight, he or she will also want to resell them by weight (I might choose to buy the biggest ones!). Thus, even in the case of a correct count, in this situation the counting model is not useful for determining the price to be paid.

Or, we can consider linear functions: We know how to work with them mathematically, which operations are permissible and which are not. But as soon as some presumed functional relationship in the real world is to be represented by a linear function, this function is a mathematical model of that relationship. For the purchase of goods, it is customary to set the price to be paid proportionately to the number or quantity of the goods to be bought. To return to our example above, if I have fixed the quantity of potatoes to be bought by weighing them ('measuring'), then I can determine the price to be paid from the results of my measurement by means of a linear (or, here, even proportional) function. The appropriateness of the model of 'proportional function' is ensured by a custom which in our society is a common practice among retailers, applying to the purchase of small amounts of goods (for larger amounts, discounts would also be a factor to be accounted for – the mathematical model would have to be changed). If I compute the price using the rule of three, in principle, I make use of the same model (or, if you will, of a variation of the model).

Taken by itself, considering these two examples with reference to the mathematical modeling behind them would, of course, be mere child's play and would unnecessarily complicate the use of mathematics in everyday situations. The perspective of modeling theory becomes productive primarily in cases in which the interconnectedness of mathematics and 'reality' is not immediately perceptible. The selection or construction of adequate mathematical models is particularly important for examining stochastic phenomena or rates of change. The usefulness of computer simulations that are increasingly being implemented for scientific analyses and prognoses is especially dependent on the models behind the simulations. Often enough, misinterpretations or exaggerating the relevance of the results of such computations are due to insufficient insight into the modeling nature of the mathematical procedures used. The internal rationale of the models used is guaranteed by the rationale of the mathematical theories invoked, by the reliable functioning of the algorithms built into the model. But how well the mathematics implemented fits the issue to be examined is a problem in its own right, precisely a problem of *modeling*. This problem cannot be solved only on the basis of mathematical expertise; rather, sufficiently subtle knowledge of the content of the problem to be modeled is also required.

The idea of modeling not only permits a better understanding of the role that mathematics plays in analyzing and solving individual concrete problems, but also generally characterizes working procedures in the exact sciences. The physicists who work with formulas and equations, measure dimensions and establish relationships among them, and develop theories for specific areas within physics construct mathematical models. An essential part of their scientific activity consists in developing mathematical models that correspond to the known empirical facts and, with the aid of such models, in making predictions which, in turn, can be checked empirically. In this sense, the idea of mathematical modeling does, in fact, prove to be universal because at various levels – from counting potatoes to relativistic quantum field theory – it illuminates the interconnections between 'mathematics' as a consistent intellectual construction and perceptible or symbolically mediated 'reality'.[94] Stating that mathematical modeling 'works' at all, is actually another way of expressing the second fundamental experience that mathematics instruction oriented to general education, according to Wittenberg, should provide.

How can we treat the idea of mathematical modeling in the classroom? I will be restricting myself here to only a few hints that seem important for avoiding the misconception that the argument behind the idea of modeling aims at letting students use the word 'model' as often as possible.

The insight that the connection between reality and mathematics is mediated by models can already have an enriching effect on mathematical learning processes even if the term 'model' is not used at all during class; in many cases, it will be sufficient if the teacher is conscious of this modeling aspect of mathematics. Yet, as the students progress through the lower grades of secondary schools, there is a marked increase in opportunities to expressly speak of 'models', e.g., within the framework of elementary geometry, in functional analysis (linear functions, rates of change), and especially in probability.

In sum, the idea of mathematical modeling establishes a firm connection between 'mathematics' as a cultural product and non-mathematical cultural aspects of our Western industrial society. One can clearly demonstrate the potential of mathematics for developing and changing civilization with the use of the idea of mathematical modeling. In addition, other main ideas of a mathematics curriculum designed for general education can be related to one another in manifold ways with the aid of this idea of mathematical modeling.

3.3 Developing an Understanding of the World in Mathematics Instruction

3.3.1 Mathematics as Part of the Real World

Developing an understanding of the world – we described this objective of schooling as: broadening the horizon of perception and discernment beyond the everyday horizon of students; systematization of thought; establishing a differentiated view of the world. Generally speaking, the subject areas of instruction represent segments of the world and, at the same time, they represent views of the world that need to be supplemented. They concentrate on one aspect of the world of sensory perception or of symbolic mediation and intensify this aspect in a way characteristic of the particular subject. Thus, contributing to an understanding of the world with or through school subjects means, first of all, establishing a connection between the subject matter's (objective) relation to the real world and the (subjective) world of the student. Students should be able to integrate that part of the world they encounter through the subject matter into their subjective idea of the world, to recognize and identify it as part of their world. This, in turn, requires that the segment of the world represented by the subject matter must not simply exist within a sphere restricted to the classroom, and it must not be totally disconnected from the students' intense everyday experiences outside school.

In the case of mathematics, even determining the 'objective relation to the world' is more difficult than for most other subjects. As a first approximation, we might formulate this problem as follows: *Mathematics is part of the real world and at the same time concealed within it*. Only to a small extent is mathematics perceptibly a part of our everyday world, namely, at the level of everyday mathematical culture. At this level it is also 'real' for children in the form of symbols encountered in widely varied situations and in the form of techniques and actions one becomes accustomed to at an early stage. Insofar as this type of mathematics is taught and learned in school, a great variety of everyday experiences outside school also corresponds to it. Its relation to the real world is still comprehensible for children and young people even if this theme is hardly touched on in the classroom. But what about the relation to the world *beyond* everyday mathematical culture?

Our original thesis that mathematics is part of the real world and at the same time concealed within it can be differentiated and formulated more sharply: Mathematics is not only part of the world, it is constitutive of our world. And this applies in a twofold sense: It is constitutive of the rational worldview of Western culture shaped by modern science and it is constitutive of the (in the broadest sense) technologically influenced life circumstances Westerners have developed since the beginning of the industrial revolution. On the other hand, the obscured nature of mathematics, its disappearance behind the phenomena, is characteristic for both spheres. I would like to comment more fully on these two aspects in turn.

With the development of modern science, the mathematics created by man proved itself as a profound means of perceptively describing (and, subsequently, mastering) the existing natural world surrounding mankind. The development of mathematics and the extension of its applications have been closely interrelated since the beginning of modern times in Western society. In comparison to its status in antiquity, mathematics has become an *epistemological tool* in a totally new sense. Mathematics as an epistemological tool is a prerequisite of a modern, scientific, and rational worldview. Our conception of the world as that of educated Westerners would be simply impossible without mathematics. But, normally, we no longer 'see' the mathematics that allows us to have this interpretation of the world. For example, we observe Jupiter in the night sky and interpret it as a planet that revolves around the sun in an elliptic orbit on the basis of our knowledge of the solar system. But we could not easily reconstruct the measurements and calculations done by Kepler and others to make their model of planetary orbits plausible. As part of our conception of the world, our knowledge is reliable even if it has removed itself from the mathematics originally implemented in the reasoning behind it. Our knowledge of natural phenomena is the result of employing mathematics as an epistemological tool to a much greater extent than is usually perceptible in the formulation of this knowledge.

Furthermore, mathematics is a constructive tool for uses related to the artificial world created by man. It is used in architecture and technology, in institutions of the private economy and of state administration; it shapes the exterior order and organization of our world – by means of calendars and the time of day; by surveying, partitioning, and registering land; by code systems from telephone numbers to zip codes and, even further, to the recording of personal data in private and government databases. Mathematics is, as it were, invisibly incorporated into many objects we use in everyday life – household appliances, means of transport, computers – without our having to explicitly reflect on this circumstance as users. Indeed, it is precisely a unique advantage of mathematics that its technological applications work regardless of whether the user is familiar with them or at least, in principle, conscious of them.

Thus, the relationship of mathematics to the 'real' world is an indirect one. The difference between mathematics and other scientific disciplines which aim directly at knowledge of the world can be characterized most clearly with reference to 'pure mathematics', for this discipline is its own object of study; it examines its own products. To the extent that school mathematics is oriented to the ideal of pure mathematics, it copies this feature – with the decisive modification that the objects studied are only rarely creatively 'produced' by students; instead, they are presented as finished products by the teacher (or by the textbook).

Of course, even the most tedious sort of classroom mathematics does leave behind traces, does effect certain changes, in the students' conception of the world. But this is not actually a broadening of their horizon, not a useful complement to their view of the world through a mathematical perspective, but only a simple appendix to such a worldview by means of mathematical objects. To use a more vivid illustration: The sort of mathematics that is learned does not become a window that offers a different, a new outlook on the world. Instead, it is just another isolated piece of furniture that perhaps even takes up too much space. The segment of the world represented by school mathematics in such a pedagogically most unsatisfactory and, measured against the claim of aiming at general education, most deficient case, consists only in itself: school mathematics. To the students, the knowledge that mathematics plays an important role in our world remains totally abstract since, apart from everyday mathematical culture, their only experience with mathematics is limited to mathematics as a school subject with ritualized problems and classroom activities corresponding to a predetermined canon of instructional material. In this way, school mathematics becomes a rather closed-out segment of the world, isolated from the rest of life.

If mathematics is constitutive of our world and yet, to a large extent, concealed behind the phenomena, then a special treatment of mathematics in the classroom is necessary to allow its relation to the world to become distinct. It seems indisputable that in this sense the application of mathematics to non-mathematical situations must play an important role. The dispute over which examples are appropriate and exactly how this application should take place is all the more intense. In the following sections, I will be analyzing to what extent some of the more common concepts of mathematics instruction oriented to applications or to making life circumstances more accessible can realize the claim of developing an understanding of the world.

3.3.2 Applications-Oriented Mathematics Instruction

In the past two decades, mathematics teaching both at the national and international level has been much concerned with the topic of 'applications', as an abundance of publications and initiatives testifies.[95] I will provide a brief overview and in doing so, I will essentially be following Schupp's presentation (1988) since he has contributed a sound tentative appraisal of the discussion, focusing on lower secondary schooling.

The term 'applications-oriented' is still relatively new. When it came into use in the 1970s, it signalized a decisive countermovement to 'New Math'. In the course of the preceding decade, the dominance of structural mathematics in the classroom and the increased orientation to mathematics as a science (which, in turn, was for the most part simply equated with 'pure mathematics') had resulted in only marginal significance of mathematical applications in syllabi, textbooks, and teaching practice.

On the other hand, the debate on the status of applications in school mathematics teaching is considerably older. They have never been totally neglected in the teaching of computation in primary schools (*Volksschulen*) since these came into existence, even though often enough the applications were made in a fixed, schematic way, and the proximity to real life being striven for remained unattained. Up to the present day, a considerable part of elementary mathematical instruction in the first grades is designated as basic computation. – Under the influence of neo-humanism, mathematical applications became largely excluded in the secondary schools of the 19th century, for they became equated with the scorned *predetermined purpose* of education and personal development. Since general education was primarily understood as formal education, it was, paradoxically, precisely the claim to the general educational qualities of mathematics instruction that reinforced the trend toward a self-complacent disciplinary field. Toward the end of the 19th century, efforts to disrupt this self-complacency finally began to increase. The greater economic significance of technological and scientific progress created an atmosphere in which scientific and technological interest groups could exert pressure on society at large. As early as the Meran Plans of 1905, explicit demands were made on mathematics instruction to develop "the ability to mathematically interpret the phenomenal world around us"[96]. After 1925, the pendulum began to swing back in the opposite direction,[97] the advocates of 'pure mathematics' and supporters of a formal educational ideal of a neo-humanist mould regained the upper hand. If we leave aside the attempts of the National Socialists to get school mathematics to serve their ideological purposes (military technology, biometrics), then topics related to application remained neglected in mathematics instruction until the 1960s. Even such educationalists as Wagenschein and Wittenberg, who at the time gained prominence as severe opponents of a specialist orientation, took recourse especially to inner-mathematical topics to demonstrate mathematics' potential for general education and personal development.[98] Later, as already mentioned above, the turn to 'New Math' actually produced a further reduction of a project that was by no means extensive to begin with. There is, indeed, mention of the necessity of "mathematically grasping reality" in the preambles to the Culture Ministers Conference's guidelines of 1968 for modernizing mathematics instruction; yet, this demand had no practical ramifications for the guidelines themselves (Damerow 1977, pp. 221ff; Schupp 1988, p. 9).

3.3.3 Mathematical Modeling as an Integrative Perspective for Applications-Oriented Mathematics Instruction

The most recent trend toward applications-oriented instruction (like 'New Math' before it, an international movement) has given the debate on the treatment of non-mathematical issues a totally new character. For all the differences in general aims and in many details to be observed among the supporters of applications-oriented mathematics teaching (I will explore such differences below), there are common elements that indicate a significant advance in comparison to the discussion before 1960. The most important new common element is the interpretation of mathematics applications from a perspective of modeling theory. Since I provided the necessary general information for the use of the concept of model and for mathematical modeling within mathematics instruction at the end of the previous section, I will restrict myself here to listing a few principles generally recognized by educationalists supporting applications (cf. Schupp 1988, pp. 10ff):

- There is no mathematics inherent in the objective situation or 'real' original problem under consideration – no mathematics which students would only have to 'discover'. Instead, a mathematical perspective is imposed on an objective situation; in other words, a situation is 'subjected to mathematical modeling'.
- All mathematical activities – deductive reasoning, geometrical constructions, computations, etc. – take place within the framework of the mathematical model.
- The (necessary) check on the mathematical model's appropriateness to the situation or to the original problem cannot be achieved by immanently mathematical means. The route from the situation to the model (How much reduction or idealization is acceptable for the problem at hand?) as well as that from the model back to the original situation (What is the significance of the mathematical results for 'reality'?) require both specialist knowledge and reflection on the purpose of and claim behind the modeling process.

For education research, there are interesting consequences that can be drawn from these principles:

- The widespread discontent with traditional 'fabricated' problems can be provided with a theoretical rationale: From the perspective of modeling theory, such problems are attempts to stimulate the use of mathematics in non-mathematical contexts without reflecting on the modeling process. If teachers wonder why students do not 'take a close look' and simply perceive what is to be computed, there seems to be an accompanying latent idea that the required mathematics is already present in the situation under consideration and only needs to be 'discovered'. From the perspective of the person constructing the problem, the image of concealed mathematics does indeed make sense: Inventing 'fabricated' problems is usually nothing more than 'hiding' a mathematical topic that currently requires some practice behind certain verbalizations. But this perspective ignores the fact that, in order to solve the problem, the students, in turn, must engage in constructive acts involving their thinking skills.
- Conversely, the perspective of modeling theory opens up possibilities for dealing with inappropriate problems in a way that is didactically sound. Students can be encouraged to make the implicit models explicit, to check their appropriateness, to replace them with alternative models, and to check whether and, if so, how the solutions are affected by this procedure (cf. Baireuther 1990, pp. 212ff). I will return to this topic in the section on word problems.
- Many topics of school mathematics take on new meaning if they are interpreted as standard models for common everyday situations, e.g., formulas for computing area and volume, proportionality and anti-proportionality, linear functions and equations (applications in physics,

in economics, etc.), trigonometric functions (for processes involving oscillation), exponential functions (rates of change), parameters and distributive functions for stochastic phenomena.

- The perspective of modeling theory provides a framework that extends from the simplest computation problems of elementary school teaching (Winter 1985, pp. 31-32) to the use of standard models in the lower grades of secondary schools, further, to challenging mathematical projections at the upper secondary level and in university-level mathematics (Reichel/Zöchling 1990, Knauer 1992), and even up to the most complex scientific applications. Didactically, at every level this offers the chance of building on insights in the relationships between mathematics and the 'reality' outside mathematics and of deepening such insights with an age-appropriate spiral 'applications curriculum' with increasing intellectual challenges.

Through active mathematical modeling, the students learn something about the object involved, about the context from which the problem is taken, and they also learn a good deal of mathematics. Conversely: A prerequisite of successful mathematically modeling is that one already understands something of both fields.

Despite the theoretical standardization that results from interpreting mathematical applications by means of modeling theory, there is no cause for euphoria. The theoretical and practical teaching problems involved in applications-oriented mathematics instruction are not at all already resolved. One superficial indication of this is that teaching practices far removed from applications are still predominant – despite convincing arguments for taking applications into account and despite an abundance of interesting examples suitably worked out for use in classroom instruction[99].

However, before we engage in a discussion of aspects related to teaching practice, the following fundamental problem remains to be addressed: To what extent do the available concepts of applications-oriented instruction meet the demand for developing an understanding of the world in accordance with the concept of general education under discussion here?

3.3.4 Applications Appropriate to Developing an Understanding of the World

If up to this point we have been primarily concerned with the views common to the supporters of mathematics instruction oriented to applications, I would now like to concentrate on the differences. These pertain to the general goals, to the assessment of specific goals within an exhaustive concept of mathematics instruction, and to plans for putting such goals into practice in classroom instruction. For this reason, the issue at hand cannot be formulated as follows: Are developing an understanding of the world (as a guideline stemming from educational theory) and an orientation toward applications (as a position of mathematics teaching) compatible in a general sense? The more discriminating formulation necessary can be expressed as: Which of the goals and concepts supported as being 'applications-oriented' ensure the attainment of the more general goal of developing an understanding of the world?

Gabriele Kaiser-Meßmer (1986, 1989) analyzed in detail the most influential didactic positions which, until then, had been presented as thoroughgoing explorations of the topic 'applications in mathematics instruction' in national and international debates. Kaiser-Meßmer distinguished five positions: a 'scientific-humanist' view (main advocate: Freudenthal) and a 'pragmatic' one (Pollack) in other countries, and, as specifically German contributions, an 'integrative' view (Winter, Wittmann, Blum, Bender, Schupp, G. Becker), a 'scientifically oriented' view (e.g., Fischer/Malle, Steiner, Steinbring), and an 'emancipatory' one (Damerow/Keitel, Münzinger, Volk). For our discussion here, these sets of goals determined by Kaiser-Meßmer can be condensed into the following three:

(1) *pragmatic-utilitarian goals*: imparting the ability to apply mathematics in real-life situations;
(2) *goals oriented to personality and to a critical view of society*: enabling the student to act as a responsible citizen;
(3) *goals oriented to mathematics as a scientific discipline*: imparting a more comprehensive concept of mathematics by taking the aspect of application into account.[100]

For our considerations here, the goals listed as (1) and (2) are especially significant. On goals of the third type, I will restrict myself to the following short remarks:

While reflecting on the topic of scientific orientation as a didactic guideline (section 2.3.1), I emphasized the necessity of imparting to young people a conception of the subject matter, of the competencies, problem-solving capacities, and specific worldviews inherent in the most important individual disciplines. Insofar as incorporating mathematical applications into instruction serves such purposes, an obviously indispensable contribution to the students' understanding of the world is made. Nevertheless, one risk involved must be kept in mind: The indispensable treatment of mathematics as an applications-oriented science can itself become distorted to specialist practice. I will go into this problem in more detail with my remarks on the goals listed as (1) above. From the perspective of general education, one question constantly to be raised in the light of concepts taking recourse to the goals listed above as (3), is as follows: How much of the *specifics* of mathematics must students understand in order to realize how mathematics as part of our scientific and technological world is of *general significance*? The orientation to main ideas (as developed in the previous section) might prove to be an appropriate guideline for the desired objective of developing an understanding of the world. – I will now proceed with some observations on the goals listed above as (1) and (2).

Applying Mathematics: Pragmatic-Utilitarian Goals

Here, we should keep in mind the fact that the 'ability to apply mathematics' includes a complete spectrum of very distinct qualifications. Located at the 'lower' and 'upper' ends of this spectrum are abilities that have very little to do with the understanding of the world aspired to here. This is because the spectrum of situations to be mastered has an enormous scope. I apply mathematics (usually in an unreflective way) whenever I cut an apple into four pieces. But I also apply mathematics (very consciously) if, as a member of an expert committee, I take part in assessing the risks of a planned nuclear power station.

Let us first consider the 'lower end' of the spectrum of qualifications. Here we find a practical sort of mathematics that I have already considered in detail with respect to preparation for later life and to the continuation of everyday mathematical culture. Essentially, applying mathematics within this context means having access to standard mathematical models and being able to adapt them to everyday situations. 'Mathematical fitness' in everyday situations emerges if the necessary mathematics is firmly associated with the particular type of situation. This connection must be established in a variety of learning situations in which mathematics is concretely applied: by counting, measuring, designating, classifying, drawing, building, estimating, and by communicating mathematical content, by writing something down and symbolizing.

As long as all of these activities remain confined to role-playing, to imagined or concrete situations involving actions, the dimension that necessitated including the objective of developing an understanding of the world into the concept of general education under discussion here is lacking. This dimension only comes into play if the child's horizon is broadened beyond the immediately perceptible, if the knowledge of the objects in the world and their interconnections attains a value in its own right. What does this mean with reference to mathematical learning processes?

It is difficult to precisely designate the point where preparation for later life in a strict sense becomes the development of an understanding of the world. Some of the transitions are not clearly defined; in practical learning situations there is often a close link between educational elements serving one or the other of these goals. Still, the following crucial difference can be emphasized: an element of reflection is added to the simple quality of being mathematically 'fit' in everyday life, to dealing with standard situations containing mathematical aspects with confidence. This theoretical, reflective element puts sensory phenomena to question, establishes more comprehensive contexts, and adapts them to an image of the world. Then, the object under consideration, including its mathematical aspects, is no longer simply given – e.g., as a mathematics problem posed by the teacher, or as the demand to present the correct amount of money at the cash register while shopping at a store, or as a practical problem of using a timetable to determine which bus I must take to get to school on time. Rather, it surpasses its own limitations in a certain sense. Thus, as a student, I can consider whether the mathematical problem presented might not be accessed by some other method than the computational procedure being practiced; at the cash register, I might contemplate why 'money' is a standard applicable both to the computer magazine and to the loaf of bread in my shopping cart; and on the bus, I could speculate on how such a detailed timetable can be set up despite traffic lights, traffic jams, and varying numbers of people getting on and off, and on what consequences this might have for the timetable's reliability.

These kinds of questioning, relativizing, and reinterpreting standard situations are characteristic of those variations of applications-oriented instruction that have committed themselves to the goal of *discovering the student's environment* in primary and early secondary education (Bender 1978, Winter 1985, Floer 1987).

Here, the pragmatics of mastering situations gives way to the goal of discovering new aspects of the objects under consideration by means of mathematics, making contexts perceptible that, without this mathematical viewpoint, would probably remain obscure. A further variation of applying mathematics that is clearly to be assigned to developing an understanding of the world and not to preparing for later life is observable in many of the examples provided by Paulos and Hofstadter (Paulos 1990). The point here is to better understand quantitative aspects of our surroundings: making estimates with large numbers; connecting large numbers to concrete ideas stemming from sensory data; relating one-, two-, and three-dimensional quantities to one another (How many letters does this book contain? How long would a freight train be that consists of tanks containing the annual worldwide extraction of crude oil? Etc.). There is a connection here that is established between simple mathematics and our knowledge of the world around us: Mathematics is used as a *means of communication* on matters about which we normally do *not* communicate in everyday life. Everyday objects or phenomena that are indirectly connected to everyday life become mathematically 'illuminated'. Whether such mathematical 'illumination' always leads to a *meaningful* broadening of the horizon is a different matter, and one that I will put aside for the moment.

In sum: At the 'lower end' of the spectrum of qualifications associated with the goal of applying mathematics, there are practical qualifications that hardly help to develop an understanding of the world in the sense proposed in our concept. At any rate, there is no clear delineation in the transition to applications also employing rather elementary mathematics, but clearly transcending pragmatic-utilitarian goals. A common element of many of the suggestions geared at *discovering the environment* is that not mathematics as such is the matter at hand, but that mathematics does actually become an *epistemological tool*.

What can be said about the 'upper end' of the spectrum? Often enough, the following ideal is behind dealing with mathematical applications in the classroom (cf. Werge 1987): A real-life, complex problem is presented; with the cautious support of the teacher, the students enter into a creative process during which the mathematical modeling of the original problem gradually takes on a concrete form; the students make use of the entire arsenal of school mathematics with which they have become familiar and, if necessary, motivated by the concrete goal, they independently supplement their mathematical competencies.

In everyday schooling, an approach to this ideal encounters considerable problems. The demand for content as realistic as possible can easily lead to a dead end. In many cases, greater proximity to reality is combined with greater complexity and incomprehensibility of the phenomena under consideration and with enormously increasing demands on mathematical skills. This can lead to a productive challenge for mathematically talented and interested students, especially at the upper grade levels. But for mathematically less sensitive young people, the hurdle abruptly becomes much too high – not only with reference to mathematical skills in a strict sense. The systematical thinking that is necessary for the analysis of complex relations within the original problem situation, for the construction of a suitable model, and, in challenging applications, for checking the relation between reality and model is also difficult for these students. In that case, the potentially motivating effect that precisely for such students could consist in engaging in mathematics not as in 'art for art's sake' is lost.

In sum: Also at the 'upper end' of the spectrum of qualifications that can be described as 'applying' mathematics there is a risk of missing the target of developing an understanding of the world. Beyond a certain degree of complexity, working on the relationships between mathematics and the 'real world' requires an expert ability that would demand too much of the majority of students. Complex applications can be meaningful as projects for especially capable students, e.g., within the framework of advanced courses in the upper grades of secondary schooling. They seem to be less suitable as a means for achieving the desired development of an understanding of the world in mathematics instruction for all students. Thus, if putting the first set of goals into practice (the ability to apply mathematics) is to help develop an understanding of the world, we must primarily be concerned with the middle of the spectrum. Beyond participation in our everyday mathematical culture, the ability to apply mathematics is only justified from the standpoint of educational theory if:

- mathematical modeling provides the students in an exemplary fashion with new perspectives on significant phenomena of their everyday lives which they experience directly (via sensory perception) and indirectly (via symbolic mediation);
- thus, their horizon is broadened in such a way that parts of the 'hidden' mathematics in their surroundings become perceptible to them, that the 'mathematical pervasion of our world' (Frey 1967) can be connected to their own experiences by means of comprehensible examples;
- the level of mathematical skills required does not exceed the intellectual potential of the majority of the students in the respective courses.

Enabling Students to Act Responsibly: Goals Oriented to Personality and to a Critical View of Society

In contrast to the proponents of the first set of goals, the advocates of this second one perceive applications-oriented mathematics instruction as embedded in an all-encompassing concept of school education with emancipation as the guiding principle. Theoretical arguments in favor of this position can be found in Damerow (1974, 1984), Münziger (1977), Keitel (1985), and Volk (1979, 1980, 1995), and, with a less pronounced attitude of social criticism, in Winter (1990). The materials offered by the German organization MUED contain a considerable number of practical suggestions for classroom work.[101] Relevant suggestions for topics and examples for classroom instruction have also been published elsewhere in no small quantity (Winter 1990, T. Jahnke 1993 and 1995).

The issue is not so much applying mathematics as such, but, rather, enabling students to think and act independently and from a critical standpoint in future, complex situations that are not yet foreseeable in detail. The issue is to enlighten the students. Thus, in comparison to the goals connected with the ability to apply mathematics, there are much more thoroughgoing transfer expectations here.

Even this brief characterization already clearly demonstrates that this second set of goals exhibits a pronounced affinity to other aspects of the concept of general education under discussion here, to *developing critical thinking* and to *enhancing the students' self-esteem*. To what extent these objectives can be taken into account by means of applications-oriented mathematics instruction will be discussed below. At the moment, I will be concentrating on the aspect of *developing an understanding of the world*.

The requirements for the selection of problem content to be made accessible and worked on with mathematical modeling are explicitly different from those in conjunction with pragmatic-utilitarian objectives. The applications should not only be closely related to everyday life, mathematically interesting, and appropriate to the students' abilities, but should also demonstrate possibilities for individual commitment and intervention, should call attention to acute or concealed problems that are of urgent significance for society (or for all of mankind). The following list of topics taken from materials presented by the organization MUED demonstrates the demands made on selection criteria (MUED 1994a, pp. 42-43, 1994b, pp. 46-47):

- risks involved in speeding (Amon);
- reducing air pollution by establishing a speed limit (Volk);
- reducing time on the job – creating new jobs (Meyer-Lerch/Volk);
- project: water (Boer);
- searching for an optimum method of thermal insulation (Volk);
- flow of traffic and speed (Volk/Meyer-Lerch);
- surveying expressway intersections (Volk/Boer).

In section 2.3.4, we maintained that public schools cannot undertake up-to-date efforts toward developing an understanding of the world without touching on major issues of the times and of the current world situation, without treating "key issues", as Klafki puts it (1985a). Suggestions for classroom instruction like those listed above demonstrate how mathematics teaching can participate in developing an understanding of the world. In doing so, mathematics certainly cannot compete with subjects like political and social science, for which these kinds of issues are of central importance. Nevertheless, traditional mathematics instruction does not even come close to exhausting its existing possibilities in this sense. Ecological and economic problems, the consumption of raw materials, population growth, estimates of risks, etc. can be fairly readily combined with standard topics of mathematics at the secondary levels. If content of this type does not simply serve as a stimulus for practicing certain mathematical techniques, but, instead, demonstrates that mathematics can help to illuminate such problems, to compute variants, to provide vivid ideas of orders of magnitude, to go through alternatives to dominant trends with mathematical models, then mathematics instruction does indeed contribute to developing an understanding of the world in the sense outlined above.

Nevertheless, in teaching practice, combining major issues of the times and of world affairs with school mathematics is not without problems: The students' motivation is not necessarily proportional to the pervading relevance of the issues to be considered, and overenthusiastic teachers committed to environmental or social causes, but lacking a certain pedagogical tact, sometimes inadvertently produce defensive reactions in students who, in accordance with expectations arising from their socialization and reinforced by their school experience, prefer to simply 'do math' during mathematics class.

In sum: If mathematical applications become part of instruction with the explicit goal of turning students into responsible, active citizens, then this can be assessed as an attempt to develop critical thinking. But since the proponents of this more or less formal goal usually treat current, socially relevant issues, mathematics instruction also takes on a nonmathematical material dimension. By dealing with topics concerning major issues and current world affairs, a substantial contribution to developing an understanding of the world can be made. There are, however, difficulties involved in putting such plans into instructional practice.

The material for problems used in instruction presents a series of didactic problems in their own right – problems of major practical relevance for applications-oriented mathematics instruction. The question is not only how problem material is to be selected and designed, but, perhaps even more importantly, how it is to be handled in the classroom.

3.3.5 Word Problems and Problem Wording

Students usually encounter 'applications' in traditional mathematics instruction in the form of word problems. Of course, problems inherent in 'pure mathematics' are also often presented as word problems. But the following remarks are not concerned with such problems; under discussion here are problems that are related to matters outside mathematics.

There are significant arguments both in favor of and against word problems. How important are word problems for a mathematics instruction designed to contribute to developing an understanding of the world? Are they indispensable? How can we distinguish between 'good' and 'bad' problems couched in words?

In traditional mathematics instruction, solving word problems is one of the least favorite activities for many of the students. This is understandable to the extent that mathematics instruction as a whole is seen as fulfilling the expectation that doing 'real' math is working out algorithms. In that case, stripping the problem of the words in which it is couched is reduced to discovering the algorithm that the teacher or the textbook has 'unfairly' concealed. Since finding the approach to the solution usually cannot be formulated as an algorithm itself, many students feel that wording a problem is a superfluous complication that has nothing to do with the sort of mathematics they expect to encounter in school. Naturally, we cannot rule out the possibility that even where word problems are simply appended to a mathematical topic currently under discussion, some insight into the usefulness of mathematics in discovering the 'real' world might be suddenly revealed to the students. Still, mathematics' potential for solving nonmathematical problems will not become perceptible; the self-isolation of school mathematics as a subject will not be disrupted. On the contrary, at the most the average student will find his or her view of the subject simply complemented by one further obscure detail: "In mathematics class there are problems to be solved that would never occur to any reasonable human being!"

Apparently, the unsatisfactory status of word problems and their unpopularity among the majority of students are closely related to general effects of socialization through mathematics instruction. For this reason, the arguments pro and contra word problems cannot be discussed in isolation, but only in connection with further qualitative aspects of instruction.

Types of Word Problems and Criteria for Evaluation

In order to arrive at a discriminating didactic evaluation of various possibilities for their use in classroom instruction, I will differentiate three types of word problems:

A. I will designate problems as 'fabricated' if they are presented for practicing a problem-solving procedure covered in class beforehand. Even elementary school mathematics instruction already relies on this kind of problem (e.g., on the topic of division): "Lars has one brother and one sister. Grandma gives him 9 EUR and says, 'Divide the money among you.'" It can also still be found at the upper level of secondary school (e.g., an extremum problem such as: "What dimensions must a cylindrical tin can with a volume of 1 lit. have if the amount of material used in producing it is to be as small as possible?"). Typically, problems of this kind are considered solved as soon as the 'solution' has been written down. At the most, the teacher might require an explicitly formulated solution; if so, the students often see this as unnecessary harassment. After all, if the 'right approach' has been found, then it is, indeed, usually obvious what the 'solution' means – why should the obvious have to be written down in a full sentence? And, at any rate, aren't we always supposed to be as concise as possible in math class? To put it briefly: To a large extent, the usual fabricated problems rob a nonmathematical problem of its (potential) context within everyday life; the problem is presented *in isolation* so that the purely mathematical procedure under discussion can be more readily implemented.

B. More challenging word problems are those that involve generating an entire set of problems. Here, a large part of the text provides nonmathematical objective information that is required for mathematical modeling, e.g., on topics such as: the population explosion, braking performance of vehicles, energy consumption and costs of 'energy-saving' lights in comparison to traditional light bulbs, etc. Such texts often include a variety of tasks to be completed: calls for independent research and the collection of data, suggestions for componential modeling that can be refined step-by-step, suggestions for consequential actions resulting from the mathematical examination of the specific problem, etc. The 'individual problems' (if, indeed, they can actually be called such) are precisely not presented in isolation, but remain related to the more comprehensive problem context. Finding a solution does not mean that the problem is 'finished'. Instead, it provides opportunity for further questions and interpretations, perhaps even an opportunity to review and reformulate the original problem.

C. A third type of word problem does not even appear in the form of 'problems' formulated for mathematics instruction, but, instead, as informative texts containing mathematics (i.e., including numbers and/or graphs). Erichson (1992, 1993) has described this approach to 'discovering textual material in our surroundings' in detail and has demonstrated its possible use with a number of examples. The informative text has its own effects, directed toward its own (nonmathematical) subject; the mathematics expressed in the text or even concealed in it originally has much the same status as it normally does in the everyday lives of young people. To paraphrase an overused educational formula about meeting the student on his or her own terms, we might say: Mathematics is met on its own terms, where it actually exists in the 'real' world. Collective examination of such texts illuminates what is meant, uncovers a variety of possibilities for interpretation, promotes familiarity with the observed use of mathematical devices, and, in addition, provides opportunities to critically inspect this use (e.g.: Is there only a pretense of exactness? Are the compared items really comparable? Can the numbers specified actually be correct?).

Word problems and informative texts that contain mathematics and initiate problem-solving activities can be classified further according to the following criteria:

(1) *Relationship to everyday life*: Are the problems to be solved relevant to the everyday lives of young people? Can the procedures required for solving the problem be used in everyday life? Obviously, the elementary school problem mentioned above is relevant to everyday life in this sense, the extremum problem is not.

(2) *Relationship to reality*: Does the problem make sense in a certain societal (economic, ecological, scientific) context? Is the information communicated in the problem realistic (i.e., not only the data in a strict sense, but also the information about the objective context)? Would experts also take recourse to the procedure to be practiced in order to solve the problem? – For older students, the relationship to reality can be more important than the relationship to everyday life. The extremum problem mentioned above under A is definitely not totally unrealistic (of course, there are cans of this type; measurements could demonstrate how much usual dimensions diverge from the values to be ascertained by solving the problem); on the other hand, one can quickly clarify that industrial designers of cans will work

differently and that there is an extremely idealized case of modeling behind the problem (cf. Rentz 1991).

(3) *Relationship to a more general problem context*: Does the wording indicate why the problem presented is significant? Particularly in the case of older students and if the problem is fairly realistic, but not related to everyday life, the following questions are relevant to the goal of developing an understanding of the world: In which societal, scientific, or professional contexts are people concerned with problems of this sort? Which more general problems are concealed behind the specific ones to be solved?

(4) *Explicit formulation of the problem*: Is the solution to be found explicitly described or only to be inferred from the problem context? The elementary school problem is only implicitly formulated; the extremum problem is of the explicit type. However: In the light of preceding instruction and prior experience with similar formulations, implicit problems also often result in unambiguous appeals for action; conversely, problems that from the perspective of the teacher are formulated explicitly can demand too much of students with less advanced mathematical understanding since the presumably explicit technical terms and symbols are only intimidating hieroglyphics to them. Thus, 'explicitness' simply designates a superficial category that only exhibits a very loose correlation to processes of understanding.

(5) *Open formulation of the problem*: Does the problem allow for various problem-solving procedures and solutions? Or does the formulation of the problem already confine it to a specific mathematical model from which the students cannot diverge without risking having their solution be assessed as false? – The general characterization of 'fabricated' problems implies that they are usually not open: Their main purpose, of course, is to activate the mathematical procedures 'covered' in class beforehand. In this sense, both of the example problems mentioned above are closed.

Didactic Evaluation

Without a doubt, word problems or informative texts of the types B and C fulfill the requirements of applications-oriented mathematics instruction aimed at developing an understanding of the world better than do traditional fabricated problems. In particular, the relationship to a more general problem context can be clarified much more easily by types B and C than with isolated individual problems whose relationship to nonmathematical material is primarily determined by the procedures to be practiced. This also has consequences for the desired proximity to everyday life or to reality: With C, this proximity is practically given in and of itself; with B, it might, indeed, be disregarded, but the risk of preparing totally unrealistic problems as a project is certainly smaller than that of contriving a totally unrealistic fabricated problem. Finally, B and C offer more opportunities for open problem formulations since 'finishing' a problem by establishing an unambiguous solution is not the focus.

Does this mean that we must refrain from using traditional word problems in applications-oriented mathematics instruction, or, more precisely, in mathematics instruction oriented to general education and aimed at developing an understanding of the world? I think not. The most important theoretical argument supporting this view is the following: A word problem does not predetermine how teachers and students deal with this problem in the classroom. In addition, there are also some practical reasons of import: Given the circumstances of limited time for preparation and for classroom instruction and in light of the fact that an enormous selection of traditional word problems is offered in the usual textbooks and in collections designed for this express purpose, such problems offer the opportunity of making a link between school mathematics and the 'real' world even in situations in which otherwise one would have to completely refrain from doing so. But, then, how can we deal with such problems so that the negative effects on our aim of developing an understanding of the world are compensated?

- Traditional word problems must not represent the sole access to mathematical applications. Sufficiently often (we might say at least once every school year), students should be given the opportunity to involve themselves more deeply in a problem context that is meaningful to them and is worked on as a project. Here, students can gain experience with step-by-step, open forms of modeling, with the usefulness, but also with the limitations of a mathematical perspective with respect to practicable problem-solving in nonmathematical situations.
- Fabricated problems cannot simply be lumped together. Using the criteria listed above, we can decide which problems are more appropriate for establishing a link between mathematics and the 'real' world than others.
- A problem context not explicitly expressed in the problem under discussion can be worked out in a collective effort. The teacher might provide additional information or encourage the students to do some research themselves.
- A lack of open problem formulations can be compensated by altering the wording (and the students can take part in the reformulating process!).
- The anticipation of certain mathematical problem-solving procedures implicitly contained in the problem wording can be reconstructed as mathematical models and critiqued in collaboration with the students. Even with 'badly' worded problems (artificially contrived, unrealistic, or overly idealized ones), a critique of the assumptions tacitly made by the authors of the problems can even permit a profitable learning process that far exceeds that of a straightforward, expected 'solution' to the problem. If the problem-solving practices cultivated in traditional

mathematics instruction are consciously disrupted now and again, then not the least result will probably be a greater immunity against notoriously unrealistic and meaningless word problems.

The activities proposed with the last point in the list above deserve a more in-depth explication. My argument is as follows: One essential specific characteristic of the process of modernization – the fact that mathematics is not inherent in the situation under consideration, but must be superimposed upon it or developed within it – can be just as readily demonstrated with 'good' as with 'bad' problems, if students are encouraged to vary the problems, to systematically 'misunderstand' them, to brush them against the grain, and to doubt the supposedly self-evident assumptions behind them (cf. Baireuther 1990, pp. 217ff). How might this occur regarding the problems cited under A, that is, for the elementary school problem and for the optimization of the tin can? For dividing 'Grandma's money', her gift to her grandchildren, the elementary school students might consider the following: Is dividing the sum into equal shares always the most just way? What if Lars's sister already has a job and earns money herself? What if his brother wants a stuffed animal that costs 3.50 EUR? What if Grandma had given Lars 10 EUR? ... – On optimizing the can, the following could be considered: Which factors might be crucial in practice (e.g., producing a minimum of waste in cutting the tin, reinforced material at the seams between the casing and the bottom/top [examination of real cans!], choice of volume so that the contents can be 1 lit., etc.)? How might corresponding models appear? Conversely: What sort of optimization model might have been used for can dimensions that groups of students have measured? ... (cf. Rentz 1991).

The amount of time involved alone will prohibit examining every word problem from so many perspectives. Yet, if such scrutiny does occasionally take place in exemplary fashion, ways of thinking and patterns of interpretation different from those of traditional mathematics instruction can already become established in this way. Within such a context, problems that are only designed to be worked on in a straightforward manner can more readily become accepted as limited exercises and as 'food for thought' even by students who are skeptical about mathematics.

In sum: The objective of developing an understanding of the world cannot be served simply by increasing the amount of traditional problems to be used. Nor is it sufficient to replace 'bad' problems with 'better' ones. Since traditional word problems are more strongly committed to the systematic structure of school mathematics than to some nonmathematical problematic sphere, there is a substantial risk of isolating the problems and neglecting their context. The process of mathematical modeling is treated inadequately if the employment of certain mathematical procedures is already predetermined. – Other types of texts presenting problems make mathematical applications more convincingly accessible. Nevertheless, there should be no general condemnation of fabricated problems: Whether they can be used for developing an understanding of the world – and thus for furthering the students' general education – depends on how they are handled in the classroom.

3.3.6 Aspects Related to Instructional Practice

In what follows I will be relating our discussion thus far to the practice of classroom instruction. What must be done in applications-oriented mathematics instruction explicitly committed to developing an understanding of the world? The underlying premise here is that mathematics instruction in its entirety cannot be designed solely from the perspective of applications or with an orientation to applications. But in view of the predominant practice, which as a whole can be characterized as deemphasizing or even neglecting applications, there is no need to fear that classroom instruction will insufficiently deal with 'pure mathematics' if, in general, the orientation to applications is intensified. For, at present, we are still far removed from a reasonable balance between nonmathematical applications and procedures inherent in mathematics.

The important thing here is: An understanding of the world cannot be developed in isolation from the other objectives of general education. If taking recourse to applications is not supported by a 'culture of instruction' that grants the students opportunities for individual activity (enhancing their self-esteem) and assigns high priority to sensible communication about the relevant objective problems and their mathematical systematization (developing critical thinking, practicing communication and cooperation), then even the best examples and most interesting suggestions for mathematical modeling will not be very effective. They will only serve to augment the abundance of instructional material.

Taking the Perspective of Modeling Theory Seriously

One requirement stemming from the psychology of learning and motivation applies equally to problems inherent in mathematics and to those involving applications: A problem posed in classroom instruction should also be a problem *for the students*; it should be comprehensible as a problem and seem worthy of a solution. Otherwise, however, there is an important distinction: A problem inherent in mathematics has reached its conclusion when the mathematical solution has been found; for problems dealing with applications, the determination of the mathematical solution is only a part of the entire solution. It is essential that *the problem is the starting point* and that *there is a return to the problem* (cf. Volk 1995, p. 45). Here, we must ask what the mathematical solution (in the simplest case, a numerical value) means for the content of the problem. For example, what sort of numerical accuracy is appropriate to the content, and how is the solution altered if the input data are varied? We can make meaningful use of computers here: If the mathematical model developed for a certain problem can be presented as a computer program, we can readily gain an overview of the scope of numerical solutions by varying the original data. While working with older students, from about grade 8 on, after determining a first solution the model itself can be reexamined. Does it really achieve what is required, can it be refined and better adapted to the real situation? Again, a computer can help to vary the model (in the form of a computer program) and to determine what effects modifications have on the solutions if the input data remain unchanged. For the upper grades in secondary schools, even more in-depth reflections are appropriate: Which aspects of the problem resist mathematical modeling? Is this due to the limited mathematical tools (and their availability at school) or are there components of the problem that, in principle, do not allow a rational solution through mathematical systematization?

Meta-Knowledge of Modeling Theory

I have already cautioned against misinterpreting my argument in favor of the idea of mathematical modeling as a recommendation to get students to use the word 'model' as early and as often as possible. Usually, it will suffice if teachers handle applications from a perspective of modeling theory without explicitly making use of the term 'model'. – On the other hand, the lower grade levels of secondary schools certainly already offer attractive opportunities to explicitly speak of 'modeling', e.g., within the framework of elementary geometry, in function theory (linear functions, processes involving oscillation, rates of change), and especially in probability theory.

The development of insights into the process of mathematical modeling at a meta-level should be accompanied by gaining much practical experience in the mathematical treatment of and reflection on problems in other fields. Making meta-knowledge of modeling theory a topic of instruction at a very early age should be regarded with a certain skepticism. Insights of this type easily become banalities if they are not supported by concrete knowledge and reflective experience. The result would be lifeless, superficially acquired information.

Mathematical Requirements

Even though experiencing how the mathematics to be learned can be applied can result in the disruption of obstacles to understanding at an 'inner-mathematical' level, nevertheless, applications-oriented instruction is generally more challenging than abstaining from applications. In addition to mastering the relevant mathematical techniques and methods of deductive reasoning, students are also required to direct their thinking to the specific fields touched upon in classroom instruction. The students must recognize structures in these fields or impose structures upon them in a constructive way; conversely, they must interpret the content of mathematical symbols, operations, terms, functions, algorithms, geometric objects, and abstract relationships. Processes of interpretation and construction are necessary in both directions, from objective contexts to mathematical formulations and then back again to the objective contexts.

For this reason, it may be necessary to make concessions regarding the mathematical content to avoid demanding too much of the students. Interesting modeling processes are also possible with the use of elementary mathematics. If through a wide variety of problems and situations students can experience how convenient many mathematical topics of the standard curriculum (e.g., linear functions) are as 'standard models', then their understanding of the world is more enriched than if they are primarily confronted with working on problems that would first require great efforts to attain the necessary mathematical know-how. An application that is interesting from the student perspective need not be 'mathematically' interesting to a mathematician. Generally speaking, the increase in motivation attainable through applications has a much greater effect if difficulties in understanding do not complicate or even prevent handling the necessary mathematical procedural skills. Such considerations are especially important for mathematically less talented students. If, for instance, mathematics instruction at lower secondary grade levels is not to refrain from an orientation to applications (and it is especially necessary here), if, on the other hand, within the framework of traditional computation exercises it is not satisfying to demand standard applications of the students as a form of mechanically memorized 'knowledge', then much didactic creativity will have to go into presenting intriguing problems that can also be modeled by means of fairly simple mathematics.

As appealing as the idea might be that the students only come to acquire the necessary mathematics in the confrontation with an objective problem or, going even one step further, that they actually discover it anew in and for this situation, this idea is hardly realistic. In practice, such simultaneous advances in modeling processes and in the acquisition of new mathematical knowledge will have to remain the exception – simply for reasons pertaining to cognitive psychology. If we refer to the theory of domains of 'subjective experience' (Bauersfeld 1983; 1993, p. 244-245), then we can convincingly argue that the modeling process is always accompanied by the construction of a new domain of subjective experience – even if the required mathematical knowledge (organized in other domains of subjective experience) is already available to the students. By applying already familiar mathematics, then, the students do not learn any new mathematics (in the form of new concepts, techniques, etc.), but they do *learn* this portion of *mathematics anew* by putting it into a new context. Their image of this part of mathematics changes; their knowledge takes on structure by connecting the new domain of subjective experience to the existing ones. Thus, reducing the mathematical complexity in favor of enhancing applicability does not at all result in less complex and less challenging learning. The converse is much more the case: active mathematical modeling is such a complex and rewarding intellectual activity that no didactic effort should be spared to allow it to also become a personal experience for less talented students. However, this involves imposing reasonable self-limitations on one's own subject area-related ambitions. For many mathematics teachers, these ambitions stem from a too confined view of mathematics.

Whenever an applications problem is presented, the mathematical knowledge and skills required for the meaningful mathematical modeling of the problem should normally be familiar to the students in principle. 'Familiar in principle' is certainly not the same as 'readily accessible': First of all, acute needs might motivate recall or repetition; secondly, behind the idea of 'readily accessible' knowledge and skills is the ideology that equates learning with accumulating stock – with the assumption that if some topic has been 'covered', then it must be in the permanent possession of the students for all time. Teachers could avoid much frustration if only they realized that this mechanistic model of learning does not at all do justice to the reality of learning![102]

Project-Oriented, Interdisciplinary Methods

In many respects, instruction in the form of a project, relying on cooperation among teachers specialized in various disciplines, would be an ideal method for learning mathematics and simultaneously developing an understanding of the world. But since this method is quite difficult to put into practice, I would like to focus more on alternatives than on adding to the detailed accounts of its advantages.[103] Under the circumstances determined by curricular organization and the division of labor at public schools, there is a risk that establishing a too close theoretical link between the goal of developing an understanding of the world and project-oriented methods overarching several subjects can discourage teachers. In the final analysis, then, this can result in a situation where applications that help develop an understanding of the world are simply further neglected. Even at experimental schools such as the *Laborschule* (Laboratory School) in Bielefeld, for which the integration of the mathematical sphere of experience with other spheres is one of their express goals, there is a constant need of special efforts and extraordinary commitments among the colleagues to realize such integrative projects. As a member of the project group 'Mathematics Instruction at the Laboratory School', I can speak from experience.

What practical alternatives are there to the ideal of project instruction, in which specialized teachers design instruction in a collaborative effort? As a result of the unsatisfactory course of many project phases strained by too high expectations, the staff of the *Laborschule* in Bielefeld turned to collaboration in developing ideas and materials on certain themes which were then given to a specialized teacher who put them into practice within the framework of specialized instruction. In this way, instructional units up to within the lower grade levels of secondary schools can be planned without specifying whether the mathematics teacher or a teacher specialized in a nonmathematical subject should be the person to put the suggestions into practice.[104]

This 'slimmed-down' project variant of the *Laborschule* also requires a considerable amount of revising one's own thinking. The radically serious assessment of nonmathematical aspects of the world can present an enormous obstacle to mathematics teachers who, in their previous applications, never went beyond fabricated problems. For this reason, small steps are a good place to start: instructional phases and lessons that only require half an hour or one or two classroom hours can establish confidence and demonstrate that the dreaded chaos does not erupt; they can arouse curiosity (not only among the students, but also in the teacher) and prepare for more comprehensive mathematical excursions into the world. But apart from units designed to thematize applications contexts within the framework of 'normal' specialized instruction, there are also many situations that provide opportunities to interpret a mathematical concept in a nonmathematical way, to refer to uses in everyday life, in technology, and in other scientific disciplines. It is crucial for the aim of developing an understanding of the world to relate the subject to everyday life, to 'dismantle the boundaries of specialization', to disrupt the systematically dictated self-isolation of the discipline in a way comprehensible to the students, or perhaps even highly motivating to them. Less significant is how much time this takes or how much explicit interdisciplinary collaboration is involved.

Culture of Instruction

Whether mathematics instruction plays a part in developing the students' understanding of the world is not simply a question of the instructional topics, of the selected segments of reality, of the objective areas to be modeled mathematically. Of course, in applications-oriented instruction, content is of great significance. Mathematical modeling of irrelevant or even imaginary problems results in a distorted view of mathematics and contributes nothing to the students' (nonmathematical) knowledge of the world. But content is not the only crucial factor. Problems that, considered in and of themselves, demonstrate some proximity to everyday life and to reality and whose modeling provides no unsurpassable obstacles to the prior knowledge of the students, do not automatically guarantee that the desired effects will actually occur. Such effects would be that students realize what part mathematics plays and can play in everyday life, that they become aware of the practical potential of mathematics as a cognitive instrument by examining cogent examples, that, through mathematical modeling, they learn something about the world and about mathematics, and about the limitations of mathematical systematization.

The following subchapters will increasingly demonstrate that the *culture of instruction* is crucial to meaningful mathematics learning. How do students and teachers behave towards one another in the classroom and how do they collectively treat mathematics? – The following premises, which seem decisive for the desired contribution of mathematics instruction to developing an understanding of the world, are to be understood as elements of a 'culture of instruction contributing to general education' that has still to be developed. They are not didactic, methodical guidelines in the strict sense (i.e., they do not suggest what teachers should do). Instead, they are descriptions of conditions of classroom communication and individual learning – conditions that are to be seen as consequential to the preceding reflections on applications and mathematical modeling:

- Ample latitude for a number of modeling variants should be provided. This might refer to decisions on which factors of the real situation can be disregarded, how much accuracy is aimed for in the result, how qualifying circumstances can be varied, how the solutions discovered can be generalized, etc.
- Students should have plenty of opportunities to become active on their own initiative, to experience their own creativity within the modeling process. Here, one should ensure that the possibility of such experiences is varied in relation to the discrete intellectual abilities of the students.
- Asking unorthodox questions and expressing doubts should be permissible. Putting the meaningfulness of a given applications problem to question should also be allowed.
- It should be possible to approach a problem from widely different angles. Problem-solving procedures (choice of aids, recourse to

illustrations) should not be predetermined by mathematical techniques just covered in the classroom.

3.4 Thinking, Understanding, and the Use of Critical Thinking in Mathematics Instruction

Learning mathematics develops intellectual faculties, especially cognitive abilities – this argument in favor of mathematics as a compulsory school subject is the one with the most extensive tradition. The roots of this argument go back as far as Plato.[105] By contrast, skeptics such as the satirist Georg Christoph Lichtenberg pointed out that also (or perhaps precisely) among mathematicians there were some very narrow-minded people.[106] At any rate, positive effects of doing mathematics on the use of critical thinking in everyday life do not seem to appear automatically – as everyday experience and empirical research on transfer achievements confirm. What, then, might today still be reclaimed of the expectation that mathematics instruction enables one to think, indeed, even to engage in critical thinking?

To call to mind some important results of the previous chapter that are related to the concepts designated in the heading above and to their context:

- The ability to engage in critical thinking is not a quality that can be acquired once and for all; critical thinking is connected to an *attitude* of wanting to get to the bottom of things and to the realization that the scope of our understanding is limited.
- As is all thinking, critical thinking is largely *specific to a certain sphere*, linked to situations and content. For this reason, training critical thinking in a way neutral to content is impossible.
- Critical thinking is based on 'understanding' certain matters under consideration. It can only be developed by learning with understanding.
- Critical thinking requires the capacity for discriminating and for logical thought. Thus, it is not conceivable without a *developed understanding*.
- Learning how to think is always *socially mediated* activity: As communicative beings, we can only learn to develop and make extensive use of our cognitive potential in exchange with others and through the feedback we receive from them.

Returning to the question posed above: How can the assumption that mathematics instruction would generally train cognitive abilities be maintained – in light of the specificity of human thought? Should we actually demand from mathematics that it should promote students' *general cognitive abilities* and *general capacities for critique*? If so, what changes must be made in traditional instruction, in which these goals have obviously remained unattainable to the majority of students? Or should we instead content ourselves with a more modest goal – that at least some of the students learn to *think mathematically*? Reduced to a pointed formulation: What can be expected as the result of doing mathematics – an *expansion* or simply a *specialization* of students' cognitive abilities? The corresponding background issue would be: Is mathematics (according to Wittenberg [1963, p. 57], "the sphere par excellence … of the laws inherent in reasoning") an especially high level of reasonable thinking or is it only a very peculiar (even if in some contexts highly effective), specific instance of human thinking?

My attempt at an answer is, on the one hand, situated between the two extreme positions outlined above. On the other hand, it goes beyond the view of the problem expressed in these positions. My main argument is as follows: *In mathematics instruction aiming at general education, the students should experience how mathematics can be employed as an 'amplification' to their everyday thinking.*

This subchapter is designed to explicate and illustrate this argument. I would like to demonstrate that this argument adequately describes what the exact contribution of mathematics instruction to the overarching goal of 'promoting critical thinking' can be – in accordance with the possibilities of mathematics as well as with the necessities of a general education.

At first, I would like to briefly explain the use of the term 'amplification' in my argument. 'Amplification' means something other than simply an 'aid'. Within the context of the instructional objective of preparation for later life, we viewed mathematics as a tool and as a means of communication. At that point, we were primarily concerned with the function of the particular mathematics in mastering everyday situations. However, the relationship of the human subject to mathematics remained unconsidered: Whether and to what extent the employed mathematics was 'understood' – in the sense of gaining insight into the reason why it works – was not the focus. I would like to stress exactly this dimension with my use of the term 'amplification'. Mathematics that I do not understand, but can routinely correctly apply, can, at best, *replace* my thinking. In that case, I can, for example, solve a practical problem without really having apprehended the problem and without knowing why the problem-solving aid works. But I can also make conscious use of mathematics to analyze and solve a problem. The difference to the first case is that, here, what I am doing seems 'reasonable' to me in a profound sense. The mathematics has been more or less integrated into my 'reasonable reflection' on the problem. Admittedly, these are somewhat vague and tentative formulations that are in need of being made more precise. At this point, I am only concerned with hinting at the direction in which my thoughts will proceed.

3.4.1 Mathematical Thinking and Everyday Thinking: Two Examples

The widespread view of mathematics as a trainer of cognitive abilities is ambiguous. On the one hand, even young people and adults who consider themselves to be mathematically untalented usually concede that mathematics has something to do with clear and logical thinking.[107] On the other hand, again and again one encounters disturbing statements that point in the opposite direction. The kind of thinking required in the mathematics classroom seems to many people who have difficulties with mathematics to be encoded in a peculiar way, to be 'distorted' and unnatural, so that 'you just don't get it' the way it is. It seems that in mathematics we need to 'think differently' than we do in everyday life.

First Example: Dividing by a Fraction with a Numerator of One

A characteristic situation involving my daughter Katharina (who was 13 at the time and not very interested in mathematics) can illustrate this phenomenon and at the same time invite tentative speculations on which misconceptions are responsible for bringing about the gap between mathematical and everyday thinking. While doing her homework, Katharina had correctly applied the rules for computing fractions and divided the number 2 by ¼. Her result, 8, surprised her so much that she then asked me how the result could possibly be larger than the dividend. After all, she had 'divided' it! I attempted to explain to her why this must be the case (for positive numbers) with division by numbers smaller than 1. Finally, she remarked, "Okay. Now I know how to compute this. But don't try to tell me that math has to do with logical thinking."

This example clearly illustrates at once several aspects that often lead to a conflict between mathematical and everyday thinking. At first, I will simply list them in consecutive order:

- Apparently, one circumstance contributing to Katharina's confusion was the fact that a word of ordinary language is used with a technical meaning. When we talk of 'dividing' in everyday situations, we almost always mean breaking something down into its components. As long as we divide by natural numbers, this everyday idea is adequate and helpful for a mathematical understanding.
- There are also 'natural', everyday situations that apply as illustrations of dividing by positive rational numbers less than one. But in such cases, we no longer speak of breaking something down into its components, but rather, of 'fitting into something': How many quarters fit into two wholes? But the cognitive route to such situations, which could provide a vivid backdrop to a considerable number of division problems, remains blocked for many students who were taught in a traditional way because they have firmly associated 'dividing' with 'breaking down into components'.
- This inevitably leads to the presumption that it is not so much different types of *thinking* that Katharina and I exhibit in the situation described, but we do seem to have different interpretations of the concepts being employed. As long as we equate mathematical 'division' to 'breaking down into components', Katharina's critique of mathematical procedure can be considered valid. For the insight that ordinary language might formulate as: "Parts are always smaller than the whole" can be repeatedly confirmed by observation in everyday life and, thus, is an extremely familiar experience. If, then, this objective condition is (apparently) negated by mathematics, this must seem strange, unreasonable, or even illogical to someone who is convinced of the validity of the observation. The apparent contradiction dissolves at the point where a student comprehends that mathematical 'dividing' is a generalizing procedure that can also be used for describing, e.g., situations of the type: "How many x fit into y?" or (more generalized and, yet, more specifically mathematical): "By which number must x be

multiplied to attain the result y?". Then, the mathematical expression seems quite natural, reasonable, and logical.

- Finally, I would like to call attention to the motivational element triggering the episode. The fact that Katharina did not simply accept the correctly computed result, but, instead, put it to question because it contradicted what she would have expected on the basis of her everyday experience is, in itself, an indication of critical thinking. Mathematics instruction that obstructs such questions or declares them to be nonsensical most certainly does not contribute to promoting critical thinking.

Second Example: Constructing and Interpreting Simple Equations

The second example that I wish to employ comes from a study by Rosnick and Clement (1980). In the last several years, it has often been presented, commented on, and – with ever new variations – replicated.[108] This example is especially interesting because it illustrates the respectively different ways of dealing with abstractions and symbolizations in mathematical and everyday thinking.

The following problem was presented to college students of various fields (in subsequent studies, also to pupils, teachers, and university professors):

Write an equation using the variables S and P to represent the following statement: "There are six times as many students as professors at this university." Use S for the number of students and P for the number of professors.

Only about 60% of the solutions were correct; among pupils, the percentage was considerably smaller.[109] Almost all of those who suggested false solutions did so in the same way: They wrote P = 6S instead of S = 6P or S = 6 · P.

My intention here is not to summarize the numerous interpretations.[110] Instead, I will be focusing on a few aspects that have to do with the relationship between mathematical and everyday thinking.

First, we can assume that almost all of the test persons would have correctly answered the question: *"How many students are there if there are 50 professors?"* (in some of the subsequent studies, this was explicitly confirmed). In connection with elementary arithmetic, everyday thinking is sufficient for finding practical solutions based on the relation specified in the problem. But formalization or dealing with symbols creates difficulties.

In the problem stated above, one source of error is the interpretation of the symbols P and S not as *numbers of* something, but, rather, as *designations* or *units*. '6 S' is interpreted as 'six students' in much the same way as '6 cm' stands for 'six centimeters'. Interestingly enough, even this misinterpretation does involve an *abstraction* that makes use of variables – as with every use of more generalized concepts in ordinary language. 'Student' (as a concept) is a variable that can represent various concrete students in an abstract way, disregarding their individual traits. In much the same way, most 'everyday thinkers' have no problem with designating this 'variable' by some other symbol ('S' or 'X'). Thus, the test persons with the incorrect solutions were by no means incapable of making abstractions or symbolizations, as is sometimes assumed of people who have problems with 'mathematical' thinking. From an algebraic perspective, these people simply abstracted from inappropriate objects – from concrete *persons* instead of from the given *numbers* of persons.

A further – and, as the empirical studies demonstrated, more significant – source of error was the interpretation of the equals sign. The 'false' solution mentioned above is completely correct if '=' is read as: 'corresponds to' or 'is confronted with'. Using *P* and *S* as designations, the result would then be the totally correct interpretation that '6 students correspond to one professor'. Malle (1993, p. 101) indicates that this perspective is very widespread in everyday thinking and, on the basis of schema theory, introduces the term *correspondence schema* for this phenomenon:

> "This schema simply represents the knowledge that certain objects of one type are confronted with or correspond to certain objects of another type. This schema is so fundamental and natural that we have trouble recognizing it as a schema at all. But it certainly is one, indeed, one that constantly directs our behavior."

The erroneous nature of their solutions usually became apparent to the test persons, who in most cases were quite capable of formally dealing with equations – they had, after all, had quite a bit of practice with them in the mathematics classroom –, if they were told to replace the alphabetical symbols with numbers. Then, they remembered that, in mathematical contexts, the equals sign stands for numerical equality and that the 'correspondence schema' would obviously be inappropriate here, but these recollections did not prevent many of them from constructing another equation with alphabetical symbols according to the correspondence schema when asked to solve a slightly altered problem.

As with the first example, here, too, there seem to be two types of reasoning. What is mathematically 'correct' seems to be 'illogical' within the framework of everyday thinking for a great portion of the test persons – even though they actually can deal with the algorithmic aspect of this area of school mathematics, with transposing terms and solving linear equations. Nevertheless, apparently they did not *understand* what, from a mathematical (or more precisely: algebraic) perspective, variables, terms, and equations actually mean.

A conflict between everyday thinking and mathematical thinking seems to be characteristic of both examples. They represent an abundance of similar learning difficulties observable in the classroom every day. Up to now, the underlying tacit assumption has been that there are such things as 'everyday thinking' and 'mathematical thinking'. In the course of our further deliberations, it will become clear that in many cases this distinction cannot be maintained and that it is psychologically and didactically only of limited explanatory value. But before I systematically examine the relationship between 'everyday' and 'mathematical' thinking (section 3.4.4), I would like to consider two related problems in more detail: What does it really mean to 'understand' mathematics (3.4.2) and what is the specific nature of mathematical abstraction (3.4.3)?

3.4.2 Understanding Mathematics

In ordinary usage, everyone knows – or at least thinks that he or she knows – what is meant by 'understanding' some matter.[111] Both teachers and students are aware that it is of supreme importance to 'understand' the mathematics presented in the classroom. Without understanding mathematical concepts, problems, procedures, rules, arguments, and theorems, mathematical thinking cannot work. On the other hand, most students and teachers are painfully aware that, precisely in mathematics instruction, understanding often does not take place. Certainly, understanding the mathematics discussed in class is a high priority for instruction committed to training cognitive abilities and to promoting critical thinking. But what does this vague demand expressed in colloquial terms actually mean, and what are the consequences?

In the last two and a half decades, an enormous amount of academic research has aimed at theoretically and empirically clarifying 'understanding mathematics'. Even a brief overview of this research, the different approaches, the theoretical perspectives, and the most important results would require an extensive article in its own right.[112] Despite this formidable amount of research, we can maintain the following: A generally accepted theory of understanding mathematics that, on the one hand, describes the mental processes of learners with sufficient accuracy and, on the other, does justice to the various meanings of the everyday concept of understanding, is still nowhere to be found. In my view, the reason why the concept of understanding resists attempts at scientific explication pertains to the fact that, in order to undertake empirical research in the first place, a phenomenon of subjective experience must be connected to 'objectively' observable process characteristics. For in everyday usage, the concept of understanding is not so much employed to described a cognitive process as it is to subjectively evaluate the result (perhaps a tentative or preliminary result) of one's own perceptive and cognitive processes. As a person comprehending something, I *experience* my understanding. It is accompanied by a more or less profound feeling of inspiration, connected to sudden insights of varying intensity. In the light of Carnap's distinction between object language and experiential language, 'understanding' would be a term of experiential language. The statement "I understand problem X" would then have the same epistemological status as the statement: "I've got a toothache".

Despite the limitations mentioned, scientific research on understanding has led to some basic insights that are largely a matter of consensus today. I will present them in condensed form and complement them with a few remarks on the controversy between representation theorists and constructivists. Then, I will distinguish three dimensions of understanding. With this distinction, I will explicitly take into account connotations of the concept of understanding in everyday usage that have been neglected in some of the research.

Understanding as a Mental Process

Cognitive psychology has provided sound arguments supporting the assumption that whenever understanding takes place or, more precisely, whenever someone has the subjective experience of 'understanding' something, a new connection linking cognitive acts to one another is established. Most of the time, this involves establishing a relationship between the new experience and existing knowledge or prior experience. Similarly, a new connection can be established between elements of knowledge that were heretofore separated from one another. As Gestalt psychologists have pointed out (Duncker 1945), such instances often involve not only adding something new to existing knowledge, but also a restructuring of prior knowledge within the process of understanding. In accordance with Piaget's terminology, we might call these two variations 'assimilating' or 'accommodating' understanding.

On the basis of these considerations, a further hypothesis that is generally accepted in cognitively oriented research on understanding seems plausible: Some issue X is that much better understood the more connections a learning individual has established or can establish between his or her idea of X (cognition of X) and other, already understood matters A, B, C, etc.

However, the explication of understanding as the 'establishment of cognitive links' does not already explain how the (subjective) experience of understanding takes place. Up to this point, only necessary, but not sufficient conditions for understanding have been described: If understanding has taken place, it is reasonable to assume that the establishment of a new connection has preceded it. Generally speaking, the converse cannot be assumed.

A further matter of consensus is the view that understanding is accompanied by the *experience of meaningfulness*, of sense, and that, for this reason, intellectual efforts directed toward understanding are motivated by a search for meaningfulness. There is an extensive tradition of characterizing the process of understanding as striving for meaningfulness both within cognitive psychology and beyond that discipline's scope. Usually, this is termed a process of 'making sense'.[113]

So much for the points shared by different researchers on understanding. Simplifying somewhat, two major positions can be distinguished that are based on distinct theoretical premises: one of *representational theory* and one of *constructivist theory*. The representational theorists (e.g., Hiebert/Carpenter 1992) assume that structures existing external to the individual are cognitively represented in an internal way. Thus, an individual would understand the concept X better the more precisely the internal representation of X (including its links to other representations) provides an image of that external conceptual structure with which X is interconnected. In this way, understanding is seen as a process of aligning an internal, cognitive structure to an external one (describable as objectified knowledge). – On the other hand, the constructivists (e.g., Glasersfeld 1987a, Cobb et al. 1992) refrain from postulating some external structure; they assume that the individual constructs his or her own knowledge in such a way as to be able to get along in life (according to the criterion of 'viability'). Knowledge that fulfills this condition is experienced as being meaningful. Constructivists maintain that the individual constructs meaningfulness in the process of understanding.[114]

In what follows I will relate the characteristics of the understanding process mentioned above to three dimensions involved in understanding.

Experience as a Dimension of Understanding

The concept of understanding has a unique status in the self-reflection of man as a creature with cognitive powers. Whoever thinks or remarks, "I understand issue X", registers, as it were, a 'mental experience', experiences himself or herself as gaining insight into something. At the same time, one evaluates the cognitive process within oneself. No one else can carry out this evaluation. Only I can really know if I have understood something – regardless of whether this understanding will prove useful in the future and regardless of whether others (someone I am talking to, a teacher, an examiner) also think that I have understood something. Of course, there is also the case where I have doubts, where I am not sure that I have understood something. But this evaluation is also something that I can only perform myself.

Let us consider a few examples of how the attained understanding of some mathematical topic X is commented on by reflective persons:

"I now know the reasons for X"; "Now I know why X is formulated as it is"; "I now know how to use X"; "Now I know how X works"; "I can now deduce X at any time"; "X now seems convincing to me"; "X now seems reasonable to me"; "X makes sense to me now".

One point in common among all of such statements is that a change within one's own person is announced. X is something postulated externally with which I am confronted and which is perhaps irritating or disturbing to me. Then, while confronting X, a change takes place *within me*, by means of which X becomes something else *for me*: I now see X from a different perspective. And this gain of a different outlook takes place consciously and has a positive emotional tinge to it.

To illustrate this with a concrete example: If as a student I am attempting to understand the rules for the addition of fractions, at first, I might just do some trial-and-error work, observe what the person next to me is doing, attempt to interpret the suggestions made by the teacher. In doing these things, I undertake more or less intense and more or less successful efforts to establish links between what I know about adding natural numbers, about the equivalence of fractions, about the specific meaning of numerator and denominator, etc. If, finally, I succeed in establishing such connections appropriately, then I suddenly (or perhaps gradually) make sense of the rule that originally seemed so strangely complicated: I *understand* it. I am rewarded for the intellectual efforts I have concentrated on the problem with an experience of meaningfulness.

To sum up the experiential dimension of understanding: In understanding something, the learning human subject experiences a confirmation of his or her own thinking, of his or her cognitive efforts by *making sense of something*. This subjective, positively tinted feeling of meaningfulness is an autonomous function of the cognitive apparatus that is not directly susceptible to willful influence.

The Object Dimension of Understanding

Acquiring knowledge involves a human subject exercising cognitive abilities and something at which these abilities are directed. Understanding can only prove useful if in some way it does justice to an object or some context which the human subject perceives outside of his or her personal self. The understanding subject appropriates part of the world in the process of understanding. Understanding cannot be imagined without this relationship between person and world. Of course, in hindsight a teacher might concede having been mistaken, that he or she only believed to have understood some issue X. If a mistake is acknowledged, the original assessment will be revoked: "I only thought I had understood", or: "My understanding was based on false assumptions".

The representational perspective on understanding takes this aspect into account by assuming that some entities (in the broadest sense) outside of the learning subject are depicted in a mental model. By constructing links among the representations of perceptible elements or elements which, themselves, have been constructed, such mental models are adapted to reality. A radical constructivist would deny the depicting function of mental models and would simply state that the mentally constructed models must be 'viable', i.e., that they ensure practical survival (von Glasersfeld 1987b). For our purposes here, this distinction is not significant. The important thing is that by adapting internal models to an external world (in the broadest sense), a relationship between the learning subject and something outside of his or her person is established. It is difficult to find an appropriate term for this dimension. Since it is a matter of a connection between the human subject and something with which he or she is confronted, I call this the 'object dimension of understanding'.

Why is understanding of such major significance especially for learning mathematics? To a great extent, appropriating mathematics can be comprehended as a (re-)construction involving understanding. The logical coherence of mathematics (to be attributed to mathematics whether one regards the discipline as a Platonic realm of ideas, as a social construction, as a conglomeration of cognitive necessities dictated by natural laws, or as an expression of action-related experiences involving some level of abstraction) prevents the learning subject from lapsing into subjective arbitrariness with his or her mathematical (re-)constructions. The appearance of contradictions (that 'make no sense') is a sort of self-corrective for the mathematical cognitive process that can only be ignored at the cost of sacrificing understanding. In other words, appropriating mathematics can only occur on the basis of mathematics that has already been understood; as a form of knowledge subject to codification, mathematics understood at a subjective level partially exhibits the same structure as (or, at least, a structure similar to) fragments of the official, 'objective' mathematics accepted at an intersubjective level. I maintain the following as a far-reaching consequence of these remarks: In contrast to empirical knowledge, understanding mathematics guarantees – in principle and on a long-term basis – the 'correctness' of what has been understood, the consistency of the conceptual construction. And if for the sake of argument we project these insights (in an admittedly speculative way) onto the domain of scientific mathematics, the following interpretation seems plausible: Formulating proofs is the method by which scientific mathematics relates what is to be understood to what has already been understood, and by which, at the same time, mathematical insights are liberated from subjective elements of understanding and thus, in a sense, become 'objectified'.

Of course, the assertion that understanding guarantees correctness is only valid in principle; it indicates a tendency. In school mathematics instruction (and, in general, in acquiring new mathematical insights), the gaps in understanding permit any number of errors. From a didactic perspective, then, it seems more urgent than ever to promote the development of understanding in its object dimension, and to permit the learning-supportive effects of the consistency inherent in mathematics through instruction aiming at understanding.

To sum up the object dimension of understanding: Understanding is a constructive act through which the human subject appropriates some 'object' that confronts him or her 'from the outside'. In this way, understanding creates a relationship between the person and the external world of experience. Within the fluid and dynamic cognitive structure corresponding to the prior knowledge of the individual, new experiences are appropriated with understanding by preparing new paths and modifying old ones, by creating new links and perhaps by severing old ones. For appropriating mathematics, which can be seen as an objectified system of what has been understood, a successively structured process of understanding is indispensable for every advance in learning and cognition.

The Social Dimension of Understanding

Ensuring that some issue X has been 'correctly' understood occurs implicitly or explicitly through social communication, especially through the collective treatment of mathematical topics currently under discussion in the classroom. Exchanging ideas with others who are also trying to understand something or have already understood to some extent is an important aid to developing one's own understanding. Understanding the mathematical topic X, then, is systematically connected to the attempt to understand other persons: If I want to understand issue X, I attempt to understand what some other person (a fellow student, a teacher, or a textbook author) means when he or she tries to explain X to me. I attempt to comprehend the train of thought expressed in words by someone who presents himself or herself to me as someone who has already understood topic X. In this perspective, the widespread practice among mathematics learners of appreciating a teacher or fellow student who can explain well is related to making understanding possible. It is true that students often interpret the ability to explain well rather naively in the sense of a model based on transportation (consider, e.g., the expression that someone can 'bring the point across'). But in light of the considerations mentioned above, it seems more plausible to interpret this ability as a complex interconnection of expertise and social competence. Someone who is able to explain succeeds in allowing his or her own process of understanding to become revitalized, to bring out thoughts into the 'public' as possible starting points for others – thoughts that have perhaps long become fairly insignificant for oneself, but indicate for others paths, points of access, connections between something new and not yet understood and something that is perhaps already familiar. This context makes clear why it is so important that the person explaining something orients his or her language to the current horizon of the learners. This horizon must constantly be determined anew by engaging in dialog: Where understanding is doubtful, points of access are always more easily established by recourse to ordinary everyday language than to technical expressions.

We can speak of the social dimension of understanding in another sense, too. Most of what is to be understood at school, and especially in the mathematics classroom, is something that has already been thought out by others. Understanding the mathematical topic X usually comes down to understanding what others have thought out: a concept, a bit of theory, a rule, a symbolization, a conventional designation. Appropriating such topics with understanding is not immediate appropriation of the world, but the appropriation of external appropriations of the world: Many others before us have interpreted the world, have imposed a network of concepts upon it. A learner, whether child, youth, or adult, is confronted with these interpretations which continually reshape his or her own spontaneous interpretations. For mathematics instruction, the focus, of course, is on *mathematical* interpretations, superficially declared as 'subject-area content'. But if understanding succeeds, then the conceptual construction devised by others does indeed become one's own. With the process of understanding, the external construction seems reasonable; with the use of this construction and through it, the world is 'grasped'. In particular, mathematical constructions distinguish themselves by being experienced as reasonable, as logically necessary, in an appropriation with understanding.

Understanding is not always understanding what has hitherto been thought of. One's own creativity is experienced with the greatest intensity if as a learner (or researcher or simply an agent), one *discovers for the first time* some connection, a solution to a problem, a possible model. A classical example of such understanding is the scientific discovery of the connection between the specific gravity (relative density) of water and floating or non-floating bodies, as commented on by Archimedes with the exclamation "Eureka!" – regardless of the historical authenticity of this scene. Archimedes's understanding was related to new knowledge, to something that in this form had never been thought out by anyone beforehand and in this sense was not socially mediated. But, again, the principle acceptance up to the present day that this originally only subjectively attributable knowledge is a valid description of physical phenomena would be impossible without social communication.

The following should also be emphasized: If someone has understood some matter in-depth and trusts his or her own thinking sufficiently, then this person is highly immunized against social challenges to the insights gained. This person will not waver in his or her views even if the insights are contradicted from within the person's social surroundings. Unfortunately, the reverse conclusion cannot be drawn: It is common knowledge that obstinately maintained views do not at all allow the assumption that they are based on valid experiences and reasonable judgments.

Summary

In understanding mathematical topics, an experiential dimension, an object dimension, and a social dimension are combined:

From a subjective perspective, understanding takes place whenever a previously unfamiliar mathematical topic 'makes sense'. The resulting understanding or meaningfulness is accompanied by more or less intense sudden insights and is experienced as a reward for the preceding intellectual efforts, sometimes also as a gift (intuition). Thus, understanding represents an emotionally positive, subjective experience of a cognitive process.

The object dimension of understanding means that understanding aims at insight, at the appropriation of 'objects' external to the person experiencing understanding. This appropriation occurs through the constructive act of interconnecting new experiences to prior knowledge. Since mathematics can be seen as an objectified network of things that have been understood, its appropriation or, more precisely, its subjective (re-)construction is necessarily dependent on understanding.

Practically, this can only be realized by taking the social dimension of understanding into account. First, mathematics as a cultural product is itself a social achievement. Furthermore, the collective way of dealing with mathematics in the classroom has a function of key importance to understanding. Whether an individual, in the last analysis, actually experiences understanding in this way does, indeed, resist social influence; but, on the other hand, the social surroundings and the extent to which they allow an exchange of experiences, interpretations, assumptions, and discoveries (i.e., the culture of instruction) are decisive factors in making understanding possible.

3.4.3 The Specific Nature of Mathematical Abstraction

In principle, the statement that some topic X has been better understood the more interconnections the learner can construct between the cognition of X and cognitive activity related to other topics already understood, also applies to learning mathematics. However, this abstract formulation remains unsatisfactory where we are concerned with describing consequences for mathematics instruction aiming at understanding:

- the content of prior knowledge relevant to understanding remains unspecified;
- the specific characteristics of mathematical knowledge that distinguish it from object-related knowledge in other domains remain ignored.

Both points are of major significance if the relationship between mathematical thinking and everyday thinking is to be clarified further.

Mathematical Abstraction and Abstraction in Other Domains

Many students who encounter difficulties in understanding mathematics complain about its abstractness.[115] Among other things, what they mean by this is that they do not succeed in making connections between important everyday experiences and the mathematics presented to them in the classroom. But, as we will see, abstractness itself is not an obstacle to understanding. In order to pick up the track of this widespread problem of understanding, it is necessary to consider how mathematics is abstract. Above all, we must pay attention to the twofold character of mathematics as an abstract referential and an abstract formal system. This twofold nature of mathematics, namely, that mathematical concepts and rules 'depict' characteristics of real objects and their relationships to one another, but that mathematics, on the other hand, is self-complacent as a formal system and only adheres to the logical necessities inherent in itself, has often been described – in various terminologies and on the basis of divergent philosophies.

We have already noted that Wittenberg (1963) declares this twofold nature to be the core of his concept of mathematical education. Here, I am concerned with his concept less from an epistemological perspective than from one of the psychology of learning. To a certain extent, I will be following the terminology and argument of Resnick et al. (1987b). For elementary algebra, these scholars have explored the effects of the twofold nature of mathematics and of its specific abstract character on learning and understanding.

With good reason, Resnick et al. emphasize that the distinctive characteristic of mathematical knowledge in comparison to conceptual knowledge in other domains is not abstraction in and of itself. Conceptual knowledge is always abstract. Concepts of ordinary language – e.g., tree, green, language, freedom – are also based on abstractions from concrete objects, processes, or characteristics. Even a concept like 'freedom', for which it is not possible to point to individual representatives of the class defined by this concept, can inductively be defined by compiling examples. By contrast, mathematical abstraction seems to be of a different sort. This can be illustrated simply by referring to such elementary mathematical objects as natural numbers: Any combination of three concrete objects is, on the one hand, a physical representative of the number 3 (interpreted as a cardinal number). Yet, the properties of the number three are not attributable to such a combination in the same way as the properties subsumed under the concept 'tree' are attributable to any concrete tree. I can point to one of the three birches in front of my house and say, "That's a tree". But it makes little sense to point to all three and say, "That's a three". Resnick et al. mean the difference demonstrated by such examples when they maintain that even an elementary mathematical concept such as that of a number must be constructed as a "strictly *cognitive* entity". Thus, from the very beginning, mathematical thinking must be concerned with objects that "exist only as mental abstractions" (ibid., p. 170).

At this point, it may seem appropriate to complement the ideas of Resick et al. by referring to a common experience: Interestingly enough, with increasing familiarity, these kinds of mental abstractions take on the status of 'semi-concreteness'. For children who have had sufficient practice in counting and computation, numbers already practically become objects that are just as 'real' as trees and houses (which, of course, as concepts of ordinary language for human subjects are originally nothing other than mental abstractions – with the difference that there actually are concrete representatives corresponding to them). This 'reification' of mental abstractions produced by practical familiarity, which seems to readily emerge from doing mathematics intensively, might, in the final instance, be the root of the Platonic interpretation of mathematics – of the interpretation of mathematics as a semi-empirical 'reality' in its own right, in which mathematical objects exist independently of man and can be studied. As Davis and Hersh (1986, pp. 334ff) maintain, mathematical Platonism is still a widespread "informal or tacit working philosophy" among mathematicians – despite the epistemological problems involved.

The peculiarities of specifically mathematical abstraction mentioned above also influence the arrangement of links or levels of abstraction and in this way cause specific learning problems. If we consider a series of abstractions in ordinary language like the following: Fido (as the name of a concrete individual), dachshund, dog, mammal, living creature, then this chain is transparent in the sense that Fido is also a representative of the most general concept. Fido is a dachshund, a dog, a mammal *and* a living creature. For the purpose of demonstrating the contrast, we might examine a characteristic (quite elementary) mathematical series of abstractions: finite set, a cardinal number as a member of the set, a variable for natural numbers. In a certain sense, the cardinal number (as an abstraction) does indeed represent various concretely given or conceivable sets, and in a certain sense the variable also represents various conceivable numbers in a generalizing way. But the set *is* not the number or an example of this number (see the remarks above), and the number *is* not a variable or an example of one. And most certainly, the set is not a variable or an example of one. What a variable is cannot be inductively defined by 'pointing' to individual representatives. At the most, we could only point to the relevant symbol (as with the explanation of what a number is). But this might cause a totally different sort of misinterpretation that often obstructs mathematical understanding: confusing symbol and concept. Thus, at this point we should go into more detail on the twofold nature of mathematics as a referential and as a formal system.

Mathematics as a Referential and Formal System

Mathematics draws its potential from the ability to symbolically represent mathematical objects and the relationships between them and to make formal use of these symbols according to predetermined rules. The scope of mathematical reasoning that can be done exclusively 'in one's head' is very limited. But such constrictions are counteracted by symbolization and formalization. Preliminary steps and tentative results can be recorded so that we can take recourse to them at any time. Similarly, the results of mathematical thinking can be made public and become a topic for intersubjective discussion. In an extreme view, mathematics seems like a formal game with no content – undefined basic concepts are manipulated according to rules established by axioms and according to purely logical rules. With reference to elementary algebra: We can formally describe what an equation is and formally work with rules of elementary algebra which ensure that from a given valid equation only valid additional equations can be attained. Since formalist interpretation strips mathematics of its content for the most part, it is arguably the worst basis for learning with understanding – cognitive networks can (officially) only be established among the cognitively represented elements of the formal 'game'. Thus, the competency attainable by a strict limitation to the formal character of mathematics is simply that of mastering complicated rules of the game.

Mathematics acquires practical significance because, in addition to its indispensable formal nature, it can also be interpreted as a referential system. Here, again, I will restrict myself to elementary algebra, which can immediately be seen as referential in a twofold sense – in a nonmathematical way and in a way internal to mathematics. Within a context internal to mathematics, the rules of algebra and algebraic expressions attain a certain meaning because they represent, as it were, operations with numbers in a generalized form. By replacing the variables with concrete numbers, this sort of meaning becomes comprehensible to anyone who has knowledge of numbers and of the permissible operations with them. For example, I can approach the meaning of

$$x - (y + z) = (x - y) - z$$

by computing

$$17 - (8 + 3) = 17 - 11 = 6$$

and

$$(17 - 8) - 3 = 9 - 3 = 6.$$

Examples of this sort do not prove anything mathematically, but they might lay the foundation for an insight such as: Whichever (natural, whole, at a later date perhaps rational) numbers I put into the equation, as long as I use the same number for the same symbol, the same result always appears on both sides of the equals sign. Now, if by contrast I compute the variation

$$17 - (8 - 3) = 17 - 5 = 12$$

corresponding to the term $x - (y - z)$, perhaps I will notice that the general rule for resolving a parenthetical expression with a minus sign before it corresponds to my experiences in computing numbers. If, in addition, I have an internal image of the number line and know how to represent addition and subtraction on the number line, then I can also visualize various operations and relate them to the general algebraic formula. In contrast to a purely formal view of algebraic rules, this sort of referring back to computing numbers and corresponding geometric operations, as references inherent in mathematics, is already an important aid to understanding.[116]

Nonmathematical situations provide a second source of meaningfulness for arithmetical and algebraic rules and expressions. Above all, inventing computational 'stories' for given expressions (thus, the 'reverse' of the usual application problem) is an interesting way to make computations accessible. To modify an example provided by Resnick et al. (ibid., p. 173): A corresponding 'story' for the arithmetical expression $17 - (8 + 3)$ could be as follows: "Maria had 17 pieces of candy. She gave 11 of them to her friends, 8 to Sandra and 3 to Thomas." To illustrate $(17 - 8) - 3$, the story might be modified in this way: "Maria had 17 pieces of candy. She gave 8 of them to Sandra. Later, she gave Thomas 3 pieces of candy." Counting would show that in both cases, Maria still had 6 pieces of candy, which would correspond to the equivalence of the two arithmetical expressions.

Before we consider which situations corresponding algebraic expressions could refer to, let us call attention to a few problems that already arise in assuming a correspondence between arithmetical expressions and computational 'stories'. Often, the appropriateness of such a story can only be judged if a basic understanding of arithmetic has already been acquired. For example: Along the lines of the preceding stories, the following computational story might be devised to illustrate *17 – (8 – 3)*: "Maria gave Sandra 8 pieces of candy; and Sandra passed 3 pieces on to Thomas." But the arithmetical result of *12* cannot be put into this story; the situation doesn't 'fit'. However, this one does: "Maria gave Sandra 8 pieces of candy. Sandra gave 3 pieces back to Maria." It is difficult to explicitly state why the second situation is appropriate and the first is not, especially in language comprehensible to school children. It is more beneficial and more promising to let children experience what works and what doesn't, using a wide variety of connections between arithmetical expressions and situations. Thus, understanding the relevant mathematics and its referential nature cannot simply be induced by coupling mathematical expressions and referential situations. At best, this develops step by step and as the result of interaction, in proceeding to and from the mathematical and the nonmathematical world.

How can the referential significance of formal mathematical expressions be demonstrated at the next level of abstraction, at the level of algebra, by considering appropriate situations? To put it somewhat pointedly, what makes elementary algebra so interesting for nonmathematical applications is its potential to allow an operative way of dealing with a lack of knowledge: I compute with numbers that I don't yet know, but would like to know, as with concrete, given numbers. And if I manage to do this with some skill, I can find out what the unknown number is.

The formula examined above

$$x - (y + z) = (x - y) - z$$

is not very suitable for demonstrating the situational reference of an algebraic expression, because as an algebraic rule for transforming equations, it is primarily related to an applicational domain inherent in mathematics. Instead, let us consider as an example the following linear equation:

$$10 - 4x = 4.$$

An appropriate situation (even if no one would compute this way in everyday life) would be: "Maria was sent to the store with $10 to buy 4 bottles of mineral water. She gave her mother $4 change. Her mother wanted to know how much one bottle of mineral water cost." If this computational story were set up as a fabricated problem, it would be a perfect example of those applications so far removed from reality that they only serve to make mathematics detestable to the students. Turned in the opposite direction, considered as a problem for finding a situation, it can help to make students more sensitive to the referential nature of algebra. If we decide to work with such problems for finding situations, there is a certain appeal to not immediately restricting oneself to 'correct' interpretations of an equation, but to consciously devising widely different interpretations that are all plausible in some way or another. Assessing which of these are compatible with the algebraic 'language-game' would then be a second step. An interesting third step would be the attempt to algebraically model those interpretations discarded as inappropriate. To illustrate this with another example: A suggestion for the equation above might be as follows: "Someone has taken unknown amounts of money four times from a cash-box that originally contained $10. Now, there are only $4 left in the cash-box." Whereas the first interpretation proved appropriate to the situation through the tacit assumption that each bottle of mineral water cost the same amount, the second interpretation violates the condition (or at least does not clearly satisfy it) that within a particular equation the same value must always be attributed to the same variable

$(4x = x + x + x + x)$.

On the other hand, the second situation could be modeled using the equation

$10 - x - y - z - u = 4$.

This interaction of problems for finding situations and for remodeling different situations, along with the relevant assessments and argumentative deliberations, allows an enormous amount of latitude for creative inventions, constructions, trials, and discussions – dimensions that usually remain totally inaccessible in traditional algebra instruction.[117]

A sensitivity to the referential nature of algebra and of other areas of school mathematics becomes indispensable as soon as there is relatively challenging mathematical modeling to be done. The problem involving the number of students and professors (3.4.1) provided shocking, but definitive evidence that technical mastery of elementary algebra by no means implies the ability to construct an algebraic model even for simple quantitative contexts. The aversion that so many students feel toward word problems cannot simply be attributed to their removal from reality – many young people take the greatest pleasure in playing the most unrealistic computer games. Rather, this aversion expresses a justified resistance toward doing something difficult that one has not learned to do. It is primarily the formal nature of algebra that is emphasized in traditional instruction, i.e., transposing terms in equations is practiced. If the referential nature of algebra does actually become a topic for instruction, it is usually limited to a demonstration of references inherent in mathematics: algebra as a generalized form of mathematics. By contrast, dealing with algebra as a system whose expressions can also be meaningfully interpreted in contexts outside mathematics usually only receives attention with respect to subsequent application problems, thus, at a point where this treatment of the subject should already have become a matter of course.

Resnick et al. (1987b, p. 173) describe the challenge for learning that stems from the twofold nature of algebra as a formal and referential system as a paradox:

> "On the one hand, formal expressions take their meaning in part from the situations to which they refer. On the other hand, algebra derives its mathematical power from divorcing itself from those situations."

In a similar way, this also holds true for other topics that are relevant to school mathematics – for elementary arithmetic (see above), for geometry with its relation to the natural experience of space and to the organization of space, for probability with its conceptualization of uncertainty and chance, and for analysis with its relation to constant changes in nature and in social and economic spheres.

In traditional instruction, this focus on the formal nature of school mathematics in comparison to its referential aspects is due to, among other things, the following circumstances: Formal rules are easier to 'teach' if we conceive of teaching in the traditional sense. They can be formulated explicitly, written down on the blackboard and in notebooks, memorized, and easily tested. The use of rules can be demonstrated and, then, imitated. Adherence to formal rules can easily be checked and evaluated according to a right-or-wrong pattern. The instructional material is unambiguously defined; the teacher can easily assess which topics have been 'covered' and which are still to be dealt with.

Paying more attention to the referential nature of school mathematics requires allowing for processes of discovery in the classroom, for varying interpretations and their explication, for the toleration of intermediate stages between 'correct' and 'incorrect'. Often, the new connections that students make when they have understood why one algebraic expression is suited to modeling a certain situation while another is not, cannot be explicated to such an extent that an immediate evaluation of learning progress is possible from an external perspective. For this reason, this kind of instruction demands more sensitivity of the teacher, a greater sense of individual differences, more flexibility and creativity. And the demand for constant evaluation of students' achievements must be, if not relinquished, at least for certain phases, deferred.

Summary

In mathematics instruction oriented to understanding, the specific qualities of mathematical abstraction are to be kept in mind as possible sources of obstacles to understanding. In contrast to other subject matter, the abstract concepts of mathematics often cannot be inductively defined by pointing to representatives assigned to a lower level of abstraction. Only manifold experiences in dealing with new concepts to be learned and their possible applications result in the necessary interconnections to prior knowledge that allow for the sort of familiarity that signalizes understanding.

Furthermore, the twofold nature of mathematics as an abstract formal and an abstract referential system is to be taken into account. The potential of mathematics, its universal applicability is, indeed, on the one hand, due to the fact that in mathematics particular aspects specific to certain situations can be ignored and one can operate with mathematical symbols in a purely formal manner. On the other hand, in addition to their formal meaning, mathematical expressions also exhibit a referential meaning that only becomes accessible in situational interpretations. *For this reason, understanding mathematical topics is not least to be identified with being able to interconnect the formal and the referential meaning.* For mathematical modeling, this connection is indispensable. Within traditional instruction, the formal aspect is usually overemphasized. Paying more attention to the referential aspect requires a culture of instruction developed in a corresponding way.

3.4.4 Mathematical Thinking and Everyday Thinking: Systematical Reflections

The difficulties presented by the examples of section 3.4.1 can readily be interpreted as demonstrations of dealing with the concepts and symbols appearing in the problems in a way corresponding to *everyday* thinking, although they were meant in a specifically mathematical way. The mathematical operation of dividing was interpreted by Katharina in the light of everyday experiences with breaking something down into smaller component parts (first example); variables for quantities were interpreted as names and the numerical equation was conceived of according to a correspondence schema (second example). What exactly is the relationship between mathematical thinking and everyday thinking?

Two Opposing Views on the Relationship Between These Two Ways of Thinking

Archetypically, this relationship can be interpreted in two diametrically opposed ways with varying didactic consequences:

Fundamental difference: Here, the assumption is that everyday thinking and mathematical thinking are fundamentally different. In this view, everyday thinking is vague, imprecise, and leads to no unequivocal results – similar to everyday language, on which it relies. One cause for mistakes is that students remain entangled in everyday thinking. But in the mathematics classroom, mathematical thinking alone is appropriate. A high-priority goal of instruction must be to replace the students' everyday thinking with mathematical thinking as far as possible. One should demonstrate the imperfections and inappropriateness of everyday thinking to students who do not succeed in 'switching' to mathematical thinking.

Continuity: Here, mathematical thinking is, in a sense, a systematical continuation of everyday thinking. Everyday thinking is made more effective for a certain class of problems (precisely for so-called 'mathematical' problems) by enhancing the precision of its concepts, by the systematic and conscious use of certain types of reasoning and strategies that, in principle, are already present in everyday thinking (but often with some inconsistency). There is an abundance of intermediate stages between everyday thinking (or everyday language) and mathematical thinking (or the technical language of mathematics) – stages that are of significant import to learning mathematics. Mathematical learning has greater chances of success, the less the learners sense a gulf between their everyday thinking and the mathematical thinking demanded of them in the classroom.[118]

There is empirical data behind both of these views. They present two different, generalized interpretations of experiences often encountered in the mathematics classroom. Yet, they are not empirical hypotheses in the strict sense, the validity of which could easily be checked by empirical studies. The two views seem to be more like 'meta-hypotheses', framework assumptions that, at the same time, represent a certain instructional paradigm. Not least, these opposing views are an expression of the difference between subject area-oriented and student-oriented conceptions of mathematics instruction.

Conventional mathematics instruction at college-preparatory secondary schools traditionally exhibits a certain affinity to the view based on fundamental difference and the related subject area-oriented conception. By contrast, the continuity view demonstrates a greater affinity to the concept of general education under discussion here. In comparison to the view based on fundamental difference, it is much more compatible with letting students experience how mathematics can be employed as an *amplification* to their everyday thinking.

Despite the problems involved in finding immediate empirical evidence for one of the two 'meta-hypotheses' (empirical findings also often depend to a large extent on the preconceptions of the researcher and the conception of the instruction analyzed), a great number of research findings from the last few years can be interpreted much more plausibly and consistently in accord with the continuity view. With reference to my remarks on 'understanding' and 'mathematical abstraction', I would like to demonstrate this in the following sections.

Mathematical Thinking and Everyday Thinking: An Attempt at Clarification

At the outset, a more precise differentiation of both ways of thinking is beneficial. My explication relies on three exterior characteristics and is compatible with the general usage related to the concepts.

- *Mathematical thinking* operates with mathematical concepts (and related rules and symbols) that function as elements of technical mathematical language and are clearly defined in a certain given mathematical context, and it operates with a diverse spectrum of other possible meanings in other contexts inherent in and outside of mathematics; by contrast, *everyday thinking* relies on concepts of everyday language that usually exhibit a broad range of meanings.
- Actions that arise in connection with *mathematical thinking* (e.g., computations, symbolizations, mathematical modeling) are characterized by an adherence to certain formal rules which themselves are part of the body of mathematical knowledge; for concrete actions that can be traced to *everyday thinking*, no explicit rules can be formulated in the first place, or, they are of a nonmathematical nature (e.g., grammatical rules for writing).
- Successful *mathematical thinking* manifests itself in successfully solving mathematical problems (including the appropriate modeling of primarily nonmathematical problems); successful *everyday thinking* manifests itself in successfully solving practical problems of everyday life without the use of specifically mathematical means.

I am consciously refraining from attributing different types of reasoning or of logic to the two ways of thinking. I am also leaving aside the use of certain principles and strategies[119] often closely related to mathematical thinking (e.g., abstraction and concretization, analogizing, generalizing and specializing, varying, classifying, systematization) as potential criteria for differentiation so that no prejudices restrict my further remarks.

Nevertheless, the question of whether the differences between mathematical thinking and everyday thinking listed above actually justify speaking of different types of 'thinking' has not yet been resolved.

The Concept of Thinking: More Precise Specifications

In ordinary language, we do not differentiate clearly between thinking as a cognitive process employing representations of content (this could be a very broad definition for cognitive psychology) on the one hand, and prerequisites, objects, and results of thinking on the other. For instance, someone says, "I think differently about this than you," and means that he or she considers other facts, ascribes a different significance to the facts, or has other value judgments as original criteria. A similar situation arises if we say that people from other cultures or subcultures 'think differently'. What we mean is, e.g., that due to other experiences, they have other biases, perspectives, and values.

By contrast, if we understand thinking in a strict sense, as a cognitive process, then all people with sound intellectual faculties who have matured in relatively beneficial surroundings think essentially the same way. Due to some form of interaction between their genetic disposition and the intellectual socialization of their everyday surroundings, they are intuitively capable of simple logical reasoning and of thinking in simple terms of cause and effect. This development of cognitive ability proceeds in a way closely related to developing linguistic abilities in one's native language (cf., e.g., Dichgans 1994). The basis described in this way corresponds – in Bruner's terminology – to reaching the linguistic-symbolic stage of intellectual development, which, in our culture, occurs around the age of six to eight. Additional elements supplementing this basis – manifest in diverse capabilities of adequately operating with very abstract concepts (thought processes involving formal operations), of taking complex causal networks into account instead of only linear causal chains, of quickly and self-assuredly coming to far-reaching conclusions, of accessing extensive spheres through association. These are, to a great extent, dependent on the content structure of what has previously been learned or, in the terms of representation theory, on the mental models available to the particular person and on the internal connections that can be established between such models. Or, if we attempt to describe perceptible differences with the theory of 'domains of subjective experience': on the diversity and extensiveness of existing domains of subjective experience, on the ability to form new domains of subjective experience, and on the facility of transition from one such domain to another (cf. Bauersfeld 1993, pp. 244ff). However, this leads us to the following stipulation: What is normally observed as different ways of thinking among different persons, can essentially be explained by the fact that, in each case, thinking relies on different knowledge bases (with respect to the specific characteristics, diversity, and degree of interconnections) that can be used with varying degrees of effectiveness.

If we use the model of human thinking roughly outlined here as a basis, there is no reason to assume that the acts of thinking themselves (in the strict sense of cognitive processes) would have to be of fundamentally different quality. This result provokes the development of a more subtle view of perceptible differences between mathematical and everyday thinking.

Mathematical Thinking and Everyday Thinking Reconsidered

Success with a newly learned mathematical subtopic means succeeding in redirecting cognitive operations available in connection with more familiar topics to elements of the new mathematical subtopic. As we have seen, to achieve this, one must have 'understood' the new elements sufficiently. They must be connected to prior knowledge in adequately diverse ways. As mathematical education advances, part of the 'familiar' knowledge is, on the one hand, mathematics that has already been understood, including the relevant procedural knowledge, knowledge of strategies, and meta-knowledge. On the other hand, part of the 'familiar' knowledge for all learners is a large body of knowledge that they have developed in connection with their everyday experiences, their knowledge of everyday life. The exact structure of this knowledge of everyday life, its division into certain spheres, and the interconnections among these spheres all vary considerably from individual to individual.

Differences between students who consider themselves and are considered by their teachers, fellow students, and perhaps parents to be 'good mathematical thinkers' or 'mathematically gifted' and those who see themselves and are seen by others as 'weak mathematical thinkers' or as 'untalented' can be archetypically described as follows:

- The mathematical knowledge of the 'good mathematical thinker' is (in those spheres where the person is successful) interconnected in a wide variety of ways: There are internal connections among the elements of this knowledge and, in addition, connections to the individual's *unique knowledge of everyday life*. This person has no difficulties with a great number of mathematical chains of thought, because he or she can establish a structural relationship between them and familiar phenomena, because he or she has access to experientially enriched images, models, analogies, and metaphors that provide interconnections to the current mathematical topic. The formal and referential aspects of mathematics are not isolated from one another. For this reason, this person is also able to mathematically model nonmathematical problems (to find approaches to problems).
- The 'weak mathematical thinker' can establish hardly any connections between his or her knowledge of school mathematics and of everyday life. Since the mathematics this person can master has been acquired solely as a formal technique to be used as a procedural formula, there is nothing to be gained from the referential aspect of mathematics. There are no useful images, models, or analogies associated with mathematics. It remains, as it were, lifeless, since no vivid experiences are connected

> to it. For this reason, this student's knowledge of school mathematics is also more weakly interconnected at the internal level. It dissolves into separate domains of experience whose interconnections are, at best, based on contexts explicitly emphasized in the classroom. Thus, a 'weak mathematical thinker's' knowledge of school mathematics does not form a network, but, rather, a fragile, linear structure that quickly dissolves if individual connections are lost by being left unused for longer periods and forgotten. The isolation of this fragmentary knowledge from the rest of personal experience is perceived as insufficient understanding, as a lack of meaning, and as a gulf between 'normal' thinking and that required in the mathematics classroom.

This distinction remains reductive in two respects: First, the broad spectrum between the 'good mathematical thinker' and the 'weak' one is populated by students with widely varying learning abilities and impediments, to whom neither of the archetypical characterizations applies. Secondly, the classifications are, of course, only relative; which students are viewed as good mathematical thinkers is largely dependent upon classroom instruction, mathematical requirements, and social milieu. A junior high school student who is a 'good mathematical thinker' and goes on to a college preparatory school, might be disappointed to find that he or she is now only at the level of the lower half of the class. And a mathematically 'gifted' student who decides to study mathematics might have a similar experience at the university.[120]

A Tentative Assessment

Talk of 'mathematical thinking' often involves the idea that this is a fundamentally different mode of thinking than is usual in other contexts. There is more empirical data supporting the idea that what in ordinary language (and often also within the technical language of mathematics education research) is called the ability to 'think mathematically' is based on a denser network of elements of mathematical knowledge – both internally denser and with reference to nonmathematical knowledge. It seems that a certain disposition to establishing such interconnections readily, quickly, and flexibly is innate. For the individual learner, the development of the corresponding cognitive abilities occurs as an interaction between this disposition and the general intellectual socialization, i.e., and the opportunities and stimuli for thought provided by the school setting and the environment outside of school, the models to which thinking can be oriented, the possibilities of engaging in active thinking in such a way that meaningfulness can be experienced. If mathematics instruction should contribute to training cognition, it can only do so by means of a corresponding intellectual socialization. Instruction will then make efforts to avoid isolating mathematical knowledge from the knowledge of everyday life. On the contrary, it will constantly make use of opportunities to establish connections.

Wherever in what follows I speak of 'mathematical thinking' or 'promoting mathematical thinking', this occurs within the framework of the preceding explication. Thus, somewhat reductively, 'mathematical thinking' is a sort of abbreviation for 'adequate cognitive treatment of elements of mathematical knowledge', taking their referential aspect into account.

However, this means that in the dispute between views asserting *fundamental difference* or *continuity*, the stand to be taken is definitely in favor of the latter: Mathematical thinking does not require 'switching' from normal thinking, much less 'switching off' everyday thinking; instead, it represents a *more systematic use* of normal thinking and its extension to new areas of knowledge – the body of mathematical knowledge – that are interconnected with everyday knowledge in a variety of ways, and it represents an effective application of the cognitive tools made available in these new areas of knowledge. In this way, the stipulation I made as a guiding proposition of this subchapter (in mathematics instruction aiming at general education, students should experience how mathematics can be employed as an 'amplification' of their everyday thinking) acquires a much more precise meaning.

Possible Objections: A Fictitious Dialog

Although the idea of the interconnection between mathematical and everyday thinking outlined up to this point (and reductively sketched in the description of the view asserting *continuity*) is much more compatible with current mainstream theories of cognitive psychology than is the assertion of *fundamental difference*, we can, nevertheless, not maintain that it has been empirically confirmed beyond any doubt. As mentioned above, this has to do with the fact that the two opposing views are not empirical hypotheses in the strict sense, but, rather, 'meta-hypotheses', principles of interpretation with empirical content. For this reason, as a final comment on this topic, I would like to examine some potential counterarguments which might be brought up against the view of continuity – less from a specifically educationalist perspective than with reference to generally accessible observations of everyday life. I will present this debate as a dialog between a proponent of the continuity view (P) and a skeptical opponent (O). Let us assume that P has presented the content of section 3.4 up to this point to O.

O: If we take both your characterization of conventional instruction and your description of mathematical thinking into account, we are inclined to ask why there are any people at all who can think mathematically. Apparently, traditional mathematics instruction with its subject-area orientation and with no particular emphasis on connecting mathematics and everyday knowledge does manage to produce students who can 'think mathematically'.

P: That's no contradiction for me. There is a good deal of evidence indicating that mathematically more talented children and youths devise their own connections to their everyday thinking in a mathematics classroom that cultivates mathematical thinking and language in a way that is primarily oriented to the subject area. It is common knowledge that many creative mathematicians operate with an abundance of visualizations, metaphors, and analogies that are by no means a component of 'official' mathematics. Applied to the classroom, this means the following: What the teacher does not explicitly mention is constructed by the more talented students on their own initiative, consciously or unconsciously. The paths that lead from a systematic approach to everyday problems to a mathematical systematization seem to be readily passable in both directions for such students.

O: Assuming that you're right, what about student-centered instruction, where the attempt is made to make the subject come alive, to take recourse to the students' everyday ideas and to promote understanding in this way? That all of this should be done is hardly a novel educational demand. And there are a great many attempts to realize just this kind of instruction. Now, is this type of instruction really more successful than that oriented to the subject matter? Everybody complains about the learning achievements of their students! It seems to me that this is just a case of one ideology vs. another: traditional, college-preparatory secondary education, oriented to science, against ostensibly more student-centered instructional practices of modern elementary and secondary schools. But neither one really works!

P: I haven't yet mentioned anything about the methodical, practical realization of a mathematics instruction that promotes cognitive abilities. I am actually convinced that more students will have success in mathematical thinking if classroom instruction shows more sensitivity in dealing with the intermediate stages between developed mathematical thinking and its preliminary forms.

O: But, in my opinion, at least this much is obvious: On the average, students who have had subject area-oriented teaching in college-preparatory secondary schools demonstrate better learning results.

P: With regard to learning achievements, first, you can only compare students with similar preconditions, thus, e.g., an average preparatory school student having subject area-oriented instruction with an average preparatory school student having student-centered instruction, and, indeed, only if the level of intellectual demands in the classroom is comparable. Secondly, we should take into account that there is often a considerable gap between intention and realization. Student-centered instruction demands much more sensitivity to differences among the students than instruction oriented to the discipline. Research on interaction in the mathematics classroom focusing on questioning has demonstrated that teachers often act contrary to their own intentions by unconsciously directing in a normative way, by using subtle means to elicit those answers from students that they consider necessary for moving along in the lesson. Thus, research has shown how teachers *and* students become the victims of a logic of interaction that remains incomprehensible to them.

O: You're digressing from the issue! The question was whether establishing connections to the students' everyday knowledge is a necessary (or at least beneficial) prerequisite for learning mathematical thinking.

P: I wasn't quite finished. Precisely this 'establishing connections to everyday knowledge' can easily deteriorate to a classroom ritual that simply precedes the 'actual' mathematics and that serves the sole purpose of defining for everyone involved what is to be considered everyday knowledge within the current context. But the everyday knowledge that is really pertinent here and that every individual teacher has to connect to new mathematical knowledge in a personal way is extremely idiosyncratic. For this reason, the decisive impulses of instruction that promotes cognitive abilities and is oriented to understanding will most likely be indirect ones. The important thing is the appropriate balance between room for one's own intellectual activity, external stimuli provoking the relevant thoughts and actions, and purposeful help and encouragement for individual students if learning processes become obstructed.

O: Tell that to the colleague who the next day has to teach the rules for transposing terms to an apathetic gang of 30 adolescents who have just reached puberty!

P: If the colleague's primary goal is to introduce such concepts formally and with absolute accuracy, then he or she has already struck out. But I can't dictate to other people what their goals should be!

O: Let's leave the ideologies aside for now. I'd rather explore a further obvious inconsistency in your conception. In essence, you said that the view of fundamental difference is part of the theory providing the backdrop to conventional mathematics instruction at college-preparatory secondary schools and that this is compatible with those schools' orientation to the subject area. On the other hand, you maintain that the mathematical and everyday thinking of mathematically talented persons is better integrated and that this is even an important reason for their mathematical superiority. And that a gulf between mathematical and everyday thinking would more likely be sensed by weak mathematical thinkers. Now, tell me: Are you assuming that the mathematics teachers at the preparatory schools tend to be untalented? Or are you supposing that they don't actually know how their own mathematical thinking works?

P: Now, don't laugh at this, but: the latter. The assumption that this gulf does not exist for mathematically more talented persons does not at all imply that this corresponds to their conscious self-perception. 'Good' mathematical thinkers have little reason to reflect on the connection between mathematical and everyday thinking that we have been discussing. Such connections are sought when they are lacking. Only if fundamental difficulties in understanding arise, i.e., only if in the course of intense personal efforts to solve a mathematical problem no meaning can be experienced, is this subjectively experienced as a distinct gap between mathematical and everyday thinking.

O: But, what is the consequence of your theory for a subject area-oriented teacher at a secondary school? Why does he or she support the view of fundamental difference if the view of continuity is the one that is actually correct? I don't understand what should motivate him or her to make the gap – which, in your opinion, is only perceived as such by the weak mathematical thinkers – the basis of a personal didactic theory.

P: Put yourself into his or her position. You experience daily how some of the children and adolescents entrusted to you have enormous difficulties with cognitive acts that are familiar and natural to you. Furthermore, you experience how other students whom you perceive as being talented readily understand and adapt the cognitive steps required in the classroom. You interpret these experiences then as follows (and seem to confirm the view stipulating fundamental difference): You and some of the 'gifted' students take the technical language of the discipline seriously and get involved in 'genuinely mathematical' thinking. By contrast, the other students cling to useless everyday concepts and balk at serious mathematical procedures. If they are not simply lacking in the necessary intelligence (in which case they really shouldn't be attending this school), one must constantly demonstrate the advantages of the technical language of mathematics and of mathematical thinking to them, using technically perfect examples. – So much for the perspective of this type of teacher. This is a consistent interpretation that is totally compatible with my theory.

O: You view the entire world from the perspective of your theory! A compatible interpretation doesn't prove anything!

I will cut short this fictitious dialog here and let the reader form his or her own judgment.

Summary

This summary refers to the preceding part of section 3.4, i.e., to the entire complex of reflections on understanding, on abstraction, and on the relationship between mathematical and everyday thinking.

Persons who are unsuccessful in learning and applying mathematics often sense a gulf between their everyday thinking and the mathematical thinking required of them. They personally experience how the thinking they are familiar with from everyday life and nonmathematical spheres does not enable them to cope with mathematical problems. They are then even confirmed in their perception of such a gulf if mathematically superior persons (e.g., teachers) tell them that it is necessary to think differently in mathematics.

Recent theories on human learning and understanding and empirical findings of the last few decades do not support the assumption that mathematics requires a qualitatively different type of thinking. Those cognitive operations that are characteristic for the developed form of everyday thinking are, in principle, also present in mathematics. Nevertheless, differences do exist: the particular concepts to which thinking takes recourse in mathematics and the interconnections among these concepts. In a certain sense, mathematics is a different language that not only adheres to a different grammar than do the natural languages, but also the concepts of which are abstract in a different way.

Since mathematics is not only a formal system with a different 'grammar', but also a referential system, there is an abundance of interconnections between it and the objects and situations to which everyday thinking refers. An understanding of mathematics can only take place if these connections (which represent the 'nonmathematical meaning' of the corresponding mathematical concepts, operations, rules, etc.) can also be learned. If these connections remain underrepresented in the mind of the learner, this will also detract from the cognitive structures which represent the internal interconnections of the mathematics to be learned. The mathematical fragments, understood only in a superficial and one-sided way, remain isolated from one another and can no longer be experienced as meaningful; understanding fails to materialize.

The specific quality of mathematics that makes it so suitable as an 'amplification' to everyday thinking cannot be recognized and used by learners if mathematics is not adequately understood. In mathematics instruction oriented toward understanding, students must be given opportunities to create connections between the mathematics currently under discussion in the classroom and their familiar prior knowledge, their deeply rooted everyday ideas.

The following sections relate more clearly than the preceding ones to the practical realization in classroom instruction of the insights developed thus far: How can instruction be designed to increase the chance of understanding mathematics? How can countermeasures be taken against the separation of mathematical and everyday reasoning? Finally: In which way can mathematics teaching help develop *critical* thinking?

3.4.5 Mathematics Teaching for Understanding (Wagenschein): Genetic and Socratic Teaching Revisited

In the German-language literature, Martin Wagenschein's efforts aiming at mathematics and physics instruction oriented to understanding merit an exceptional status. Although his approach can now be considered from an historical perspective and some of his ideas require modification, complementary remarks, and reevaluation, a considerable number of insights on learning that have only found a broad theoretical and empirical basis in the last two and a half decades seem intuitively anticipated in his work. Because of the reemergence of interest in Wagenschein's ideas in the last few years, a discussion of his didactic concept also seems advisable in the light of current trends.[121]

A major issue in Wagenschein's train of thought is the demand for assigning a high priority to understanding (1974).[122] He emphasizes that he is not concerned with understanding 'how something is done', i.e., not with the mastery of (mental) abilities, algorithms, technical procedures, but, instead, with understanding in the sense of "realizing how something can be" (1975, p. 100). As major didactic principles, he proposes the genetic, Socratic, and exemplary principles. He is, indeed, not the first theorist to advocate these principles (cf. Schubring 1978). Yet, Wagenschein interprets them in an especially impressive way and illustrates them with vivid descriptions of examples for classroom practice. In the way Wagenschein employs these principles, they aim more immediately than do many other seemingly more 'modern' principles (at least with respect to their terminology: "problem orientation",[123] "discovery learning",[124] "communicative learning") at unleashing students' critical thinking.

A unique fascination that many researchers on mathematics education and other educationalists find irresistible emanates from Wagenschein's works. At the same time, semi-official mathematics education research and educational science have some difficulty taking his position seriously because, at first glance, he seems to advocate an anachronistic idea of instructional practice. Wagenschein's ideas on teaching are neither a didactic concept in the usual sense (a theoretical conception of how instruction should generally be carried out) nor are they a comprehensive concept for specialized instruction as part of a general education. When in doubt, he always takes an example rather than a theoretical context as the starting point of his argument. In doing so, he transfers what he demands of mathematics and science teaching to his pedagogical, didactic argumentation. I will discuss his ideas in order to provide a major aspect of my own concept of mathematics instruction – *making understanding possible and promoting critical thinking* – with a more distinct profile.

A thorough-going assessment of Wagenschein's approach would be out of proportion here.[125] I will restrict myself to depicting and interrogating the main ideas involved in genetic and Socratic teaching since the exemplary principle is more concerned with structuring the selection of instructional material, instead of aiming at understanding and reasonable thinking in the strict sense.[126]

Genetic Teaching

The main idea behind genetic teaching is as follows: Theoretical knowledge should not be taught as a 'finished product', without reference to a question or problem requiring a solution and to which it provides an 'answer'; the learner should be able to clearly recognize the emergence of theoretical, conceptually abstract knowledge from a problem that is originally unresolved, but plausible.[127] Wagenschein emphasizes the following points (here, I am condensing his ideas in my own words): Much theoretical knowledge is rooted in everyday phenomena that is obvious to anyone who observes his or her surroundings consciously – in phenomena about which we can communicate in everyday language and which entail questionable, astounding, and seemingly contradictory aspects. Theoretical, conceptual knowledge arises from interrogating such phenomena, from examining what is apparently already familiar, from examining common judgments and prejudices. To mention only two examples: How can we be so sure that the earth actually rotates on its axis or that if we set a compass to the radius of a circle and mark off the radius along the circumference exactly six times we will then arrive back at our starting point? Such interrogation and examination often opens up a new perspective on the original phenomenon. We then engage in speculations that could prove to be false, and, within a zigzag learning process that necessarily also includes mistakes, new insights gradually emerge – insights that we do not have to believe because authorities maintain them, but, instead, because our own considerations, examinations, actions, comments, and renewed questioning have convinced us that these insights are correct. By overcoming seemingly self-evident prejudices in this painstaking, sometimes time-consuming process, we become more keenly receptive to essentials; the insights acquired become part and parcel of our intellectual power.

Genetic instruction is concerned with grounding the knowledge of what something is on the insight of why it is this way and not otherwise. Its 'method' consists in letting students experience the context of the origin of this knowledge. Heinrich Roth puts this into a pedagogical prose that seems almost poetic:

> "The art of all method consists in transforming lifeless matters back into the living actions from which they have arisen: objects into inventions and discoveries, works into creations, plans into concerns, contracts into decisions, solutions into problems, phenomena into prime phenomena."[128]

Not everything worth knowing can be linked back to phenomena and problems in the way preferred by Wagenschein and other proponents of genetic instruction such that understanding is awarded to the students as a result of their own efforts, their own intellectual reconstruction, their critical inquiry into and sensible examination of these phenomena and problems. Wagenschein's examples (not by chance only a limited number, even if this number – also with regard to other subjects – could certainly be considerably increased) are actually combinations of fortunate circumstances which, in turn, can also be called exemplary for the possibility of a far-reaching, often surprising degree of understanding instead of superficial adoption. Without a doubt, their realization in the classroom demands much didactic skill of the teacher.

Genetic teaching in Wagenschein's use of the term is certainly not the only way to have students participate in instruction with understanding. But this reservation does not invalidate genetic teaching. Instead, it helps to determine its status more precisely and to prevent exaggerated expectations.

Socratic Teaching

The main idea of Socratic teaching can be grasped with the following maxims: Take your partner seriously as someone who is capable of thinking independently and critically; don't make use of your experience and perhaps superior knowledge as an authority; don't advise your partner from an elevated position; don't persuade, instead, give impulses – especially by asking questions – for becoming conscious of the reasons behind one's own judgments.

Thus, the subject matter of Socratic teaching cannot be facts or empirical knowledge in the broadest sense. Socratic teaching constitutes a pedagogical situation, because the partners as such do, indeed, have equal rights, but they are not equally well-informed. This distinguishes Socratic teaching from "rational discourse" (Alexy 1978), from "undistorted communication" (Habermas), and from other procedures of rational communication that aim at a consensus between persons who, ideally, have equal rights in every respect.

If the idea of genetic teaching is primarily concerned with how the students' encounter of the content to be imparted can lead to an 'understanding' appropriation, then the idea of Socratic teaching encompasses an ideal image of the relationship to be developed in the dialog between the teacher as someone who is 'already knowledgeable' and the students as persons who are 'not yet knowledgeable'. Thus, here the focus is not on the relationship 'student-topic', but, instead, on the relationship 'teacher-student', and thus, in the broadest sense, on the social dimension of instruction. Both genetic and Socratic teaching share the goal of making understanding possible.

The idea of Socratic teaching has repeatedly been subject to misconceptions. The traditional Platonic dialogs between Socrates and his contemporaries can by no means be considered as perfect examples of a 'Socratic' dialog suitable to instruction. Precisely that dialog most interesting for teaching mathematics, *Meno*, (in which Socrates lets a slave discover on his own that, for a given square, a new one with twice the area can be constructed by using a diagonal of the original square as one side of the new one) can be criticized in a number of ways: Socrates makes exaggerated use of suggestive questioning and drives his 'student' into a corner, leaving him no other choice but to give the expected answer (Struve/Voigt 1988;

Bodenhemer 1984, pp. 25ff). To put it briefly: The original Socratic dialogs allow us to recognize many of the constraints and dictates so typical of questioning instruction as it is practiced in our schools on a day-to-day basis (cf. Bauersfeld 1978, Voigt 1984a/b, Bauersfeld et al. 1988).

The justified critique of the shortcomings in the communicative surface structure of Plato's dialogs often lets the critics forget what sort of instructional practices the idea of Socratic teaching is directed against: against one-way communication, one-sided teaching that drums the finished product, the teacher's knowledge, into the student; against the practice of spelling out even those things to a student that he or she would readily discover with slight intellectual efforts; against the disqualification expressed by not discussing students' ideas that would lead to other results than the officially valid teacher's opinion, but simply classifying such ideas as false. To put it pointedly: all of the types of questioning instruction with which teachers attempt to elicit from students what they could simply tell them, with which only the superficial properties of not advising from a position of superiority are realized, with which suggestive questions subtly drive students exactly into that corner where the teacher wants them to be – all of these types of instruction have as much in common with the idea of Socratic teaching as the image of a window painted on a cement wall does with a real window.

As with the genetic principle, a meaningful application of the Socratic method also requires didactic competence, experience, and pedagogical sensitivity. In the concrete situation, it is up to the teacher to make the appropriate decision: When should students simply be informed about empirical data or common conventions, and when is it better to approach answers to questions under discussion by collectively considering reasons for such answers? When is it more important to insist on discipline in the classroom or on a rigorous advance in subject matter, and when does the classroom situation allow for venturing into a dialog among people with equal rights?

With all of this said, it seems self-evident that a successful Socratic dialog more strongly elicits and releases the students' critical thinking than many other forms of instruction, that with this kind of 'instruction' (if, indeed, any kind at all should take place) the (presumed or aspired to) responsible judgment of the students is more greatly respected than with advising, prescribing, simply informing, or 'enlightening' measures from a position of superiority.

Evaluation

There is no need for a detailed argument stipulating that with respect to the general objectives, there is a high degree of compatibility between Wagenschein's ideas and the concept of general education under discussion here, especially regarding the goal of promoting critical thinking. Many of cognitive psychology's findings on developing cognitive abilities, as recapitulated in section 2.4.5, as well as the consequences for mathematics instruction specified in the preceding sections are also included in Wagenschein's work, if not explicitly, then at least implicitly. Here is a keyword list of the most important areas of agreement:

- Understanding the instructional material currently under discussion has a higher priority than its technical mastery. Although Wagenschein does not explicitly formulate his concept of understanding, his examples demonstrate clearly that the dimensions of understanding worked out in section 3.4.2 (the experiential, the object, and the social dimensions) are inherent in it.
- With Wagenschein's concept of Socratic teaching, major significance is attributed to social exchange, to mutual consultations about the phenomena to be understood; in addition to efforts focused on a more stringent rationale, there should be ample opportunity for speculation, subjective interpretation, and doubts, and for the social evaluation of all of these activities. In this way, the significance of the social surroundings for developing cognitive abilities is (at least to a certain extent) taken into account.
- Issues to be treated in the classroom should be developed from phenomena or ideas that are rooted in the students' everyday knowledge or are accessible to their everyday thinking. Here, Wagenschein's approach proves to be compatible with the view stressing continuity explicated in the preceding section. Also related to this context: Wagenschein's concept relies on the assumption that, for arguments initiating the processes of understanding, everyday language is to be preferred to an early use of technical language: "One's native language is the language of understanding; technical language puts the seal on this, as the language of what is understood" (Wagenschein 1980, p. 137).
- Finally, Wagenschein's examples for instructional practice demonstrate his awareness of the important connection between cognitive abilities and the motivation to employ them – by using concrete phenomena and content to constantly encourage students to engage in independent thinking (cf. section 2.4.5).

These areas of agreement, which reflect Wagenschein's significance for the current debate, are accompanied by certain features that – from the perspective of the concept of general education proposed here and in light of more recent research – signalize the need to supplement his train of thought:

- The content of Wagenschein's mathematical examples is confined to a few areas of 'pure' mathematics: Euclidean geometry and elementary number theory (this limitation also applies to Wittenberg); there is no discussion of the possibilities of utilizing applications-oriented mathematics to make connections to everyday experiences and to augment the re-discovery of mathematics with the re-invention of the mathematical modeling of practical problems.
- Wagenschein's examples demonstrate that, on the whole, his concept is primarily oriented to a level of mathematical understanding conceivable

at college-preparatory secondary schools (this also applies even more directly to Wittenberg).

- No attention is paid to the informal level of instruction, to implicit learning and the indirect effects of classroom socialization; Wagenschein's focus is solely on the level of conscious, didactic and methodical actions.[129]
- Wagenschein exhibits an inclination often noticeable in more dated didactic positions by equating the learning progress of the class as a whole (in the sense of a collective group the teacher encounters and communicates with in the classroom) to the advances of the individual class members.[130] What might perhaps be nearly practicable for small groups (circumstances ensuring that all of the students have comparable chances at every level of understanding and problem solving) is hardly feasible for the usual class size in our public education system.

Wagenschein himself admitted that genetic instruction would have to be complemented by phases of instruction geared more toward informing and providing systematic overviews. According to Wagenschein, such phases should be conceived of as bridges ("lofty arches") between "piers" or "pillars" ("islands", "platforms"), where one struggles to reach a more in-depth understanding by intensive treatment of topics dealt with in exemplary fashion (Wagenschein 1975, pp. 10-11).

More soberly, regarding Socratic and genetic teaching, we can maintain that they can only rarely be realized in ideal form in everyday classroom instruction. In fact, their realization would not even be desirable in many situations and with reference to much instructional content. The ideas of Socratic or genetic teaching represent primarily background knowledge (providing orientation), and only to a lesser extent technical knowledge.[131] Attempts to transform them into didactic, methodical recipes for action that could guarantee 'understanding' run the risk of totally missing the mark.

Taking these reservations into account, we can summarize as follows: Mathematics instruction oriented to developing independent thinking (and thus to a treatment of mathematics focusing on understanding) cannot materialize without certain elements of genetic and Socratic teaching as proposed by Wagenschein – adjusted, of course, to the particular personal and curricular circumstances.

3.4.6 Developing Cognitive Abilities, Critical Thinking, and Instructional Content

In conclusion, I would like to treat in a more systematic way two aspects that have only been mentioned in passing thus far. The following can serve as guiding inquiries:

- Which *requirements* are to be *placed on content* in mathematics instruction that is especially concerned with developing cognitive ability and critical thinking?
- Which *requirements* are to *placed on* the way of *dealing with such content* in this type of mathematics instruction?

The present section is concerned with the first question. In the following section, the second question will not yet be examined in detail, but at least a provisional discussion of its systematic context will be provided.

We have already emphasized several times that cognitive abilities are developed with respect to content or, more precisely, that cognitive abilities and skills cannot be developed without some connection to particular content. Mathematical thinking – as an adequate way of thinking about mathematical phenomena – can, in principle, be learned by employing various sorts of mathematical content. Yet, the results will also vary greatly, depending on the peculiarities of the content and the possibilities for making connections to other content both inherent in mathematics and outside of mathematics; the learner will have varying possibilities for application, varying chances for transfer, and these will thus be of varying meaningfulness to him or her. Content and the formal qualifications acquired with it remain more or less closely bound to one another. This explains the significance of our guiding inquiry. A satisfactory answer can only be found if we refer back to the other objectives for our concept of general education.

In consecutive order, I will discuss three propositions which allow a differentiation (admittedly, only a rough one) of content relevant to learning mathematical thinking (in the sense defined above):

First Proposition: Mathematical thinking can be learned in conjunction with *any* topic, as long as this content is taught in such a way that it is understood.

Second Proposition: Among such instructional topics there are some that are more promising for the promotion of *general* cognitive abilities than others.

Third Proposition: Among these topics, there are some that not only promote general cognitive training, but, in addition, offer an impulse and specific opportunities for *critical* thinking.

First Proposition: Topics for a Mathematics Instruction For Understanding

The first proposition is a consequence of the train of thought expounded in sections 3.4.2 to 3.4.5: Sensible thinking in and with mathematics has no basis if the mathematics has not been understood, but, instead, acquired superficially as a technique or as knowledge of rote memorization. At first glance, this would seem to allow for a very broad range of topics corresponding to this criterion. Yet, the structural conditions must be kept in mind: We are concerned with topics for mathematics instruction aiming at general education. For this reason, the first proposition implicitly includes the demand that the topics must be able to be taught under everyday classroom conditions in such a way that the greater majority of the specific group of learners can understand them. Or, formulated as a criterion of exclusion: Topics that cannot be taught under everyday circumstances such that the majority of the students understand them are ill-suited for a mathematics instruction designed to promote cognitive ability.

Unfortunately (or, perhaps: fortunately), we cannot use this criterion to 'mechanically' sort out mathematical topics according to their appropriateness or inappropriateness. Since understanding a new mathematical topic is dependent on which topics have been understood beforehand and to what extent they are readily accessible and interconnected to one another, the decision on whether the criterion contained in the first proposition actually applies to a certain topic depends, in the final analysis, on the entire curriculum. And to mention a further contributing factor: The comprehensibility of a topic correlates – *ceteris paribus* – to the time spent on it in class. For this reason, the abundance of instructional material is certainly an inhibiting factor (it reduces the time available for each individual topic) for teaching with understanding.

Of course, the dependence of the comprehensibility of a topic on the topic's status within the entire curriculum also means that taking an internal mathematical systematization into account is no trivial matter. Among other things, comprehensibility also means that new knowledge must be interpreted on the basis of prior, familiar knowledge and must become interconnected to it. Often, important concepts cannot be understood if not beforehand a basic understanding of other, logically prior and less complex concepts has been acquired. For example, someone who has not understood how we can use natural numbers aside from counting, how we can compute with them, will have no chance to understand what fractions mean and how we can compute with *them*; much less will this person find an understanding approach to variables. Thus, in any case, the interconnections inherent in mathematics are reflected in the networks we wish to establish.

But from the stipulation that a systematization should be made possible for the acquisition of mathematical concepts, it does not necessarily follow that this must be the conventional systematization of school mathematics, even less, that this would be the best or only useful systematization.[132] Mathematics is not a strictly hierarchical system, but, rather, offers opportunities for many distinct and yet, in and of themselves, quite logical systematizations, even for fairly small branches of the discipline. Furthermore, a systematization need not at all be imposed from the outside. Instead, students themselves, with the support of their teacher, can *construct* it or *discover* it (depending on the epistemological point of view) as a meaningful possibility. Freudenthal (1973, pp. 126ff, p. 142) argues in favor of such a process, which he refers to as local organizing. The syllabus introduced by Wittenberg (1968, pp. 68ff) and developed according to a method of grouping themes for 'rediscovering' geometry in the classroom is an instructive example of such local organizing, of acquiring a systematization from the process of a step-by-step understanding approach to mathematics.

In the final analysis, the criterion contained in the first proposition proves to be rather vague regarding a better determination of topics promoting cognitive ability, as many factors that do not primarily depend on the topics as such also play a role here. The use of the criterion only makes sense if we take recourse to empirical experiences. In any case, the deficits in treating mathematics reported at the beginning of this subchapter (which were only representative for a list that could be extended almost indefinitely) do indicate that many topics of current school mathematics require a different form of integration into the curriculum, which presumably cannot be accomplished without considerable changes in priorities and even deletions in the syllabi.

Second Proposition: Topics for a Mathematics Instruction that Trains Cognitive Ability in a General Way

The second proposition is also based on the comments in the sections 3.4.2 to 3.4.5: If the communicability of cognitive skills depends on the content with which they are developed in the classroom, the topics must be selected in light of this fact. In particular, this means that if a connection is to be established between mathematical and everyday thinking, this connection must already be potentially available in the network of the themes discussed. For this reason, a mathematics that, from the students' perspective, can be connected to the rest of their everyday lives is much more promising for the development of general cognitive ability than is one that is primarily of interest for reasons inherent in mathematics. At this point, all of the arguments of the previous sections in favor of going through the intermediate stages between everyday thinking and mathematical thinking could be presented again. But discovering such intermediate stages is not equally feasible for all mathematical topics. In case of doubt, those mathematical subjects, themes, and representational techniques with which the referential nature of mathematics can be made apparent without enormous efforts or complicated tricks are definitely to be preferred.

To provide a further concretization, we can take recourse here to the concept of general education at the basis of our discussion. For the other six goals of schooling described in that concept help specify how a mathematics more promising for developing cognitive abilities is one that can be readily connected to the rest of the students' everyday lives. Every one of these goals represents a particular type of interconnection between content-related qualifications, instructional material, cultural accomplishments, etc. on the one hand and the life circumstances of the students on the other. The legitimate demand that content (not only in the mathematics classroom) be 'more promising for training cognitive ability generally' coincides to a great extent with the seemingly much more general demand of being 'significant for a general education'. One direct consequence of this is that the content treated in mathematics instruction designed to promote cognitive ability (in addition to its comprehensibility as outlined by the first proposition and insofar as this can be combined with a sensible internal mathematical systematization) should fulfill at least one of the following conditions. These conditions can apply to a topic or subject matter under consideration in any combination or alternatively.

Topics of mathematics instruction aiming at general education and not primarily concerned with promoting specialist cognitive abilities (only useful with regard to mathematical subjects), but, instead, general ones should:

- be of immediate practical usefulness for everyday life, i.e., for the private and professional everyday life of the majority (cf. 3.1); and/or
- be representative for mathematics as a cultural achievement, e.g., for one or several of the main ideas that mark the intersecting points between mathematical culture and that outside of mathematics (cf. 3.2); and/or

- provide many opportunities for applications or mathematical modeling in contexts outside of mathematics or at least demonstrate characteristic features of the science of mathematics as a part of our world (cf. 3.3).

And if I might anticipate sections 3.5 and 3.6 and the subjective and social elements of learning mathematics to be treated there, it is absolutely necessary that these conditions are augmented by the following stipulation. In addition, such topics should:

- provide possibilities for dynamic instruction: for object-oriented communication, cooperative work, practical work, self-responsible action, sensory experience, creative invention, and animated problem solving.

Conversely, this means: *Content that does not meet at least one of these requirements is inadequate for a mathematics curriculum oriented to general education.* The stipulation that other mathematical content could also help develop cognitive ability is perhaps not incorrect, but, from the perspective of general education, it is misleading at any rate. The assumption that any treatment of (any sort of) mathematics would per se have educating effects and promote cognitive abilities is, in light of the requirements of mathematical education developed here, oblivious of the needs of *the majority* of children and adolescents and opens the way to the sort of specialist and esoteric predisposition already severely criticized by Whitehead (1913 [1967]). In any case, there is indeed, for various levels of intellectual development, enough mathematics available that exhibits manifold interconnections, can be taught in a way aimed at understanding, and fulfills the requirements specified above. If we take recourse to such mathematical topics, we can make good use of the time allotted to mathematics instruction in our public school system.

Transforming these thoughts into practicable curricular suggestions requires a great number of detailed considerations and coordinated efforts that exceed the capacity of a single writer. For this reason, I will simply hint at a few tendencies that could function as the starting point of further discussion: In light of a 'mathematics for all', which topics warrant more intense treatment in the classroom and which could be downgraded? If we examine the current standard curriculum of secondary schools, then, e.g., more statistics, data analysis, probability topics related to everyday life, and geometry related to the perception of space would be desirable. Reductions would be justified for: mere exercises for transposing terms and similar topics for drills (certain computations with fractions), quadratic equations, powers and logarithms, trigonometry (unless closely related to applications), geometric topics without any connection to the experience of space (unending constructions of triangles in the lower grades of secondary schools, abstract linear algebra in the upper grades), and analysis (refraining from systematic treatment in basic courses).

Third Proposition: Topics for a Mathematic Instruction that Promotes Critical Thinking

A developed cognitive ability and a developed understanding are necessary, but not sufficient conditions for the use of critical reason. Similarly, regarding instructional content, we can stipulate the following: A topic of instruction that is well-suited to developing cognitive potential does not necessarily provide an opportunity for *critical* thinking in the sense intended by the idea of enlightenment. Thus, we should inquire whether some of the mathematical topics which, according to the second proposition, seem appropriate for mathematics instruction oriented to promoting cognitive abilities are also specifically suitable for stimulating *critical* thinking. Strictly speaking, this involves a distinction of whether mathematics functions primarily as a *means* ('amplification') or as the *object* of critical thinking or if both aspects appear in combination:

- First of all, mathematics can help to illuminate and rationally clarify assertions, issues, natural phenomena that influence our life circumstances and social relations (in the broadest sense). It has been sufficiently demonstrated that this occurs by means of the construction or reflective application of appropriate mathematical models. Here, mathematics is a mediating force ('amplification') for critical thinking.
- Secondly, the use of certain mathematical models that are involved in presenting information and in decision-making processes in our society can be critically analyzed. The clarification then consists in revealing the widespread abuse of mathematical models (i.e., their unreflective use or even consciously deceptive application) and in dispelling the belief that information cloaked in mathematical form would be more reliable than other information. Here, mathematics is the means and object of critical thinking.[133]
- Thirdly, we can reflect on mathematics itself, on its scope, on the limitations of modeling, or, to put it briefly, on the relation between mathematics and the world outside of mathematics. Here, mathematics is the object of (philosophical) critical thinking.[134]

All of these three aspects were mentioned in section 3.3.4, where I indicated that they can be considered both with respect to 'developing an understanding of the world' and to 'promoting critical thinking'. Possible topics and themes of mathematics instruction were also mentioned. I will refrain from repetition here and restrict myself to two examples and a few additional remarks. The examples should demonstrate that promoting critical thinking can also permeate everyday mathematics instruction. In principle, this requires neither complicated projects nor particularly advanced mathematical ability on the part of the students.[135]

First Example

The third quotation that serves as a motto at the beginning of chapter 3 can be presented without a commentary to students who can already compute with percentages and simple fractions for them to express their opinions on it: "If, some years ago, one out of ten drivers still drove too fast, now 'only' one out of five does this. But even five per cent is too much, and thus, the controls persist, and the fast drivers will continue to pay fines."[136] The analysis of this text corresponds to the second of the three categories named above. Knowledge of the world around us does not figure into the analysis in any pronounced way since it is only necessary to trace the conflict between the mathematical models used to discover a contradiction. The important thing here, then, is the (implicit) knowledge of the referential meaning of segments, fractions, and percentages.

Second Example

World population growth is the subject of this diagram. It lists the number of births worldwide (from top to bottom) as follows: 3 per second; 180 per minute; 10,800 per hour; 260,000 per day; 7.8 million per month; and over 90 million per year.

In a regional daily newspaper[137], I came across this diagram with a commentary on a conference on the world's population. I would have enjoyed immediately presenting both of them to students from about the 9th grade on. The commentary begins as follows: "Every second, three children are born on Earth – most of them in the southern hemisphere. But, every second, a new car comes from the assembly line, almost always in the countries of the North. The not quite 920 million people of the industrial countries consume 70% of the world's energy… ."

I will list some questions that can be posed in this connection. Of course, it would be desirable that the students ask at least some of these questions themselves:

- Can some information such as 'number of births worldwide' actually be correct? How do the numbers specified correspond to what we know about the total population of the Earth and its increase within the last few decades (the newspaper article numbers the current population as being 5.7 billion, the annual increase as 94 million)? What about the people who die? (surplus of births vs. number of births)
- Does the cone in the diagram really help us to get a better idea of the numbers or numerical relationships specified? What is the relationship

between the numbers and the corresponding sections of the surface of the cone (volume of the cone)? (demands placed on meaningful geometric illustrations)

- How long does a birth last? Does it really make sense to speak of so many children being born within a second? How does such numerical information as 'three births every second' come into being? (computation for the purpose of (questionable) illustration; the model supercedes real processes)
- Which more appropriate illustrations can be found both for the relative increase and for the absolute values? (computing the relative increase for one's own community, etc.)
- Is there really the same average increase in population for the same amounts of time? (linear vs. exponential growth; limitations of 'exact' modeling)

And if we also consider the commentary:

- How many new cars are there for every newborn baby? Does the number of cars also only mean the surplus (here, of course, in comparison to scrapped cars)?
- What is the relation between new cars and newborn babies if we only take the population in the industrialized countries into account?
- What is the relation between the per capita energy consumption of the average inhabitant of an industrial nation and that of an inhabitant of the Third/Fourth World? Which differences are not taken into account with such a computation?

And if one's own concrete life circumstances are to play some role, a more or less comprehensive discovery project can develop in connection with the topical excerpt from the newspaper:

- How large is the surplus of births within the families of the students if computed for the last 30 years? How large was that surplus a generation before that?
- How large is my personal per capita energy consumption (as a student, as a teacher) in a year or on a single day (direct and indirect consumption – e.g., taking the bus, heating system at school –; estimate and approximate what cannot be computed exactly)? How large is our collective consumption on the average? What energy resources are available to an individual in the Third World (based on a rough relation between 'industrial nations – Third World')? Which actions or enjoyable activities involving energy consumption might this individual be able to afford on the average on a single day?
- If everyone were to consume as much as I do: Which effects would this have on worldwide energy consumption and on the corresponding resources of raw materials?

At this point, I will discontinue this list of possible questions. The example – the concrete treatment of which in the classroom was not our topic since context information was missing (group of learners, prior knowledge, etc.) – illustrates the fact that the demands made on content here can sometimes be met by dealing with a *single* topic. On the one hand, the example offers an abundance of opportunities for critical thinking in relationship to mathematical considerations. Yet, in addition, a practical, useful treatment of mathematics can be learned more or less in passing (estimating, approximating, imagining and visualizing large numbers); several of the main ideas discussed in section 3.2.4 can be treated (number, measuring, mathematical modeling, perhaps functional relations); and, in light of the inevitable discussion of the relevant *key issues* (cf. 3.3.4), there is certainly a contribution to developing an understanding of the world.

Mathematics teachers and educators primarily oriented to the subject matter as such will probably object that the example and the related questions contain too little 'demanding' mathematics for 9th graders and older students, especially for those at college-preparatory secondary schools. At best, they might be inclined to accept the newspaper clipping as a 'motivating' prompt for introducing exponential function. But both of these reactions (are they really only hypothetical?) would be fatal. Mathematics instruction aiming at general education cannot be designed solely on the basis of such examples and themes as those listed above. But if it does not repeatedly motivate connecting everyday reasoning with mathematical thinking, it runs the risk of imposing upon the majority of students a type of thinking with which they cannot identify – and precisely in this respect, there is an unfortunate tradition in mathematics teaching. To put it pointedly, mathematics teaching then tends to dull critical faculties instead of stimulating them.

3.4.7 Cognitive Ability, Critical Thinking, and a Culture of Instruction

Although it would be quite mistaken to neglect specific topics in light of the goal of promoting critical thinking, the decisive factor is how these topics are treated in the classroom. One aspect cannot be played off against the other since both aspects have effects on classroom instruction at different levels: A catalog of topics (or, more precisely, a specialized curriculum developed in however much detail) does not already determine what becomes of it in the classroom. Specialized content specified in the curriculum is only established as actual content for learning *within* the instructional process.

Above (subchapter 2.4), I have demonstrated (at that point in the argument still independently of the specifics of the mathematical discipline) how significant a culture of instruction is for developing cognitive ability, especially for promoting critical thinking and a certain reflectiveness – a culture of instruction in which the subjective cognitive efforts of the students are permissible and also have social repercussions. I will not repeat this in detail here. Instead, I will simply list some keywords to recall the main points:

- Students must be able to 'observe' the ways others engage in thinking.
- They rely on the fact that teachers and fellow students socially evaluate both the finished products of their thinking and their tentative lines of thought.
- The sort of promotion of critical thinking aimed at here cannot be realized solely by direct pedagogical and didactic measures; the culture

of instruction to be achieved must also encompass the informal (socializing) level of instruction. 'Sensible behavior' in the context of mutual interaction must be cultivated.

- The way of treating the topics of instruction must be interconnected to this behavior: Learning social behavior and learning subject manner cannot be separated from one another. How content is treated within the framework of social interaction – what is accepted as correct, which interpretations, associations, and digressions are permissible, whether and, if so, how the teacher exercises a sole right to formulate definitions – also influences the content-related concepts of students and, among other things, their conception of mathematics.

This would be the point where we might discuss how these general demands can be put into practice in relation to mathematical topics. Since I will systematically examine this entire complex of developing a culture of mathematics instruction contributing to general education in section 3.6, the preceding provisional remarks will suffice for the moment.

3.4.8 Summary and Concluding Remarks

In and of itself, doing mathematics does not lead to an improvement of general cognitive abilities. Even less can we assume that mathematics instruction automatically enables students to make use of critical reasoning. From their experiences at school, too many children, adolescents, and adults get the impression that there is an enormous gulf between their everyday reasoning and the 'mathematical thinking' expected in the classroom. Empirical data have shown that even for many persons who obviously were not unsuccessful at school mathematics, mathematical and everyday concepts are at conflict.

Mathematics instruction that promotes cognitive abilities must be oriented to understanding. From a subjective perspective, understanding takes place whenever some previously unfamiliar content is experienced as being meaningful. Understanding cannot be imposed on someone from the outside. But conditions that promote understanding can be established. Some of these conditions were described by Wagenschein with his concept of genetic and Socratic teaching.

One aspect of understanding a new mathematical topic is the establishment of interconnections with prior knowledge. Mathematics instruction aiming at understanding has a greater chance of being realized if cognitive strategies, heuristic devices, images, and metaphors of everyday thinking are made productive for mathematics; if, in addition to the formal nature of mathematics, its referential aspect also receives adequate attention; if we repeatedly attempt to make connections between mathematical and everyday thinking. In this way, students have the chance to become familiar with the distinctive features of mathematical abstraction step by step. Only on the basis of sufficiently understood mathematics can students experience how mathematical concepts and techniques are useful as 'amplifications' of their everyday thinking in many situations. And only on this basis can mathematics also be experienced as a means to enlightening students or as a subject for critical inquiry.

Understanding the mathematical topics currently under discussion in the classroom is a necessary, but not sufficient condition for promoting critical thinking. Since thinking is always content-related, promoting critical thinking is also dependent on the choice of content. Certainly, astute thinking can be trained by means of mathematics. But if this astuteness is not to be restricted to mathematics, the desired transfer must constantly be practiced using topics rich with interconnections in which mathematics and the non-mathematical world are related to one another. Yet, it seems even more significant than the choice of topics that children and adolescents can experience the treatment of mathematics and interesting mathematical applications within a living, social praxis involving reasonable contention, inquiry, doubt, and substantiation. The demand for the development of a culture of mathematics instruction contributing to general education refers precisely to the cultivation of such a praxis.

Thus, whether mathematics can justifiably be considered a sort of training for cognitive abilities or even for critical thinking depends on a number of conditions that are often neglected in conventional instruction at all types of schools and at all grade levels. Mathematics instruction that focuses on routine use of the usual standard solutions for traditional school mathematics instead of on understanding, on conscious efforts at transfer, on explicit challenges to the learners' ability to engage in criticism, serves to dull critical faculties rather than to mobilize them.

3.5 Mathematics Instruction With Socio-ethical and Student-Related Objectives: Responsibility, Communication and Cooperation, Enhancing Students' Self-Esteem

3.5.1 Learning Specialized Content and Learning Social Skills

Within the current chapter, I have dealt with the first four of the seven objectives of public schools outlined in my concept of general education in individual subchapters. Briefly, I would like to mention a few reasons for treating the last three objectives as a group, for setting aside to a certain extent the systematic structure that I have been employing thus far.

The relationship of the mathematical discipline to the objectives of general education dealt with up to this point demonstrated apparent content-related elements:

- Mathematics makes a specific contribution to mastering situations of everyday life.
- As a high-level cultural achievement, mathematics is a part of our culture.
- By means of its applications, mathematics is interconnected to our world in a variety of ways.
- In a sense, reasonable thinking is the means by which mathematics can materialize as a creative achievement of human beings in the first place.

By contrast, mathematics displays no immediate connection to the socio-ethical and student-related objectives designated by the heading (3.5) above. Obviously, this distinguishes mathematics from a number of other school subjects: These three objectives or, at least, certain aspects thereof can readily be discussed in terms of content in subjects such as political science, social sciences, civics, religion, or even in literature, art, and foreign languages. Generally speaking, they can, indeed, be the object of certain discussions or reflections inherent in the subject matter.

The fact that such a content-related interconnection between mathematics and these three objectives of general education is lacking might help explain why there is often a perceptible tension between pedagogical and subject area-oriented goals in mathematics instruction.[138] Apparently, as soon as teachers begin with mathematics instruction, they find it more difficult to contribute to schooling's aim of general education than with most other subjects. Even if they are in general agreement with this obligation, in many cases their specifically pedagogical activity remains more strongly separated from their specialized instructional activity than in other subjects. Instruction and education are not in accord.

Certain characteristics of school mathematics, as they are usually perceived by both teachers and students, seem to potentially conflict with all three of these objectives of general education: with inspiring a sense of responsibility, with training communication and cooperation, as well as with enhancing the students' self-esteem. This does not mean that mathematics teachers (who often also teach other subjects) do not generally strive for or reach these goals. But even if special efforts in this sense and a certain success are noticeable, sober observation from an external perspective usually results in the impression that success has not been achieved *by means of* specialized instruction, but, rather, *despite* it.

The major argument of this section can be formulated pointedly as follows: If socio-ethical and personality-related objectives such as developing a sense of responsibility, practicing communication and cooperation, and enhancing the students' self-esteem are to be taken seriously in mathematics instruction, this also has considerable effects on the collective treatment of mathematics in the classroom. For we learn *social skills* primarily by learning *specialized material* in commune with others.

The previous sections have clearly demonstrated how greatly cognitive development also depends on an appropriate social arrangement of the surroundings for learning – both in a specialized mathematical and in a general education perspective. For this reason, I have repeatedly stipulated that a sufficiently developed culture of instruction is an important prerequisite for successful specialized learning.

The issue here is the reverse: How can we design specialized learning in such a way as to promote a willingness to assume responsibility, to further communication and cooperation among the students and, at the same time, to enhance their self-esteem? Which characteristics of the traditional culture of mathematics instruction are more of a hindrance to achieving this goal? Considering these questions allows us to give the desired 'culture of instruction contributing to general education' a more distinct profile.

Any instruction – thus, of course, also mathematics instruction – not only relates to a certain subject matter, but is also an interactive process. This interactive process not only influences the students' view of the subject matter that develops through instruction (or, at least, on the occasion of classroom instruction) in a specific way. The students also – inevitably and simultaneously – learn something about their fellow students, their teacher, and themselves. They learn which social forms of conduct and which social norms apply in the particular group; they learn (with more or less success) how to treat the topics discussed within the social situation of the classroom; and they learn how well or badly received their own behavior is, which strengths and weaknesses others perceive in them, which of their own ideas, questions, doubts, and fears they must conceal and which of these they can bring out into the open and still 'survive' in the classroom. Thus, the classroom situation *always* has the function of socialization, regardless of whether the participants are aware of this and regardless of whether the teacher intends this or attempts to avoid it.

Proponents of the sort of teaching exclusively aiming at informative instruction tend to neglect learning that is not immediately related to the subject matter or to consider it as of marginal importance. By abstaining from pedagogical intentions, they allow the socialization effects of classroom instruction to develop in a fully arbitrary manner.

Conversely, pedagogical goodwill alone does not suffice. In fact, one characteristic of pedagogical processes is that the results corresponding to the pedagogical intentions of the teachers often do not come about. In "instruction oriented to personal development"[139], instruction and personal education should not simply be joined to one another by addition. Significant to the realization of such instruction is not only the teachers' sensibility to what happens in the classroom, their situational alertness and spontaneity, their perception of the unique traits of individual students, and their knowledge of their own strengths and weaknesses, but also their knowledge of interaction structures and of the constraints imposed on action in everyday classroom situations. There is, indeed, no simple formula which generally helps prevent unintended side effects of didactic and pedagogical action. Yet, if teachers are aware of which undesired side effects their actions in the mathematics classroom are likely to have, they can prepare themselves and take conscious counteractions in the particular situations.

At this point, I wish to examine in more detail three potential complexes of causes for tension between pedagogical and subject area-oriented goals. Subsequently, I will briefly consider each of the three general education objectives under discussion here and indicate factors inhibiting and promoting these objectives in mathematics instruction.

3.5.2 Three Causal Complexes of Tension Between Pedagogical and Subject Area-Oriented Goals

The causal complexes to be considered in what follows are situated at three different levels, but there is interaction among them and they are mutually reinforced by their effects. In consecutive order, I will be discussing distinct features of the usual interaction structure in the mathematics classroom, the specific socialization of mathematics teachers, and the conception of mathematics held by mathematical nonprofessionals.

Interaction Structure in the Conventional Teacher-Centered Mathematics Classroom

At all grade levels, mathematics instruction is characterized by the dominance of teacher-centered instruction, mostly in the form of questioning instruction[140] (Hopf 1980, Maier 1986, Maier/Voigt 1992). Empirical studies have demonstrated that even in the last few decades there has hardly been any change in the dominance of this form of instruction.[141] In particular, this predominance is observable in phases in which new mathematical content is introduced.

We can describe the interaction structure in this type of instruction as a sequence of three steps as follows (Voigt 1984, Maier/Voigt 1992, Maier 1994):

- 1st step: question or impulse formulating a request from the teacher;
- 2nd step: contribution(s) of the students as response(s) to the preceding question or impulse;
- 3rd step: teacher's comments on the students' contributions, aiming at evaluation, correction, or providing supplementary information.

Of course, with respect to content, instruction with this formal structure can be of widely varying quality in individual cases. The quality of interaction can also vary considerably within this formal structure: from a more rigidly schematic sequence of individual steps with very restricted expectations of the students' contributions to an immense variety in the teacher's formulation of impulses and questions with an extensive scope of student responses considered acceptable by the teacher. Nevertheless, if this interaction structure dominates to the extent repeatedly indicated by empirical surveys, this inevitably leads to limitations and restrictions that, if they do not indeed prevent reaching the socio-ethical and student-related goals, at least make these enormously difficult.[142] The most important individual factor in this complex seems to be the extremely dominant role of the teacher. The teacher exercises nearly exclusive direction of the 'official' process of interaction; practically all of the official, verbal and nonverbal contributions of the students are directed exclusively at the teacher. As an authority on the subject, the teacher decides what is correct or incorrect, what is mathematically relevant or irrelevant. The teacher 'embodies' in his or her own person and instructional activity the discipline of mathematics. For this reason, the students' subjective ideas on what mathematics is – their conceptions of mathematics – are strongly influenced by the interaction experiences, which, in turn, are very much related to the particular teacher. The more pronounced and uniform the classroom interactions with various mathematics teachers whom children and adolescents encounter during their school years demonstrate characteristics of the interaction structure described above, the more restricted – *ceteris paribus* – the resulting conception of mathematics will be.

Socialization Specific to the Discipline of Mathematics

During the 1970s, several promising approaches made use of theoretical concepts of socialization with the aim of determining more precisely the connection between the university education and personality structure of mathematics teachers on the one hand and the culture of conventional mathematics instruction[143] on the other (Bürmann 1977, Reiß 1975 and 1979, Heymann 1984). Since then, the theory of socialization specific to a discipline has become more elaborate, and there are now also comparative empirical studies of various disciplinary cultures at postsecondary institutions.[144] Yet, regarding the academic discipline of mathematics and especially the effects of the university study of mathematics on the behavior of teachers in the mathematics classroom, there have been no substantial new contributions to the research since Reiß's survey.[145] With this in mind, the following remarks are to be conceived of as plausible generalizations of the findings contained in the existing empirical studies. They seem to be well supported by everyday experience. I would like to call attention to the fact that these remarks refer to teachers and aspiring teachers for secondary schools, primarily for college preparatory schools.

One major difference between the humanities and social sciences and the natural sciences and mathematics seems to be as follows: With the former group, there is a

> "practically unlimited interpretative dominance regarding the objects of study, issues for research, and the content of professional training. ... criteria for views on certain content and for advances in the processes of conceptualization and problem solving are primarily only negotiated in the actual interactive situations. ... For the mathematical and natural sciences, social interaction and communication not only have much less significance, but also one that is qualitatively different. ... There is a higher degree of prior agreement among those involved concerning criteria for views on certain content and for advances in the process of conceptualization; thus, communication on these topics is relatively sparse." (Reiß 1979, p. 282)

Mathematics students intending to become teachers at secondary schools have more success with their studies if they adapt their communication practices to the prescribed and habitual rules. Refraining from speaking about mathematics and the widespread exclusion of reflection on the significance and meaning of the topics covered in seminars are internalized as adequate ways of dealing with the discipline and its content.[146] Even students who in addition to mathematics also study a humanities or social science subject tend more toward 'switching' between the opposed disciplinary cultures than toward a reciprocal transfer of elements of the two different styles.

Since the accompanying educational science courses fall into the category of humanities and social sciences, it is not surprising that teachers of mathematical and natural sciences often cannot identify with pedagogical content and with the disciplinary culture specific to pedagogy. This would also explain why mathematics and science teachers, confronted with the opposition 'subject-matter orientation vs. pedagogical orientation', exhibited a stronger predisposition to the first alternative than did teachers of other subjects (according to a somewhat dated study, cf. Koch 1972, p. 126; cited in: Reiß 1975, p. 302). The disciplinary culture of pedagogy, with its less rigid interpretations and negotiations on values and goals, is experienced as alien, contrary to one's own understanding of science.[147] In that case, the inclination of mathematics and natural science teachers towards a conservative stand on pedagogical issues, demonstrated by somewhat dated empirical studies (cf. Reiß 1975, p. 303), is hardly a surprise. This general tendency could probably also be substantiated by current studies.

The disciplinary socialization of mathematics teachers does not at all commence with their academic studies: Usually, only those students decide to study mathematics who had no considerable problems with mathematics at school, i.e., who already successfully appropriated certain elements of the usual treatment of mathematics during their years at school. In addition, there are indications that, in terms of personality structure, this group has a predisposition for a 'purely objective' style of working, i.e., for one that is limited with respect to its communicative variety. With such predispositions already at hand, university studies have a largely stabilizing effect. The typical structure of interaction in conventional mathematics instruction, the pronounced subject-matter orientation of mathematics teachers (especially of those at college-preparatory secondary schools), their reluctance to consider genuinely pedagogical perspectives, their tendency to exclude questions on the significance and meaning of mathematical topics from their classrooms – all of this could thus to a large extent be explained by a self-stabilizing vicious circle of socialization specific to the discipline both at school and at the university.

The dichotomy 'humanities & social sciences vs. mathematics & natural sciences' described above exhibits a number of parallels to the dichotomy 'female vs. male', as it is often perceived in our society. The inadequate representation of women in mathematical and scientific fields still apparent today could be an indication that women's opportunities are more strongly restricted than those of men by the disciplinary cultures of mathematics and natural sciences.[148] Conversely, it might be easier for female mathematics teachers than for their male colleagues to develop the type of mathematical culture of instruction informed by general education under discussion here.

A further aspect of university socialization specific to mathematics, relevant for at least a considerable number of future teachers, should not go unmentioned. Many prospective teachers who chose mathematics as a major not least because they were good mathematics students at school suddenly feel confronted with extremely high demands on their expertise, sometimes with demands that are too high.[149] It remains largely the exception for someone to adequately understand the instructional material, which is presented in an abstract, deductive, and formalist way, *during* the actual lecture or seminar. Very often, there is backlog to be taken care of. The individual is confronted with university mathematics as with a dangerous mountainous terrain to be crossed without the necessary equipment. Certainly, every rock wall scaled incites a feeling of achievement. Yet, on the whole, a general impression that maintains during their course of studies for many non-gifted students is that they, as vulnerable human subjects, are confronted with a science from which all subjective elements have been banished, which is threatening with its 'objectivity', and which seems to demand personal sacrifices rather than to provide opportunities for developing one's creativity. Thus, in the course of their studies, this group is belatedly subject to an experience they were spared during their schooling.

It is quite probable that such experiences contribute to the tendency – also exhibited by many mathematics teachers – to treat mathematics essentially as a predetermined 'finished product'. The one-sidedness of such a conception of mathematics is something these teachers share with most mathematical nonprofessionals – despite the considerable differences in expertise.

Mathematical Nonprofessionals' Conception of Mathematics

Whereas many creative mathematicians see mathematics as an entirely new world full of questions, secrets, and surprises which, the further one delves into them, the more elaborate they appear, many nonprofessionals view mathematics as a solid, unalterable block of knowledge.[150] In our society, the fundamental encounter with mathematics takes place at school. The conception of mathematics that remains dominant for most people during their entire lives is determined by their experiences during this time at school.

What does the majority of students experience in the mathematics classroom? To make things clearer, I will provide a few somewhat pointed answers (cf. Brückner et al. 1983): Mathematical knowledge appears to the students, indeed much more so than the knowledge involved in most of the other school subjects, as complete and final knowledge. It is firmly established in the textbooks. For every mathematically permissible question, there is precisely one single correct answer. Opinions are of no import.[151] Mathematics seems to be uncompromising, strict, and objectively true. Correctly solved problems indicate a successful appropriation of the 'finished' knowledge available; falsely solved ones mercilessly disclose one's inadequacy.

This conception of mathematics is certainly compatible with the dominant structure of interaction in the conventional mathematics classroom as described above. This structure is the 'form' that corresponds to that 'content'; it is the 'frame' within which this 'image' of mathematics gains recognition. Form and content, image and frame mutually stabilize each other. During the course of their years at school, students' expectations adapt to this image to such an extent that any deviations only cause irritation. Teachers who have dared to occasionally develop their instructional content in totally different ways (letting students work on unsolved problems, philosophize on mathematics, treat mathematical issues experimentally) have repeatedly reported that the students are very quick to ask, "Is this really math?"[152]. Similar observations can be made if unusual methods are employed: if group work is a major component, if discussions among the students are stimulated, if students are required to write compositions about the mathematical topic currently under discussion in class. A great deal of patience is necessary to convince students with the usual prior experiences and correspondingly firm expectations that a totally different way of approaching mathematics is possible, that there is more to mathematics than just those formal and algorithmic traits that are usually overemphasized. Steps in this direction are much the same as developing a new culture of instruction. This process requires much time and patience, for a new culture of instruction can only develop if new norms and assessments both for collective action in the classroom and for individual learning are accepted.

Thus, students also contribute to the stability of the predominant form of classroom instruction with their expectations and their conception of mathematics, which is to be seen as a repercussion of their classroom experiences – regardless of whether they personally hold the subject in high esteem, whether they are among the mathematically capable and successful students or among those who tend more to have difficulties with mathematics. The former can get along quite well with the 'social character' of the subject as described above. There is no reason for them to suspect a problem here. For the members of the latter group, who attribute a lack of mathematical talent to themselves, the 'social character' is firmly attached to the subject and to their primarily negative experiences.

Summary and Concluding Remarks

Conventional mathematics instruction takes place primarily as teacher-centered, questioning instruction, with little direct communication among the students and a marked adherence to the dominant role of the teacher. In many ways, this predominant structure of interaction is in accord with the quite one-sided conception of mathematics on the part of most students and many mathematics teachers, for whom mathematical knowledge is predetermined, beyond doubt, and objectively true. This structure of interaction is an element of the 'normal' culture of instruction and an expression of the 'social character' of mathematics. The expectations the large majority of the students have regarding mathematics have been profoundly influenced by their experience of this social character.

If the social character of the discipline and the predominant conception of mathematics among the students already stabilize each other, then, in secondary schools, the specific socialization of the mathematics teachers within their discipline acts as a further stabilizing factor. Experiences at postsecondary institutions, particularly the way of dealing with college-level mathematics there, might reinforce personality traits by which the mathematics students already tend to distinguish themselves from their peers. Such traits include the following: an inclination toward primarily impersonal communication, a far-reaching exclusion of personal aspects from work, refraining from questioning disciplinary norms, an inclination toward conservatism, and a skepticism of reform ideas inspired by pedagogy or by the social sciences.

Of course, in every individual case, for any concrete constellation of teacher and class, the three factors mentioned above take on particular form and by no means predetermine classroom events. Nevertheless, the cumulative effects of these factors can to a large extent explain why mathematics instruction seems to be more immune to pedagogically inspired reforms than most of the other school subjects.

I will consider individually each of the three objectives of the public school listed in the main heading above. Since in the preceding chapter I have already examined these objectives in a general sense and those considerations also apply, in principle, to mathematics instruction, I will confine myself to a few brief remarks.

3.5.3 Factors Inhibiting or Promoting a Willingness to Assume Responsibility

Within the framework of schooling, there are, above all, three aspects that are significant for developing a willingness to assume responsibility:

- The first aspect concerns the function of adult teachers as models: Setting an example with an unassuming and credible attitude of a sense of responsibility is of greater significance for moral education than any amount of verbal proclamations. Since this applies to all subjects and to school situations outside the scope of instruction, I will not go into further detail here.

- The second one concerns instructional topics: Do these, at least occasionally, provide opportunity for responsible reflections? As, generally speaking, those topics examined in the preceding subchapter in connection with promoting critical thinking are the ones that present themselves for mathematics instruction relevant to this aspect, here too, I will refrain from going into further detail once more.
- The third aspect concerns the issue of whether the school classroom or, more specifically, the mathematics classroom, actually provides the students with opportunities to take responsible action to an extent that is not negligible. I will be concentrating on this aspect here.

Within classroom instruction, taking responsible action relates primarily to fellow students and to one's own learning process. Which opportunities to actively assume responsibility does mathematics instruction provide for the students?

Let us begin with the responsibility for one's own learning. The structure of traditional instructional dialog with its small steps suggests to the students that they should only assume responsibility at certain specific points. Presuming that a student is actively participating, this responsibility extends from the particular question or problem posed by the teacher to the proposed solution (addressed to the teacher), which the student then expects the teacher to evaluate as correct or incorrect.

But if students are really to assume responsibility for their own learning – indeed, to an increasing extent as they get older –, then, supported by their teachers, they would have to be able to form age-appropriate ideas about the objectives of their own learning. What is important for learning mathematics – both in a general sense and in reference to the topic currently under discussion in class? And along with this, they should be allowed to develop their own criteria for determining – even independently of the teacher – whether and, if so, to what extent they have come closer to their objectives. Finally, and this is perhaps most important, 'havens' of self-determined learning must be provided where an age-appropriate sense of responsibility for the learning process can be put into practice. Individualized phases of instruction can also take place under very diverse preconditions: Traditional individual and partner activities are primarily intended for working out specified sequences of problems in order to consolidate the material covered beforehand. If students experience how they can develop their own learning in a self-determined way, they must also have opportunities to make their own decisions: which points are to be emphasized, where more in-depth work is necessary, how long one should solitarily 'slave away' at problems, and when it would be better to consult fellow students or the teacher.

It can hardly be denied that mathematics instruction, within the framework of its conventional structure of interaction, offers little opportunity to assume responsibility for fellow students. Certainly, there is no lack of teachers who consider it their pedagogical goal to motivate students to responsible action toward their fellow students, for example, toward a general willingness to help if someone is ill, disabled, or at a disadvantage. But, normally, this sort of effort is not integrated into specialized instruction; instead, it takes place more or less as an external event: In that case, the gulf between learning specialized content and learning social skills mentioned above, between specialized and general pedagogical objectives, also presents itself here.

As a student taking part in specialized instruction, one fundamental prerequisite for my ability to demonstrate responsibility for a fellow student is that, within the scope of that instruction, I must be allowed to directly interact with my fellow student. But precisely in this respect, conventional mathematics instruction – owing to the focus on the teacher, who mediates the entire communicative process and assesses every statement, – provides few opportunities. Thus, under the conditions of the predominant structure of communication, one major prerequisite for assuming responsibility for fellow students – as individuals or as a group – is simply not met.

There are forms of instruction that provide considerably better preconditions. If students engage in group work the results of which are to be justified collectively, each member of the group also assumes responsibility for what the others are doing. But this requires the assignment of suitable tasks, which, e.g., could relate to collective work on unresolved issues and problems: explicitly mathematical inquiries into the properties of numbers or into geometrical matters, the modeling of nonmathematical situations – not to be done by employing 'tricks', but by using various approaches, etc. There is no need to emphasize the fact that an ill-suited group composition can neutralize the social advantages of this type of work – if, e.g., mathematically more capable students or more eloquent ones seize the entire responsibility themselves. A further possibility, of course, is having certain students function as 'tutors' who, in this way, assume part of the responsibility for the learning process of their fellow students who are experiencing difficulties.

These suggestions are only a small part of the many forms of cooperative work that creative teachers and educationalists have already developed, tried, and reported on extensively (cf., e.g., Davidson 1990). Hopefully, though, it has become apparent that any efforts toward teaching responsibility also entail consequences for learning specialized content. Learning social skills and learning specialized content cannot be separated from one another. Placing other emphases regarding social behavior also means changing the way one deals with mathematical subject matter – and, thus, there are also effects on what students learn of and about mathematics.

3.5.4 Factors Inhibiting or Promoting Communication and Cooperation

Here, I will first examine opportunities for cooperation in conventional mathematics instruction; subsequently, I will consider the more general question of how and to what extent the type of communication striven for is possible.

It is obvious that the predominant structure of communication during teacher-centered phases of mathematics instruction hardly gives students opportunities for content-related cooperation – possibilities could be, e.g., mutual support while becoming familiar with new content or a cooperative effort (perhaps with assigning separate tasks to different persons) at examining and solving mathematical problems. Usually, students can only cooperate with one another in a way that is undesirable to the teacher – at the 'unofficial' level of classroom activity, by telling someone the answer, cheating, playing tick-tack-toe, writing letters, and engaging in similar activities with which students evade the official focus on competition. As intended here, cooperative learning depends on an opportunity for direct communication among the students – an opportunity which teacher-centered instruction only provides as an exception to the rule. For this reason, social forms of learning such as group work or pair work are a *conditio sine qua non* for more intensive phases of cooperative work.

Let us dwell on teacher-centered instruction for a moment and inquire about the opportunities it provides for the participants in the classroom to practice communication, in particular, communication related to the mathematical topics currently on the agenda. The answer is not so simple here, since the quality of communication within the class as a group is a decisive factor. According to the usual classification of social forms of learning (cf., e.g., Meyer 1987, pp. 136ff), 'teacher-centered instruction', considered formally and without prejudging it, simply means that there is only *one* official strand of communication related to the entire group of learners. Nevertheless, the patterns of communication observable in actual teacher-centered instruction – of which the participants are often not even conscious – frequently disturb the communication on the mathematical topics at hand and, as a result, also the comprehension of this mathematics.

We have already ascertained (section 2.6.4) the enormous significance of communication between experts and nonprofessionals in a modern, democratic society. A further consideration in this connection was that one important characteristic of general education, both for experts and for nonprofessionals, is the capacity for such communication. With a view to possible realizations, I indicated that the 'teacher–student' constellation could be interpreted as a special case of the more general constellation 'expert–nonprofessional'. The way the teacher communicates as an expert with the students offers a model by which the students might learn a great deal about the possible communication between experts and nonprofessionals.

There can hardly be any doubt that if the teacher appears as an authority on the subject – as is often the case in mathematics instruction – and hides behind his or her far-reaching knowledge, this is exactly the reverse of the 'well-educated' expert. In this way, the students get an emphatic demonstration of the supposed fact that (at least) mathematical expertise cannot reasonably be put to question. In that case, mathematics instruction has only a negative effect on the desirable practice in communication between experts and nonprofessionals.

The same applies to practice in communication and cooperation as to teaching a sense of responsibility: Different social forms of learning and working could be a first step toward better achieving this goal. I have noted some examples above. But the realization of other forms of work should be accompanied by a different way of treating mathematics. Distinctive features of a new culture of instruction would have to include the following: students can express even incomplete thoughts and genuine questions without risking losing face; together with others, students can get involved in learning mathematics as a process of discovery and, with their teacher, – of course, at various levels of intellectual sophistication – they can reflect on the concrete mathematical activities in class.

3.5.5 Factors Inhibiting or Promoting Enhancement of Students' Self-Esteem

At the outset, I would like to mention a concrete inquiry. For several semesters, I have been posing the following question to education students with various majors: Which situation from your years at school has been inscribed in your memory as especially unpleasant, distressing, humiliating, or embarrassing?

More often than anything else, the students described, with a certain amount of variation, a situation from the mathematics classroom: A problem was to have been worked out at the blackboard, but the student had experienced a mental blackout and had failed to make the right computations. Even if the student had not additionally been humiliated by the teacher in front of the entire class – and this sort of thing still happens today –, a strong feeling of having been exposed was usually the result. In such situations, one's own person seems to be reduced to performing some intellectual achievement. If the expectations remain unfulfilled, one's own cognitive abilities seem diminished, indeed, one's entire person seems deprecated. There are only few comparable situations in the classroom in which so inescapably – as many students have experienced it – personal inability is put on display.

At first glance, it might seem that such scenes are avoidable if the teacher responsible developed more pedagogical sensitivity, more pedagogical tact. Regardless of the particular subject matter, this sort of situation might give rise to a general pedagogical appeal such as: Respect the personality of the student, never expose anyone's weaknesses. But this would fail to recognize the fact that precisely the treatment of errors is part of the social character of the discipline of mathematics. Baruk (1989) has collected an impressive number of examples that demonstrate how widespread in mathematics instruction the inclination is to simply interpret errors as indications of failure.[153] For this reason, I will begin to develop my ideas on enhancing the students' self-esteem in the mathematics classroom with reference to the issue of dealing with errors.

A new way of treating errors requires above all that they are interpreted in a different way – no longer simply as indications of failure, but, rather, as necessary attendant circumstances of any learning process (cf. Führer 1984). Teachers can better understand what is going on inside the heads of children and adolescents if they explore the reasons for errors. For students it is also often more helpful if, with the support of their teacher or fellow students, they can make sense of their own mistakes, instead of just having the 'correct' solution presented to them. A different way of treating students' errors requires a different conception of the learning process – mathematics is something that every individual must construct for himself or herself anew, not some finished complex of objective knowledge that simply must be transmitted from the outside to inside the students' heads. If students can repeatedly experience this in the everyday mathematics classroom, their capabilities will not automatically increase. Yet, there is a chance that the feelings of failure and humiliation experienced so often will give way to a more realistic treatment of one's own weaknesses, that students will develop more confidence in their own cognitive abilities and, as a result, experience more success – an enhancement of the students' self-esteem.

To be sure, this new way of treating errors also requires structures of interaction different from the usual ones. Some errors can certainly be touched upon within a dialog involving the whole class and discussed in a way profitable to the entire group of learners. Yet, in many cases this would have little effect and could even be distressing to the person who made the error. Individualizing phases of instruction offer better opportunities for responding to individual students and their problems. Group work, with its collective solving of problems and working out of new concepts and interconnections, offers many more opportunities for direct communication among the students, for 'genuine' questions (as opposed to testing questions put to the students by the already knowledgeable teacher), and for mutual explication of certain topics than does instruction with the class as a whole. All of these conditions are characteristic of a culture of instruction that tends to do more justice to the claim of bolstering students' personalities.

Briefly, certain other conditions and elements can be mentioned that many students experience as inhibiting in the mathematics classroom – although they often accept them as a natural part of this subject matter:

- Only rarely are there opportunities to develop one's own imagination and creativity. Most of the time, instruction proceeds along predictable paths and provides few occasions for personal initiative.
- Students' contributions are often experienced as essentially being directed from the outside, as responses to the demands of the teacher.
- Personal questions about the significance and meaning of what must be learned in the mathematics classroom are not a topic of instruction.
- Students in the upper grades encounter the virtual impossibility of critically questioning the norms and conventions of the discipline.

Such inhibiting conditions can, indeed, be counteracted by the following measures: using more open-ended problems, allowing various individual solutions, providing opportunities for more playful ways of dealing with mathematics, tackling unusual ideas, and – in the upper grades – opening instruction for issues stemming from general philosophy and from philosophy of science.

Thus, the tension between disciplinary and pedagogical demands so often observable in the mathematics classroom is by no means irreducible. We can certainly imagine a way of treating mathematics in which these demands are not strictly contradictory. Those mathematics teachers whose primary orientation stems from the discipline itself, especially those at college-preparatory secondary schools, tend to impart mathematics in the same way as they themselves received it. These teachers seem to fear bringing too much pliancy into the rigid discipline of mathematics (which they venerate because they proved themselves able to measure up to it during the course of their studies) and thus denigrating the discipline if they become responsive to pedagogical ideas. The counterargument to this standpoint is as follows: A culture of instruction informed by general education, in which the pedagogical objectives formulated here are taken seriously and in which learning specialized content and learning social skills are coordinated in the way described, also gives those students for whom the discipline in its strict form would otherwise remain obscure for their entire lives an opportunity to become familiar with mathematics.

3.6 *Elements of a New Culture of Instruction*

The preceding sections have continually made one thing obvious: Whether school mathematics can appreciably contribute to general education as this is understood here is only partially a question of the instructional material, of mathematical topics that are dealt with in the classroom. In the final analysis, the decisive factor is the way teachers and students treat these topics in the classroom and how they treat each other. In other words: General education in the mathematics classroom is a question of the *culture of instruction*.

3.6.1 *Possibilities for Change in the Culture of Instruction*

What I call a 'culture of instruction' was explained for the first time above in section 2.4.6 and then in more detail especially in sections 3.3.6, 3.4.7, and 3.5. I will list some of the main points here once again so that they can be more readily called to mind:

- As with the general concept of culture, the concept of a culture of instruction can be used in a *descriptive* and in a *normative* sense.
- In a *descriptive* sense, every sort of instruction has its own specific culture of instruction: the characteristic arrangement of customary patterns of action and interaction, corresponding value systems, perspectives, and expectations of the participants – regardless of the content-related, didactic, or pedagogical quality of that instruction and regardless of how conscious the agents are of the particular patterns, values, perspectives, and expectations.
- In a *normative* (or prescriptive) sense, the concept of a culture of instruction represents the cultivated quality of instruction aimed at here. In that case, it is connected to a particular conception of instruction including ideas on how instruction *should* be. If we speak of the development of a 'culture of (mathematical) instruction contributing to general education', the context implies that, here, we are dealing with the normative variation. This type of culture of instruction, then, represents an arrangement of patterns of action and interaction, and the corresponding expectations, perspectives, and values of which we have reason to assume that they are beneficial to learning mathematics in accordance with the concept of general education at the basis of our argument.
- This means that the concept of culture of instruction includes – in both its descriptive and normative variations – the official (formal) as well as the unofficial (informal) level of activity in the classroom. It includes conscious didactic and methodical actions as well as the general atmosphere, e.g., the emotional climate in which classroom activity takes place.
- Finally, the culture of instruction refers to learning specialized content and learning social skills. If these elements are separated from one another, this is (in a descriptive sense) *one* characteristic of the culture

of the instruction under consideration – just as, in the reverse case, their integration would also be one such element. In any case, overcoming this widespread separation is certainly an important characteristic of the 'culture of instruction contributing to general education' postulated here.

To preclude a false impression: If the concept 'culture of instruction' is used here in connection with the phrase 'contributing to general education' in a normative sense, this does not at all stipulate that such a culture of instruction must in some way itself be *standardized* or *promote standardization*. For this would ultimately lead to replacing the outdated, fossilized structures of 'conventional' mathematics instruction with new, rigorous directives. Instead, the concept of a culture of mathematical instruction contributing to general education represents making

KEY TO TABLE 1 (ON THE FOLLOWING PAGES):
PL: preparation for later life in a strict sense; **CC**: promoting cultural competence, including orientation to main ideas; **UW**: developing an understanding of the world, including applications; **CT**: promoting understanding and critical thinking; **WR**: developing a willingness to assume responsibility; **PC**: practice in communication and cooperation; **ES**: enhancing students' self-esteem; **CM**: developing a conception of mathematics. – The symbol ■ indicates a marked relevance, the symbol ▫ only a partial relevance of the specific characteristic for the corresponding objective of general education.

	Characteristics of a culture of mathematics instruction contributing to general education	**PL**	**CC**	**UW**	**CT**
GE1	Students communicate directly with one another.	▫			
GE2	Students pose 'real' questions to teachers and fellow students and provide 'real' answers (not only with an eye to getting the teacher's attention); the students discuss arguments among themselves.	▫			■
GE3	Understanding mathematical topics has higher priority than procedural skills. The teacher gauges understanding by observing how much students reflect on their mathematical activity.				■
GE4	Formalization is never an end in itself; instead, in certain phases of instruction, it aids in making content that has been thought through and understood available for further use and operations.				■

GE5	Errors are welcome opportunities to reflect on the reasons behind the errors.	□			■
GE6	There are various levels of approach to insights, hypotheses, partial solutions, etc.	■		□	■
GE7	Errors are accepted as necessary attending circumstances of learning processes.	■			■
GE8	There is ample latitude for indirect paths, unconventional ideas, an openness for diversified classroom activities.				■
GE9	Individually distinct solutions are not only tolerated, but are encouraged as discrete ways of accessing the tasks at hand.				□
GE10	Learning mathematics is often experienced as a discovery process that can be engaged in individually or in the course of an intense exchange of arguments and ideas with others.			□	
GE11	Students and teachers do not hesitate to communicate thoughts that are not yet 'finished products'.	□			■
GE12	There are unresolved problems and problems that allow for widely varied approaches and more than a single 'reasonable' solution.	□		■	■
GE13	Students playfully explore their own creativity in mathematics-related situations.				■

WR	**PC**	**ES**	**CM**		**Characteristics of a 'traditional' culture of mathematics instruction**
□	■			T1	Students communicate primarily with the teacher or through the teacher with one another.
□	■	■		T2	The predominant pattern of interaction is composed of three steps: impulse from the teacher – response(s) from the students – teacher's comments.
		□	■	T3	The teacher assesses mastery of a particular mathematical topic by demanding correct solutions to predetermined problems.
			■	T4	High priority is accorded to scientific, abstract language and to a pronounced degree of formalization.

□		□	■	T5	Errors are corrected immediately.
		□	■	T6	There are only correct and incorrect answers.
		■	□	T7	Errors are only interpreted as indications of failure.
	□	■	□	T8	Students' thoughts which in the teacher's view digress from the official topic are not followed through.
□		■	■	T9	Only a single solution is permissible.
	■	■	■	T10	Students experience learning mathematics as following routes specified by the teacher.
	■	□	□	T11	Essentially, only the results of cognitive processes are communicated and made 'public' for other participants in class.
			■	T12	The mathematical problems dealt with in class are unambiguous and conducive to single solutions.
		■	■	T13	Dealing with mathematics is primarily experienced as strenuous, serious, dull activity that stifles one's imagination.

	Characteristics of a culture of math. instruct. contributing to general education (continued)	**PL**	**CC**	**UW**	**CT**
GE14	Curiosity, excitement, engaged interest, surprise, enjoying implementing thinking skills and doing the mathematical activity are not unusual aspects of instruction.				■
GE15	Mathematical activities are subjects for collective reflection.			□	■
GE16	The meaning and significance of mathematical topics currently on the agenda are themes of discussion in class.		■	■	■
GE17	Interconnections among mathematical subject areas are emphasized; the connection to main ideas is made clear.		■	□	■
GE18	One substantial component of instruction is discussing the meaningfulness of applications (or of the assumptions of the mathematical modeling	■	□	■	■

	Characteristics of a culture of math. instruct. contributing to general education (continued)	**PL**	**CC**	**UW**	**CT**
	behind them).				
GE19	The ostensible self-evidence and the conventions of school mathematics are subject to doubt and to critical analysis.			□	■
GE20	Students assume responsibility for their own learning.	■			
GE21	Students often have opportunities to become active and to identify with problems.				□
GE22	As a matter of course, one assists fellow students in furthering their understanding and also allows oneself to be assisted in this way.				□
GE23	There are constant opportunities to discuss problems with others, to clarify goals and strategies with them, to mutually compensate for weaknesses, to collectively enhance strengths (pair work and group work).	■			□
GE24	One's partner – whether student or teacher – is taken seriously as a human being; his or her ability to reason is presupposed.	□			□
GE25	The quality of an argument is more significant than the social status of the proponent of the argument.	□			■

WR	**PC**	**ES**	**CM**		**Characteristics of a 'traditional' culture of mathematics instruction** (continued)
		■	■	T14	Instruction proceeds in foreseeable ways; upcoming topics will hardly be matters of emotional concern.
	■		■	T15	Reflecting on mathematics is of no particular importance.
		□	■	T16	Questions on the meaning and significance of mathematics are not topics for classroom instruction.
			■	T17	Essentially, every mathematical subtopic is an isolated subject.

	□	□	■	T18	Applications problems are subject to critical analysis just as rarely as are purely mathematical problems.
	□	□	■	T19	The difference between mathematical necessities and conventions is not a topic of instruction.
■		■		T20	The teacher is responsible for students' learning.
		■	□	T21	Students experience classroom activity primarily as something directed from the outside, as a response to what the teacher demands.
■	■			T22	Fellow students are primarily perceived as competitors.
□	■			T23	Students experience themselves as being left on their own. 'Cooperation' takes place only at the informal level of instruction (cheating).
□	■	■		T24	The teacher makes use of his or her institutional power to direct the students in a preconceived way.
□		■	■	T25	In case of conflicting opinions, the teacher is always right as a result of his or her position.

mathematics instruction more accessible – to less standardization concerning permissible actions and verbal contributions, to detaching oneself from too restricted conceptions of mathematics, to consciously allowing for more subjectivity on the part of learners and teachers, to a greater variety of individual approaches to mathematics, to more opportunities for self-directed discovery, to a more constructive way of treating errors, to more intensive involvement with what and how others think, to greater sensitivity to the individual cognitive acts and the accompanying feelings of the students, in short, to more vitality.

However, we should still take the following into account: Such a list of desirable changes can lead to feelings of inadequacy or, indeed, of incapability even among teachers who, in principle, are willing to accept innovations. For this reason, a second possible misconception should be dealt with: An 'antiquated' culture of instruction, established and stabilized over many years, cannot be replaced with a 'new' one from one day to the next. This can only be an evolutional process. Often, there will be intermediary or transitory stages between what – for a teacher, but also for a student involved in such a process – has long been a matter of habit and those changes considered desirable. The important thing for practical realization will be to consciously undertake small steps and to productively make use of the tension between customary everyday practice and high ideals. I will return to this consideration presently (3.6.3).

3.6.2 'Antiquated' and 'New' Cultures of Instruction: A Comparison of Ideal Types

While examining the comparison presented in Table 1, one should keep in mind that in actual classroom activity the elements listed will only rarely appear in their pure form. More subtle descriptions of observable instruction will often have to take recourse to elements that can be unequivocally assigned neither to the column on the left nor to that on the right. Nor does this list make any claim to being exhaustive. My main concern was that the most important characteristics of a culture of mathematical instruction contributing to general education, already discussed in the text, should reappear within the table.

The elements of the column on the left point to the culture of instruction aimed at here; the sum of these individual characteristics is not to be equated with this culture of instruction. In particular, these characteristics cannot be transformed into simple recipes, whose use would necessarily result in the realization of the targeted culture of instruction. But the reverse conclusion is fairly plausible: A type of instruction in which mutual respect on the part of the participants and sensible treatment of the subject matter are cultivated, in which teachers are sensitive to the mathematical learning processes of children and adolescents, to the need to understand and make sense of classroom activity – such instruction would tend to display more characteristics from the column on the left than from that on the right.

The overview in tabular form serves a further purpose: It systematically illustrates the fundamental significance of the culture of instruction for mathematics teaching oriented to general education (a significance I have referred to repeatedly). The eight columns in the middle of the table facilitate this systematic presentation. Seven of these columns are assigned to the fundamental objectives of general education with regard to mathematics instruction (cf. the caption to table 1); the eighth one represents the aspect of acquiring an appropriate conception of mathematics.

This final category, which transcends the others to a certain extent, has been mentioned several times in the course of this study – in connection with preparing for later life, promoting cultural competence, developing an understanding of the world, and stimulating critical thinking. Thus, for a reason mentioned in the preceding subchapter, I wish to call attention once more to this aspect: The conception of mathematics that children and adolescents develop during their years at school is influenced by the concrete experience of how mathematics is treated during class. If many adults view mathematics as a closed, inflexible system of abstract rules, peculiar symbols, and frightening formalisms, this is not because someone has explicitly taught them to do so, but, instead, because they have tacitly surmised this from their experience of how mathematics was treated in the classroom. We can interpret their subjective conception of mathematics as, not least, a repercussion of the culture of instruction in which they participated.

To conclude this section, here are a few remarks on how to read the table:

- The pages of the table are to be considered as a unit. The (seven plus one) categorical columns have only been separated into four columns on the left and four on the right for reasons related to the printing of this book.
- The squares demonstrate the general relevance of the specific characteristic for the corresponding objective of general education.
- 'GE' (general education) characteristics are to be evaluated positively, 'T' (traditional) characteristics are to be evaluated negatively for the particular objective.

To illustrate this in relation to the characteristics GE11 and T11:

- There are large squares (■) in the columns 'CT' and 'PC'. Expressed in words, this means that, generally speaking, being allowed to 'publicize' thoughts that are not totally worked out, has positive effects on developing cognitive skills and on the communication among the participants in classroom instruction. Limiting the permissible 'public' statements to thoughts as 'finished products' (T11) obstructs both of these goals.
- There are small squares (▫) in the columns 'PL', 'ES', and 'CM'. Expressed in words, this means that being allowed to 'publicize' thoughts that are not totally worked out, can also have positive effects on preparation for later life and on the development of the student's personality. And with respect to the conception of mathematics, it might help to experience mathematics as a 'process', not as a 'finished product'. By contrast, limiting the permissible 'public' statements to the final results of cognitive processes can have inhibiting effects on both of these goals.

3.6.3 A Culture of Instruction Oriented to General Education as a Combination of a Variety of Such Cultures

There is a wide variety of possible routes that can be taken to realize the culture of mathematical instruction outlined here. With an exemplary sketch of some of these routes, I would like to point out that the demand for the development of such a culture of instruction does not at all signify a standardization. On the contrary, extensive room becomes available for competing concepts of instruction that are very distinct with regard to their particulars. Yet, each one, in its own way, is quite interesting and stimulating.

Basic Ideas and Travel Journals (Gallin/Ruf)

I will start with a short description of a concept of instruction in sharp contrast to conventional mathematics teaching at the lower grade levels of secondary schools. It was developed in the course of a twenty-year collaboration between two creative Swiss teachers, the mathematician Peter Gallin and the German teacher Urs Ruf (Gallin/Ruf 1990, 1993a, 1993b, 1995), and has been tested at different types of schools (*Volksschule* as well as *Gymnasium*).

One main idea of these two educationalists is to make concepts of native language instruction – narrative, expressing experiences and feelings in writing, exact descriptions of concrete contexts in everyday language – productive for learning mathematics. In this way, they hope to counteract the impression of many students that mathematics is something totally unconnected to their personality. Like Wagenschein, they attempt to make "the singular language of understanding into the starting point for the long route to the normal language of what has been understood" (Gallin/Ruf 1995, p. 65). They consider the "external perspective" on the discipline – in this case, that of the German teacher – to be a "prerequisite for determining the aspects of mathematics instruction suitable to general education" (ibid., p. 58).

I will confine myself here to a brief sketch of the essential features of the practical procedures used by Gallin and Ruf. The innovational potential of their concept of instruction can be most readily understood in this way. For further theoretical arguments, details of instructional practice (e.g., assessment), and concrete examples of lessons, the reader can consult the publications of these two authors.

In the mathematics classroom, each student is to maintain a single notebook in which individual learning experiences are recorded – a 'travel journal'.

> "Everything that goes on in class is continuously noted down in the travel journal: specific inquiries, exercises, homework, the teacher's comments, and overarching theoretical considerations. The travel journal is the special place where the singular language of the learner finds expression." (ibid., p. 73)

Thus, every student writes down individual thoughts on the instructional topics encountered – including emotionally charged assessments and associations transcending mathematics. At regular intervals, the teacher reads each student's travel journal and offers feedback; the students themselves also read each other's entries. However, the material itself is not simply predetermined; rather, it emanates from 'basic ideas' with which the students can identify. The basic idea should be "attractive, authentic, and motivating"; at the same time, it is allowed to be "tentative and somewhat imprecise"; it should focus attention and provide an overview; it should "arouse enthusiasm" and motivate learners to content-related action (p. 67).

> "At this first encounter with content, and this is especially important, things are not explained, but, instead, narrated. Here, too, mathematics instruction can learn from literature. Explanation generates pressure; narration liberates" (p. 67).

Gallin and Ruf refer to "possible worlds of understanding" through which the teacher as a narrator leads his or her students, taking recourse to his or her own personal experiences. Once the basic idea has been established in the classroom, it condenses into an assignment. The writers demand that this assignment "take the range of talent within the entire class into account and allow for intensive efforts at various levels of ability" (p. 67). The assignments often relate to topics specified by the syllabus, but they intensify the connection to the students' experiences. Furthermore, they should reveal an entire complex that is worthwhile and make long-term individual work on the travel journal possible. For example, the first acquaintance with quadratic equations can be arranged with a worksheet that allows the students themselves to discover the theory behind this topic – from quadratic residue to the quadratic formula (p. 69).

The writers had notable success with a two-year school experiment involving a total of seven grade levels. Apparently, they managed to increase interest in mathematics, understanding, and achievements especially on the part of weak and mediocre students (cf. Gallin/Ruf 1991).

It is difficult to judge solely on the grounds of the description of an instructional concept how much the concept actually does justice to the appeal to a culture of instruction contributing to general education. (Gallin and Ruf, by the way, do not make explicit use of the term 'culture of instruction'.) In any case, many of the elements of Table 1 can be found in a concept of instruction that has a great emphasis on individualization. In particular, almost all of the "GE" elements connected to enhancing the students' self-esteem (GE5-11, GE13, GE14, GE20, GE21, GE24, and GE25) or to developing critical thinking (in addition, GE3, GE4, GE12, GE15-17) are of fundamental importance for Gallin and Ruf's concept of instruction. Even if one refrains from a painstaking examination of every single element, it is still evident that this concept includes much of the spirit of a culture of instruction oriented to general education – and, indeed, in conjunction with a very original outlook.

A critique can be made of Gallin and Ruf's curricular conservatism, as they are primarily concerned with officially recognized syllabi. In Switzerland, these also offer only one variation of the internationally widespread standard curriculum for the lower grade levels of secondary schools. For this reason, certain emphases that would be indispensable to a mathematics instruction oriented to general education for realizing objectives such as preparing for later life or developing an understanding of the world (in the sense of an orientation to applications) are missing from this concept, or, at least, they are not immediately recognizable in the writers' remarks. Nevertheless, the culture of instruction described by Gallin and Ruf does allow for an inclusion of these aspects. On the other hand, it is inevitable that the appropriateness of this culture of instruction for mathematics teaching oriented to general education be questioned in another respect: Might the heavy emphasis on individual work and written presentation too categorically limit communicative exchange among the learners? Which role is assigned to verbal arguments addressed to fellow students, what significance does classroom dialog have for dealing with the mathematical topics under discussion? Gallin and Ruf are aware of this possible objection to their concept and point out that their emphasis on writings must not at all be misinterpreted as a "devaluation of oral discourse". On the contrary, they maintain that on the basis of individual reflection and notes, oral discourse would gain "variety and dynamism" (1995, p. 73).

Here, I will not try to resolve the question of whether this alternative concept of mathematics instruction aiming at general education – with its approaches to a highly individualized, nonetheless fascinating culture of instruction – might provide a model for other teachers. Its development is intricately interwoven with the personal lives of the two writers and with a form of collaboration that is practically unrepeatable. Attempts at direct imitation might even risk simply being reduced to establishing the practice of using 'travel journals', to that methodical element of the conception that can be most readily employed in a strict, concrete sense. But, certainly, such a reduction would be unable to capture the spirit behind Gallin and Ruf's operations; the attractive quality of their teaching efforts would surely be lost. In this respect, too, one is reminded of Wagenschein.

Within the present context, the question of duplication is not really significant. I was more directly interested in demonstrating in which extraordinary ways – compared to the usual ideas of how mathematics instruction should take place – one can approach the realization of a 'culture of instruction contributing to general education'.

A Mild Form of Mathematics Instruction (Andelfinger)

One of the first German mathematics educationalists who consciously made use of the concept of culture of instruction to emphasize the singularity of his ideas on instruction was Bernhard Andelfinger (1989, 1993). His concept of a "mild form" of mathematics instruction was developed in cooperation with teachers who were dissatisfied with conventional mathematics instruction and were searching for alternatives. Of major significance for this concept is the idea that teachers and students take each other seriously and have mutually enlightening effects on each other.

In contrast to Gallin and Ruf, Andelfinger does not describe concrete procedures in the classroom. He first designates in an abstract way a series of elements of the "mild form" of mathematics instruction he aims at and then he looks for effective concretizations in the classroom. In this respect (i.e., regarding the level of abstraction), his concept is more comparable to the concept of 'mathematics instruction as part of a general education' under discussion here than to Gallin and Ruf's concept.

Andelfinger's starting point is an opposition between a "Cartesian" and a "gaiatic" (Gk. "gaia": Earth) view of the world. Conventional mathematics instruction is seen as a manifestation of the former; the "mild form" of mathematics teaching that Andelfinger aims at should correspond to the latter. The theoretical and ideological standpoints behind this position stem from ecologically oriented, alternative lines of thought and from a fundamental critique of civilization inspired by such sources.[154] Andelfinger contrasts the "gaiatic" paradigm with the "Cartesian paradigm, which (also) serves as the basis of technological and industrial domination of the world" (Andelfinger 1989, p. 28; 1993, pp. 4 and 9) by assigning to them such polar "categories, methods, and elements" as:

- interconnected/in fluid equilibrium	vs.	linear/causal;
- discovering/developing	vs.	concluding;
- dialogical/sensitive	vs.	technological/instrumental;

- enriched/diverse vs. reductionist;
- receptive to error vs. rejecting error;
- subjective/involved vs. reserved;
- ambiguous/diverse vs. unambiguous/clear; (etc.)

Summing up, Andelfinger outlines his position as follows:

> "A mild form of mathematics instruction is a culture in which gaiatic thinking can engage in an open dispute with Cartesian/Baconian thought. This dispute must take place in such a way that it concurs with the principles of 'peace', 'justice', and 'preservation of the Earth'." (1993, p. 2)

Several aspects of the theoretical foundation of this concept are open to criticism. For instance, it would be worth considering whether arguments to be maintained against the more radical forms of ecological pedagogy (that they reduce the public school to an instrument of establishing new social processes of thought and thus actually demand too much of schooling) are not also pertinent here. Furthermore, we can question whether the polarization of perspectives and the radicalness of the reorientation demanded do not obstruct possibilities for altering everyday classroom instruction that teachers could and should be encouraged to undertake with step-by-step procedures. Yet, regardless of these critiques, one might ask which concrete ideas of a new culture of instruction are manifest in Andelfinger's train of thought. The answer, then, is that a great deal is compatible with what I have described in section 3.6.2 as a 'culture of instruction contributing to general education'. Even though not all of the "GE" elements in Table 1 appear explicitly and with the same significance in Andelfinger's remarks, not one of these elements would be contrary to the characterization of a culture of instruction of the "mild form" of mathematics teaching. In this sense, this concept can be mentioned as a further example – and one that is quite different from that of Gallin and Ruf – of how the demand for a culture of instruction fostering general education can actually be realized.

Additional Concepts

There is certainly no lack of further examples. Without making claim to presenting a comprehensive discussion, I would at least like to mention a few. To make the considerable scope of possibilities more apparent, I will consciously combine examples of varying theoretical sophistication and with varying consequences. Elements of a culture of instruction oriented to general education as I have described can be found (explicitly or implicitly) in the work of the following scholars:

- Heinrich Winter (1989), who has worked out the concept of "discovery learning" in more detail for mathematics instruction. Even his introductory opposition between "learning by letting students discover" and "learning by presenting information" (ibid., p. 4) has a number of features in common with my comparison in Table 1;
- Dieter Volk (1995), who is committed to an emancipatory concept with practical, real-life applications and employs many examples to describe the culture of instruction of an 'enlightening' mathematics teaching;

- Hartmut Spiegel (1989), who reports astounding experiences from his Socratic dialogs with adults on mathematical topics: how unexpected possibilities for understanding mathematics can be discovered by treating mathematics in a new way if the partners of the dialog are definitely committed to taking each other seriously;
- Martin Winter (1989, 1994, 1995), who, with stimulating lesson plans, has repeatedly demonstrated that the culture of mathematical instruction aspired to here can also be developed under the conditions prevailing in normal public schools; above all, his examples from everyday classroom activity convincingly demonstrate how much the benefits students personally gain from mathematics instruction depend on the culture of instruction realized in class.

Individual elements of a culture of instruction as it has been described here are also to be found in less recent works of scholars such as Wagenschein, Wittenberg, and Freudenthal. With reference to these 'classical' authors of mathematics education research, we must, indeed, acknowledge that they do not treat the social aspects, especially the informal aspect, of instruction systematically. Yet, this is hardly surprising because most of the knowledge we now have about individual learning being interwoven with the structure of interaction in the classroom is based on research from cognitive psychology and sociology that was conducted within the last two and a half decades. For this reason, these authors deserve even more acknowledgement for having opposed the reductive behaviorist models of learning so prevalent in their time and for having intuitively anticipated with their didactic concepts much of what today's enlightened pedagogues and mathematical educationalists consider to be the 'state of the art'.

Chances of Realization

The examples mentioned above might serve as an indication that a new culture of instruction more strongly oriented to general education along the lines of the concept developed here is possible in a variety of ways. The idea of a culture of instruction contributing to general education does not promote standardization; instead, it provides options for a number of cultures of instruction which exhibit considerable differences in detail and which can be adjusted to various guiding principles of mathematics teaching and to various personal preferences and strengths of teachers and students.

The outline of a culture of instruction contributing to general education is a *pedagogical frame of reference* and, in this respect, is similar to the entire concept of general education at the basis of our discussion here and to that concept's concretization as a concept of 'mathematics instruction aiming at general education'. The characterizations of this new culture of instruction listed above complement and concretize this concept of mathematics instruction at the level of classroom action. As elements of a frame of reference, these characterizations, on the whole, clearly specify a direction in which the culture of instruction is to be developed. But with regard to individual details, they provide a great deal of freedom for actions in the classroom. For this reason, individual teachers or, even better, groups of teachers can begin to take steps in the direction of a new culture of instruction from the standpoint of their own personal situation. Most of the concepts of instruction listed above were originally developed within personal constellations and were only subsequently put into writing for the purpose of relating experiences to others.

Many of the "GE" elements listed in Table 1 are well suited for use as a compass for the small steps to be taken in everyday classroom instruction. Good intentions expressed by wanting to do everything differently starting the next day usually end up in resignation and in a return to the accustomed practices. There must be overlapping stages between the old culture of instruction and the targeted new one – a structural element of all evolutionary processes which, in the final analysis, do indeed lead to something qualitatively new. Actual instruction – even if it essentially takes place according to conventional, long-established patterns – always contains the seeds of possible change. Such seeds must be detected; then, they can be developed and, quite literally, 'cultivated'.

Of course, even small steps are not necessarily easy steps to take. Good intentions alone are not sufficient. A teacher who decides to change his or her way of treating the students and the mathematical subject matter in the direction of a culture of instruction contributing to general education must be sufficiently sensitive to what goes on in class, to what goes on in the minds of the students and is not accessible to direct observation. The teacher must have a well developed faculty for judging classroom interaction, a sensor whose development and practical implementation remain ignored in conventional teacher training as long as orientation to the discipline and the focus on a product predominate. In addition, the teacher requires a certain repertoire of actions for the development of which conventional teacher education only provides very little stimulus and assistance. Certainly, from the very start some teachers have little difficulty in developing the necessary sensitivity and sufficient empathy for identifying themselves with the problems students encounter in learning mathematics and for creating a learning atmosphere in which a treatment of mathematics oriented to understanding can develop for almost all of the students. Still, taking recourse to natural pedagogical talents is of little help to the students and to the majority of present and future teachers.

Here, I cannot go into detail on possible changes in teacher education programs (cf. Biehler et al. 1995), but one comment does seem appropriate. Integrating methods of interpretative research on instruction into teacher education and further training seems to give prospective teachers a chance to arrive at a new view of their own actions and to become sensitive to processes that are often ignored under the pressures of everyday teaching (cf. Bauersfeld 1986, Harten/Steinbring 1991, Jungwirth et al. 1994). Self-perception from the vantage point of others – made possible by video feedback and the interpretation of transcripts of lessons – provides initial opportunities for revising even long-established delusions about one's own behavior in the classroom. I have been able to make similar observations in connection with research on 'subjective theories' of mathematics teachers. The increased sensitivity and altered self-perception of the teachers who participated seemed, at least in some instances, more noteworthy than the more general results to be deduced from my empirical findings (Heymann 1982, pp. 162ff; 1984, pp. 108-109).

3.6.4 Summary and Conclusion

To which extent mathematics instruction actually helps to effect general education can only be determined at the level of classroom activity. What is learned in and about mathematics and its connection to the non-mathematical world greatly depends on how mathematics is treated within the framework of classroom interaction. Mathematics instruction aiming at general education requires a corresponding culture of instruction in which learning subject matter and learning social skills are not separated from one another. Such a culture of instruction provides opportunities for the students' subjective perspectives, for questions about significance and meaning, for indirect paths, for alternative interpretations, for an exchange of ideas, for experiential treatment of mathematics, and for self-responsible action. Studies from cognitive psychology and sociology which have led to more subtle ideas of learning in general and of learning processes in mathematics classrooms in particular in the last two and a half decades highlight the significance of the emphases mentioned above.

In many of the more recent concepts of mathematics teaching there are elements that are compatible with the ideas developed here. Like the more abstract concept of general education at the basis of our discussion here, the concept of a culture of instruction contributing to general education is a pedagogical frame of reference. As such, it does not standardize instruction. The framework can be filled by a wide variety of cultures of instruction in which specific situational and institutional circumstances as well as the individuality of the participants are reflected.

The culture of instruction still predominant in conventional mathematics teaching with its often inflexible patterns of interaction, its usual one-sided conception of mathematics focused on products, its ignorance of individual differences among the learners (with the exception of differences in achievement) cannot be transformed into a culture of instruction contributing to general education from one day to the next. And, of course, in the actual mathematics classroom there is an abundance of transitional stages. Changing the culture of instruction in everyday schooling relies on a large number of small steps, on an evolutionary process of innovation. In addition to enlightening research, especially in the areas of pedagogy and teaching practice, and to teachers' identification with the relevant goals, what is needed on a long-term scale are better institutional preconditions and different emphases in teacher education and further training programs.

CHAPTER 4

A PROFILE OF MATHEMATICS TEACHING AS PART OF A GENERAL EDUCATION

"... Our teachers use a very wrong theory of how to understand things, when they shape our children's mathematics, not into robust networks of ideas, but into those long, thin, fragile chains or shaky towers of professional mathematics. A chain breaks whenever there's just one single weak link, just as a slender tower falls whenever we disturb it just a little. ... Perhaps this helps explain how our society arranges to make most children terrified of mathematics. We think we're making things easier for them to find what's right, by managing to make things go all wrong almost all the time! So when our children learn about numbers (or about anything else) I would prefer that they build meshy networks in their minds, not slender chains or flimsy towers."[155]

Marvin Minsky, 1985

The starting point for this book was the following proposition: Conventional mathematics instruction at public schools does justice neither to society's requirements for the future nor to the individual needs and qualification interests of the majority of students. This stipulation is highly plausible since it is supported by a variety of observable phenomena.

The necessity of compulsory mathematics instruction is not, in principle, questioned in this book. Yet, it cannot simply be justified by referring to the value and significance of mathematics in and of itself. In particular, mathematics as a scientific discipline can provide no answer to the question of which mathematics should be learned in which way by students at school. To engage in debate on this complex issue, it is necessary to argue from a definitive pedagogical viewpoint, from a viewpoint outside the discipline of mathematics.

In order to gain such a vantage point, the examination of traditional and more recent approaches to educational theory led to the development of a modern concept of general education that can provide a more elaborate idea of what might be important for all children and adolescents within the framework of school learning. The guiding principle was to use this concept as a pedagogical measure of the individual school subject, i.e., of mathematics instruction. For this permits a more in-depth study guided by the following question: To what extent does conventional mathematics instruction promote a general education in the sense specified above, and how could mathematics teaching be designed to do more justice to the claim of furthering general education?

The concept of general education was discussed using descriptions of and arguments in favor of seven major objectives of schools. Briefly, these objectives can be listed once more as follows: preparing for later life, promoting cultural competence, developing an understanding of the world, guiding critical thinking, developing a willingness to assume responsibility, practicing communication and cooperation, and enhancing the students' self-esteem.

With reference to each of these objectives of general education, we discussed how conventional mathematics teaching can contribute to their realization, which elements tend more to inhibit this realization, and, finally, which innovational elements might help mathematics teaching to attain these objectives. Proceeding from existing instructional practice and taking current conceptions of mathematics teaching into account, we looked for possibilities for change. For this search, the concept of general education at the basis of our discussion served as a guiding principle. In this way, in a step-by-step process we were able to develop a profile of mathematics instruction promoting general education and personal development in a strict sense.

Eventually, the following emphases of mathematics instruction oriented to general education in contrast to those of conventional teaching resulted (here, I present these points in strongly condensed form and list their order according to that of the objectives of schools at the basis of our discussion):

- Directly practical everyday activities such as estimating, approximating, interpreting and designing graphic representations, and an intelligent use of technological aids should become topics of instruction, should be mathematically reflected upon and practiced more often and with greater intensity – at all levels of mathematics instruction and with a continuous increase in the demands placed on learners. In general, the obligatory canon of instructional topics should show more consideration for those students who will not be practicing a mathematics-related profession (preparation for later life).
- Main ideas that exemplify the connection between mathematics and non-mathematical culture should serve as central themes arching over individual topics and explicitly become subjects of discussion – with a view to their historical origins and in connection with intensive mathematical activity. Main ideas of this type could be the following: number, measuring, structuring space, functional relationship, algorithm, mathematical modeling (promotion of cultural competence).
- In a wide variety of ways, it should be possible to experience how mathematics is used for interpreting and modeling, for better understanding and mastering phenomena that are primarily non-mathematical. The confines of conventional applications of school mathematics, which are particularly obvious in the traditional, 'fabricated' problems, should be counteracted by a more reflective way

of treating the problems under discussion (development of an understanding of the world).

- Students should have sufficient time and opportunity to actively use their own understanding in constructive and analytical ways to better understand mathematics and implement it to clarify questionable phenomena – as a sort of 'amplification' of their everyday thinking. Teaching should pay significantly more attention to the peculiarities of mathematical abstraction and the resulting problems in learning mathematics. Teachers should be conscious of the fact that students often perceive new mathematical topics as something unfamiliar and alien that can only become familiar by active practice, as a form of resistance that must be overcome, as something not yet extant that must first be constructed. Furthermore, mathematics should be experienced in its enlightening function, as a means of critical thinking (promotion of understanding, cognitive skills, and critical thinking).
- How mathematics is treated in the classroom is in many respects a more decisive factor for effects on general education and personal development than the actual content discussed. With a view to both the socio-ethical and personality-centered objectives of schooling and to fostering cognitive goals, a culture of instruction should be developed that provides better opportunities for the following: students' subjective perspectives, mutual communication about current mathematical topics, productive treatment of errors, indirect paths and alternative interpretations, a lively exchange of ideas, more animated and creative ways of treating mathematics, and self-responsible action. A more pronounced internal differentiation and a variety of forms of classroom work can help to disrupt too one-sided ritual procedures and to do more justice to the variation of individual approaches to mathematics (willingness to assume responsibility, communication and cooperation, enhancing the students' self-esteem).

Considered individually, none of these demands is totally original. Nevertheless, if their many-faceted interconnections, which have been emphasized in this book, are taken into account and constantly related to the idea of an up-to-date general education, they could provide school mathematics instruction with a new profile.

Since antiquity, one characteristic of Western culture has been that a reflective and enlightened form of existence is more highly valued than a non-reflective one. This value judgment also has repercussions for ideas of schooling, general education, and personal development. If today's schools are expected to impart fundamental abilities and skills to the younger generation, this expectation also applies to the development of an ability for reflection and of a mental horizon that goes beyond the immediate matters of everyday life. In our society, so deeply influenced as it is by mathematics, these two goals are inconceivable if young people are not at least able to experience how their own thinking can be amplified by the formal techniques developed in mathematics (by taking advantage of the laws inherent in that discipline) and how these formal techniques can be related to the world of concrete, practical experience. If such experiences are made possible in a way that learners can appropriate them and integrate them into their personal images of the world (at varying levels of intellectual sophistication), then school mathematics does indeed make a significant contribution to general education and personal development.

In order to realize a type of mathematics instruction promoting general education in a strict sense, teachers must not only be enthusiastic about their subject matter, they must also have a certain modesty – and realize that mathematics cannot have the same importance for all students. It will be helpful in this regard to occasionally remind oneself of the fact that hardly a single unresolved problem with which we are confronted on a global scale (or, indeed, within our private spheres) has to do with the condition that too many of us have too little mathematical ability.

In accordance with the stipulation that the idea of general education mediates between the individual's right to personal development on the one hand and the general culture and societal necessities on the other, the concept developed here presents a horizon encompassing new possibilities on the basis of sound pedagogical arguments. It outlines a sphere of activity in which possible steps in the direction of a more pedagogically sound mathematics instruction can be realized – towards classroom instruction that not only fulfills legitimate demands of society (in fact, to a greater extent than is currently the case), but also does more justice to various types of students and can be a meaningful experience for the majority of them.

NOTES

1 Whitehead (1967 [1913], p. 78).

2 From a questionnaire addressed to novice students in elementary-school teacher education programs (Spiegel 1988, p. 5).

3 For an example, see the findings of the British CSMS Project discussed by Howson (1984). A more recent prominent example from a representative survey on the general knowledge of Germans conducted by the Emnid Institute in Bielefeld (in *"Der Spiegel"*, Dec. 19, 1994, p. 100): Only 38% of those questioned were able to answer the following question correctly: "Which box of laundry detergent is less expensive – the 10 kg box for 13.99 Marks or the 3 kg box for 3.99 Marks, or are both boxes the same price?"

4 During the last two decades, there were also debates on the issues of general education in schools (with a number of parallels to the more recent German educational debates) in other 'Western' industrial countries. To list a few examples: Kirk (1986) [Great Britain and Australia], Hirsch (1987), Westbury/Purves (1988) [USA], College de France (1987) [France]. In addition – and, similar to the situation in Germany, largely independently of the more general debate –, in many countries there has been debate on a fundamental reform of mathematics instruction, indicating that the orientation crisis of mathematics instruction is an international phenomenon. Again, to mention only a few select publications: Cockroft et al. (1982) [Great Britain], Howson/Wilson (1986) [international], National Council of Teachers of Mathematics (1989), The Journal of Mathematical Behavior (Vol. 1/1994, especially Davis 1994) [USA].

5 This statement is generally correct even though, of course, there has been no lack of exceptions to the rule (to name some of the more recent ones: Plöger 1989, Lohmann 1990, Keck/Köhnlein/Sandfuchs 1990, Arbeitskreis Mathematik und Bildung 1993, Meyer/Plöger 1994).

6 This basic attitude (conservatively relying on values embodied in the idea of enlightenment) can perhaps explain why I do not explicitly deal with the debate on postmodernism, in which many educators have intensively participated during the preceding years (cf., e.g., Vol. 1/87 of Zeitschrift für Pädagogik). In my opinion, this debate suffers from a fashionable exaggeration of certain symptoms of crisis in the (dialectically progressing) modernization processes of Western societies. Unfortunately, this stylization also has a significant impact on reflections that, in their own right, could be quite interesting. Aware that the following statement does not do justice to certain individual standpoints, I maintain: Confronted with the urgent issues and problems of modern times, the postmodern theorist persists in the role of the intellectually witty, aesthetically inclined, glass-bead game player, who, in the final analysis, only purports to 'know' that one cannot know anything (cf. for a pedagogic perspective: Göstemeyer 1993; for a sociological perspective: Alheit 1994, especially pp. 127ff.).

7 Pascal (1987, p. 26).

8 Goethe (1977, p. 282).

9 Russell (1994 [1961], p. 149).

10 This is the method employed by, e.g., Sühl-Strohmenger (1984). This scholar's presentation of the most important approaches to educational theory since World War II is followed by his own attempt at a synthesis. This attempt remains somewhat pallid because it claims to result from the theories presented, some of which, however, emanate from contradictory standpoints, and because, in his efforts to compensate for and counterbalance these contradictions, Sühl-Strohmenger arbitrarily aligns the individual insights of widely diverse authors, from W. Flitner to H. v. Hentig.

11 The history of alphabetization is one impressive example of how qualifications imparted by schools had an impact (within the processes of social interaction) on social change and on shaping the future.

12 A (small) sample of the relevant literature: Bundesminister für Bildung und Wissenschaft (1986a), Meisel et al. (1989), Calchera/Weber (1990), Scheilke (1991), Wilsdorf (1991), Projektgruppe Schlüsselqualifikationen in der beruflichen Bildung (1992).

13 Such culturally specific misconceptions include the belief in the fundamental superiority of one's own culture, worldview, or race – ranging from moderate forms of Eurocentrism to such extreme positions as that of Nazi ideology.

14 In more succinct form than in many scholarly publications, this critique can be found in J. Abner Peddiwell's (pseudonym of the American educational theorist Harold H. Benjamin) satire: "The Saber-Tooth Curriculum" (1972 [1938]).

15 This idea has been repeatedly expressed in various forms by philosophers and educators. For example, Martin Buber (1969 [1935], p. 42) notes, "A … contemporary concept of education and personal development must be grounded in the insight that in order to get somewhere it is not enough to move toward some place, but that one must also start from somewhere. And conditions are such that we can determine 'where to' ourselves, but not 'where from'.

16 Mead (1973, p.27) describes a "postfigurative" culture as a society in which transition takes place only very gradually and in which authority is derived from the past. The adults' past is the standard for the new generation's future. By contrast, modern societies represent a "cofigurative" culture, are characterized by learning from peers, and exhibit more pronounced techniques for incorporating change. The new generation expects to differ from the generation(s) preceding it (cf. ibid., p. 61).

17 From his studies on the history of schooling, Paulsen (1885) concluded that schools generally lag one generation behind social development, i.e., according to the definition put forth here, they would not impart any current content at all.

18 To put the proposed classification into operation, a number of details would have to be clarified: Is content still the same if it is didactically and methodically prepared in a different ('more modern') way? Should different learning objectives be taken into account? Does not, e.g., Goethe's ballad "Elf King" represent totally different content depending on whether the poem is to be recited by memory or to be interpreted?

19 One exterior reason for continuity in school curricula can be seen in the fact that, in cases of doubt, the committees responsible for establishing syllabi tend to opt for conservative solutions: If there are no significant objections to doing so, any specific content is also maintained for the future.

20 Löwich's (1989, esp. pp. 51ff) concept of *cultural maturity* includes many elements similar to the concept of *reflective cultural identity* proposed here.

21 An alternative expression of this idea is contained in the quotation from Russell at the beginning of this chapter.

22 One typical proponent of such a scientific orientation is Wilhelm (1982, pp. 66ff; 1985, pp. 143ff).

23 A 1985 survey of the Allensbach Institute demonstrated this. (Bundesmin. f. Bild. u. Wiss. 1986c, pp. 7ff).

24 Many of the arguments critical of scientific orientation mentioned below are similar to those proposed by Menze (1980b) and Klafki (1985b).

25 For an overview, cf. Gerner (1970). Prominent categories of earlier debates such as those of the 'classical', the 'fundamental', the 'representative', and the 'elementary' (cf. Klafki 1964, 1965b), can also be assigned to the concept of the exemplary principle.

26 Klafki also argues in favor of a renaissance of the exemplary principle (1985c).

27 This phrase coined by Wilhelm (1969) expresses the goal to be aimed at fairly adequately and can be upheld even if one does not agree with Wilhelm's own preference for an interpretation stipulating a 'scientific school'.

28 With reference to 'general education as a way of crossing borders', Schulze (1990a, pp. 30ff) discusses transitions that lead "from something more specialized to something more general" as a distinctive feature of a modern general education.

29 Conservative educators are especially inclined to warn of the danger that pluralism might change into a lack of social bonds and of well-defined values. Cf. the proposals of the forum "Mut zur Erziehung" (ZfPäd 24 (1978), pp. 235-240). But the suggested therapy, the return to clearly defined worldviews, is certainly an inadequate response to the problem.

30 This issue, listed first by Klafki in 1985, demonstrates clearly that key issues of this sort cannot be canonized. The dramatic political developments in Eastern Europe have led to a shift in priorities that would have been inconceivable at the time. Currently (1995), one would

be inclined to discuss the (still topical) issue of peace primarily in connection with the wars in Bosnia, in the Caucasus, and in various regions in Africa.

31 The tendencies contrary to the Enlightenment that accompanied the gradual establishment of compulsory education in Germany in the 19th century and aimed at the actual elimination of critical thinking in schools culminated in Ferdinand Stiehl's Regulations in Prussia in 1854 (cf. Reble 1971a, pp. 262-3; 1971b, p. 472).

32 Immanuel Kant, "What is Enlightenment?" in *Perpetual Peace and Other Essays*, tr. Ted Humphrey (Indianapolis: Hackett Publishing Company, 1983), p. 41. The differentiation between understanding and reason (also to be traced to Kant) is not significant in this citation, but will be discussed below.

33 For Wittenberg's (1963, p. 7 and pp. 16ff) idea of general education, 'love of the truth' and correspondingly, a "commitment to the truth" are still major concepts.

34 Cf. Heydorn's publications. The concept of emancipation also acquires a significance far superior to anything else in Blankertz' assertion that the distinctive structure of education is emancipatory and that education is therefore practically identical to emancipation (pointedly formulated, e.g., in Blankertz/Born 1978, pp. 42ff).

35 Others also emphasize that in view of the material and of the formal elements of general education in schools there is a complementary relationship between an understanding of the world and critical thinking. For Hardörfer (1978), e.g., the mutual complementariness of these two poles with reference to the topics 'learning to think' and 'all-encompassing orientation" is the basis of his concept of general education.

36 A quarter of a century ago, many educators were convinced that Marxist social criticism should be imparted in this way (one example of many: Gamm 1972). Fortunately, the conviction (inspired by the tradition of the Enlightenment!) that it is also a case of dictating to and indoctrinating someone even if the indoctrinating person believes to be in possession of a higher truth has regained ground.

37 Hentig (1989, p. 316). For a more thorough pedagogical discussion of the themes "enlightenment" and "reason", see Hentig's book "Die Menschen stärken, die Sachen klären" (Hentig 1985a).

38 Nüse/Goeben/Freitag/Schreier (1991) describe the complications that result from the intermingling of these theoretical levels in radical constructivism.

39 For mathematics instruction, among the first studies of this sort were the following: Bauersfeld (1978) and Voigt (1984b); Maier/Voigt (1994) is an anthology containing more recent studies. These research studies have implemented primarily 'micro-analytic' procedures: videotaped lessons are transcribed and subjected to a qualitative, interpretive analysis. For an introduction and an overview, cf. Maier/Voigt (1991).

40 Within the present context, it is not possible to explore in detail the debate on the importance or necessity of imparting morals or value systems through schooling, a debate that emerged once again during the previous decade. Terhart (1989) provides a concise overview of the issues and standpoints involved. – Since in the present section I refrain from explicit references for the most part, I would like to mention that my ideas on schools' promotion of a willingness to assume responsibility would, as a whole, be compatible with Kohlberg's theory on the development of moral judgment (cf. Kohlberg 1981, 1984, 1986).

41 Cf. Merten (1994), who refutes the assertion (so often carelessly made in the general discussion of the matter) that today's children and adolescents no longer have any values.

42 Picht (1969, p. 331) assigns the status of a fundamental ethical category to the concept of responsibility. "Because there is responsibility, there is also morality and law. But it would be false to assume that, conversely, morality and law would be able to establish responsibility in their own right. In principle, the scope of responsibility is greater than the scope of any possible morality and of any possible law." (Cited in Danner 1985, p. 351). – Jonas (1979) attempts to establish a modern "ethics for our technological age" in strict coherence to the "imperative of responsibility".

43 Erich Weniger called education and personal development (*Bildung*) a "condition, in which one can assume responsibility" (1953, p. 138). Klafki (1965, p. 46) supports the notion of responsibility as an educational category with an impressive list of contemporary pedagogical scholars. – To name a few representative views in the more recent educational discussion: Wilhelm (1985), Danner (1985), and E. E. Geißler (1977, pp. 58ff; 1984, p. 276).

44 In accordance with Max Weber's distinction: From the perspective of an "ethics of conviction" actions are to be ethically judged with reference to how they were 'intended' by the agents; for an 'ethics of responsibility' they are to be judged with reference to their actual effects.

45 Kant's categorical imperative can still be seen as the valid formulation of our moral commitment toward fellow humans.

46 Cf., for instance, Jonas (1979, esp. pp. 245ff) and Meyer-Abich (1984), which contains a critique of Kant's view ("Metaphysics of Morals") that, judged by simple reason, a human would have no other duty than that toward humanity (ibid., pp. 70ff).

47 Suggestions for the practical design of an education oriented to responsibility are contained in many concepts of reform educators, e.g., in that of Kurt Hahn (1958).

48 Handschuh (1982) provides a fundamental examination of problems dealing with the social responsibility of scientists and academics.

49 By contrast to the objections mentioned beforehand, this one is more often proposed by school practitioners than in the specialized literature. In discussions following lectures I have given, I have often encountered this objection.

50 This twofold aspect of avoiding disappointment by apparently abandoning any claim to morality and secretly regretting that one has given up former educational principles over the years of professional practice finds a rather drastic expression in the following self-characterization of a mathematics teacher: "I drum mathematical content into the heads of little devils!"

51 This is the reason why it is such a serious matter if teachers are compelled to such untruthfulness by the political system or believe to be compelled in such a way.

52 For v. Hentig's attempt to "form a new conception of schooling", this is a major demand, one which he numbers among the "minima paedagogica" (Hentig 1993, esp. pp. 216ff).

53 A harmonistic image of society occasionally flashes up in the numerous traditional conceptions of an "education promoting community" devised by the German pedagogical reformist movement (cf., e.g., Scheibe 1969). Unfortunately, history has demonstrated that such concepts can readily be adapted to a nationalist and racist educational ideology. – Conversely, during the 1960s and 70s, when the demand for 'social competence' achieved by learning social skills was high, the emphasis was on furthering one's own interests, on resolving conflicts, and on discussions connected to emancipatory aims. In the midst of efforts stemming from such an attitude of social critique, it was sometimes forgotten that also within the sphere of educational institutions the democratization aimed at requires complementary factors such as tolerance and a willingness to compromise and to form a consensus. The classical textbook on learning social skills edited by Prior (1976) reveals the one-sidedness that resulted from making excessive demands on certain Marxist theories, a practice that was fashionable at the time.

54 Here, it is again easy to perceive the significance of differentiating between (societal) functions and (pedagogical) objectives of schooling, between actual prevailing circumstances and the projected goals of general education in school. It is possible that a socially necessary *function* reemerges in a theoretical concept as an educational *objective* – as, e.g., with the qualifying function that is developed in nearly all educational theories and is postulated here as the objective of preparing for later life. On the other hand, there are objectives (such as the demand for practice in communication and cooperation) that are expressly designed as counterweights to an inevitable societal function of schools, that 'deflect' the respective function, humanize it, and countermand it pedagogically.

55 For more detail on the possibilities and issues involved in intercultural or multicultural education, see Auernheimer (1990) and Nitzschke (1982).

56 There are many and diverse reasons for this – ranging from institutional and bureaucratic limitations of current public schools to the indubitably high demands on the competence, time, and emotional commitments of teachers. On these issues, cf. Kalb/Petry/Sitte (1990).

57 Even today, no significant qualification must be made to the stipulation formulated by Weinert/Treiber (1982, p. 8) two decades ago: "A modern transfer theory, in the sense of a central psychological theme of the research on teaching and learning, is nowhere to be seen."

58 Ulrich Beck (1986, p. 156) succinctly characterizes and caricatures modern deformities of 'self-realization' and of the search for one's own identity as follows: "The result is that people

are more and more forcefully caught up in the labyrinth of self-doubt, of self-questioning, and of establishing self-assurance. The (infinite) regress of questions such as 'Am I really happy?', 'Am I really self-fulfilled?', 'Who am 'I' really?' constantly leads to new fads as responses that, in a variety of ways, can be converted into markets for experts, industries, and religious movements. In search of self-fulfillment, people go on cataloged tours to all corners of the earth. They destroy the best marriages and establish new relationships at an ever-increasing pace. They undergo retraining; they fast; they jog; they change from one therapy group to the next. Obsessed with the aim of self-realization, they uproot themselves so that they might determine whether their roots are actually healthy."

59 A few references from many: Mead (1968 [1934]), Nipkow (1960), Erikson (1966), Krappmann (1971), Loevinger (1976), Döbert et al. (1977), Schweitzer (1985), Kegan (1986), Riedel (1989), Geulen (1989). Lapsley/Power (1988) provide an overview of more recent developments in the Anglo-American countries, including the 'cognitive revolution'.

60 Yet, the conventional literature on Rousseau that only takes that aspect of his thought critical of civilization into account is one-sided. In addition to 'natural' education, Rousseau also conceived of public education promoting responsible citizenship. These competing, sometimes contradictory educational ideals confront each other within Rousseau's writings; and he himself expressed no unequivocal preference for either one (cf. Blankertz 1982, p. 72).

61 An international empirical study by Czerwenka et al. (1990) indicates that disengagement and apathy are particularly pronounced among German school students.

62 Cf. Cohen/Taylor (1977), Rumpf (1976, pp. 143ff). This concept indicates the danger of a reification of traditional concepts such as 'person' and 'individuality'. If something is to be 'worked on', then it is at one's disposal in a way open to manipulation. 'Work' connotes rolling up one's shirtsleeves and clearing things up. – Interactionist identity theories have generally been criticized because of their tendency to dissolve the self socially by overstressing role activity, to consider the self a sort of 'phantom', and to adhere to a 'relativistic view' in the assessment of biographical events (cf. Brumlik 1989, p. 775 and his critique of Krappmann 1971).

63 Cf. Rumpf (1981, 1983, 1987), Fauser/Fintelmann/Flitner (1990).

64 Precisely this issue was one of the major problems in the instructive dispute over a modern form of general education between Wilhelm (1985) and Hentig (1985b). Whereas Wilhelm vigorously defends the necessity of a categorization (if at all possible, along the lines of his own suggestions) of human imagination, Hentig contradicts him by justifiably stipulating: "Therefore, we should not start with the world of science (which would not be a solution to the problem of content) and then assign each discipline a particular educating function, but we should specify the functional objectives and then consider how they can best be achieved" (Hentig 1985b, p. 70).

65 Even a pedagogically committed mathematics teacher like Freudenthal encounters difficulties in appropriately distancing himself from his own subject matter when he emphasizes "that the others who will never apply mathematics should also learn mathematics because they require it in order to become fully human" (Freudenthal 1973, p. 70).

66 Goethe (1977, p. 308).

67 According to notes taken by the author of the present volume.

68 Cited in "Der Spiegel" 41, 1991, p. 352.

69 For an illustration, please refer to note 3 above.

70 These common findings provide a significant argument against justified critiques of the studies to be mentioned: Each of these studies exhibits certain methodological deficiencies which would certainly be substantial if each one were to be considered individually. E.g., consumer and workplace surveys are questionable because statements about the qualification requirements in certain professions are influenced by the questioned persons' experiences with the traditional educational system, by estimations of social relevance, and by subjective opinions; and these influences cannot readily be kept under control.

71 This survey was not published. The people questioned had the following occupations: bookstore owner, lawyer, specialized physician, special educator, librarian, employee with a degree in business management, minister, curate, engineer.

72 This list ignores helping one's own children with their homework. Otherwise, the argument would become circular.

73 This is particularly noteworthy because within a broad range of vocational practice, the use of mathematics as a tool is only rarely of any significance. Usually, if any sort of complicated mathematics were needed for problem solving on the job, this mathematics would be relegated to other functional levels (e.g., to mathematizing machines). – For checking these levels, mathematics is then used more as a *means of communication* than as a *tool*.

74 The aspects that Bardy (1985, pp. 45-46) mentions as relevant to mathematical curricula of vocational schools point in the same direction.

75 Here, we can ignore the question of whether conventional music lessons actually succeed in this respect. On the whole, conceptualizing and realizing meaningful instruction oriented to general education is probably no easier for the subject of music than for the subject of mathematics.

76 This conception of a basic course is very similar to that of the Dutch "Wiskunde A" program. Whereas "Wiskunde B" corresponds to advanced courses in German college-preparatory secondary schools, "Wiskunde A" is designed "for the pre-university students (16-18) in The Netherlands. This curriculum is meant for those students who are not heading for the study of exact sciences like math or physics but who need mathematics as a tool, for instance in social sciences, psychology, economics, biology , etc." (Kindt/Lange 1986, p. 14 and Lange/Kindt 1984). Incidentally, here, the term 'tool' includes mathematics as a 'means of communication'.

77 Damerow (1980, p. 78) parodies this deficit by examining the possible realization of a learning goal set by the overall guidelines for mathematics teaching in the German state of Hesse ("The ability to apply the concepts single-member set and empty set"). He asks, "Is there really a meaningful application of these concepts apart from a mathematical system strictly designed for set theory? Does the word 'apply' in this case mean the application in such comical problems like the one I found in a secondary school textbook: 'Name the members of the set of students in your class who are taller than 3 m!' Should I apply the concepts someday by going into a store and saying, 'I'd like a single-member set of batteries. At home I've only got an empty set of batteries!'"

78 The literature on mathematics teaching offers a wide variety of concepts and examples that cannot be explored here. H. N. Jahnke (1995), e.g., pursues an interesting approach that also includes the history of mathematics. But because of the considerable intellectual demands involved, I suspect that his approach could only become productive for a certain group of students.

79 Schweiger's article was published after I had just finished the first draft of this chapter. Since his line of questioning is not primarily oriented to educational theory and since he attempts no synthesis of the approaches presented, I have not made reference to his article systematically, but only sporadically.

80 These claims exhibit obvious similarities to three criteria which Schreiber (1983, p. 69) presents as the result of a discussion of the topic: The ideas, which he calls "universal", are to be characterized by: "(1) Comprehensiveness (logical universality)"; "(2) profusion (wide-ranging applicability and relevance in individual areas of mathematics)"; "(3) meaningfulness (grounded in everyday thinking, significance for life circumstances)".

81 Gay/Cole (1967) did a study on children of the Kpelle tribe in Liberia, Lancy (1983) and Lean (1986) on Papuans in New Guinea, Harris (1980) on Australian aborigines, Closs (1986) and Pinxten (1983) on Native Americans.

82 Naturally, mathematical or mathematics-related achievements of non-Western cultures can be thematized in mathematics instruction. But, in the final analysis, they only serve to highlight contrasts, to illustrate, or to demonstrate certain features of Western mathematics.

83 Whitehead's article "The Mathematical Curriculum" was originally published in a collection of essays in 1929, but can actually be traced to a speech Whitehead made in 1913.

84 It is quite remarkable that Whitehead makes this distinction, for "Principia Mathematica", that work he coauthored with Bertrand Russell and published in 1913, is one of the most in-depth studies of extensive mathematical literature and is very challenging to read – even for mathematicians.

85 Here, there are obvious similarities to the reforms formulated by Felix Klein, who assigned major significance to "functional thinking", spoke out in favor of physics applications in the mathematics classroom, and campaigned for the firm establishment of infinitesimal calculus in

syllabi for college-preparatory secondary schools. It is possible that Whitehead was familiar with Klein's ideas, which had repercussions on the Meran Plans of 1905.

86 Yet, we can see how progressive Whitehead's thoughts are when he remarks, "In fact a slight study of statistical methods and their application to social phenomena affords one of the simplest examples of the application of algebraic ideas" (ibid., p. 84).

87 Intellectual proximity and mutual respect characterize the association between Wagenschein and Wittenberg. Elements they have in common include: supporting exemplary and genetic teaching, objecting to a strictly specialized position and to imposing on students a systematization primarily stemming from a subject-area orientation, and a skepticism toward the beginning reform of mathematics instruction with a mathematically structural emphasis ("New Mathematics"). Due to these affinities, Lenné (1969, pp. 54ff) felt justified in speaking of an opposition between the "educational concept of Wittenberg and Wagenschein" (as a sufficiently homogeneous type of "traditional" mathematics teaching) and "New Math". Nevertheless, the differences should not be totally neglected: in comparison to Wagenschein, Wittenberg makes more extensive use of systematic considerations of educational theory as a rationale for his didactic insights and suggestions.

88 "If in this book we have primarily spoken of fulfilling this obligation with reference to the most talented students, this must not at all imply that the other children are to be neglected. On the contrary, not least can our study also be assigned this exemplary significance: the example of instruction for the most talented should demonstrate the care with which the goals and realization of any instruction are to be reflected on." (Wittenberg 1963, p. 271)

89 Enumerating the technical terms of the mathematics schoolbooks for the 5th and 6th grades in use at the time, Lörcher (1976) discovered that the number of 'vocabulary items' to be acquired for mathematics learning was on the same scale as for beginning levels of foreign language instruction.

90 It seems that the non-mathematician Bruner was put on the wrong track here by the structurally oriented mathematicians whom he consulted for their expert opinions. This is demonstrated by the way he refers to his algebra example once again in "The Process of Education": "To decide that the elementary ideas of algebra depend upon the fundamentals of the commutative, distributive, and associative laws, one must be a mathematician in a position to appreciate and understand the fundamentals of mathematics" (p. 19). The idea that immediately presents itself to today's reader, namely, that the cognitive psychologist Bruner ("To learn structure, in short, is to learn how things are related", p. 7) and the structural mathematicians consulted might have had very different conceptions of 'structure', seems not to have occurred to Bruner – a misunderstanding that, in various forms, was an attending circumstance during the entire history of 'New Mathematics' in schools.

91 Schweiger (1992, p. 207) uses the term 'archetype' to characterize fundamental ideas. One of his suggested criteria is as follows: "A fundamental idea is a combination of actions, strategies, or techniques that have a corresponding linguistic or action-related archetype in the language and thought of everyday life."

92 This is a publication of the National Research Council of the Mathematical Sciences Education Board, published in a series of books designed to be greatly influential among the general public, e.g.,: "Curriculum and Evaluation Standards" (National Council of Teachers of Mathematics 1989) and "Everybody Counts" (Mathematical Sciences Education Board 1989).

93 I use the concept of model in much the same way as Stachowiak, who takes both everyday and scientific usage into consideration and generally assigns three characteristics to models (1973, p. 131ff): (1) "Models are always models of something, that is, illustrations or representations of natural or artificial originals which, themselves, can also be models" (illustrative feature); (2) "in general, models do not include all of the attributes of the original they represent, but only those which seem relevant to the respective model designers and/or users" (abbreviative feature); (3) "models are not per se unequivocally assigned to their originals. They carry out their replacement function a) for certain human subjects engaged in thinking and/or acting, in employing the model, b) within certain time intervals, and c) limited to certain cognitive or actual operations" (pragmatic feature).

94 Within mathematics, the concept of model can also prove very useful: The "number line" is a geometric model of whole, rational, or real numbers; "pie charts" can be used for computations involving fractions; the vector space of position vectors can be interpreted as a

model of Euclidean space, etc. In many cases, the connections that can be established across topics within school mathematics represent interconnections among the main ideas examined above.

95 Some examples: Becker et al. (1979, 1983), Weber (1980), Burscheid (1980), Blum (1985), Winter (1985), Kaiser-Meßmer (1986, 1989), Schupp (1988), Blum et al. (1989), Blum/Niss/Huntley (1989), Austin (1991), Mason/Davis (1991), Niss et al. (1991), Blum (1993), Volk (1995).

96 Cited in Schupp (1988, p. 5). According to Schupp, even as late as 1922, the German committee on mathematics and science education supported greater emphasis on mathematics' "value for reality" and demanded that "students acquire the overall impression of a science applicable in many other branches of knowledge and under the conditions of everyday life" (cited in Schupp 1988, p. 6).

97 In 1925, Richert's guidelines for secondary schools in Prussia warned that actual computation should by no means be neglected by placing more emphasis on concretizations (cf. Schupp 1988, p. 6).

98 E.g., understanding that there are infinitely many prime numbers (Wagenschein 1965, pp. 102ff), Pythagoras's theorem (Wittenberg 1963, pp. 128ff).

99 Apart from the relevant articles in specialist journals (*Der Mathematikunterricht*, *mathematik lehren*, *mathematica didactica*, *Der Mathematisch-Naturwissenschaftliche Unterricht*, *Didaktik der Mathematik*, *Mathematik in der Schule*), the "MUED" (*Mathematik-Unterrichts-Einheiten-Datei*) also deserves to be mentioned. This is an initiative of dedicated mathematics teachers who are especially concerned with treating socially relevant applications in the classroom (MUED 1994a/b. The organization can be contacted at the following address: MUED e. V., Bahnhofstr. 72, 48301 Appelhülsen).

100 More and more scholars are inclined to take an integrative standpoint. For example, Schupp (1988, p. 13) declares applications orientation to be a necessary component of any mathematics instruction. For "students should feel comfortable with fundamental, often recurring and important situations of everyday life in which mathematical models provide help (material goal); they should acquire abilities to behave in a meaningful way even in complex situations (formal goal); finally, they should have a coherent view of the essence, the potential, and the limitations of mathematical methods (a goal pertinent to philosophy of science)". These three goals can be assigned to the three sets of goals listed above.

101 Cf. the reference to the MUED above.

102 Most likely, the belief in this model of learning persists among mathematics teachers because their own experience seems to substantiate it. The content of the standard mathematical curriculum has become a firm 'possession' of theirs, not susceptible to loss. They often do not reflect on the fact that their subjective impression of having this knowledge readily at their disposal depends on their constantly refreshing this content during the course of their professional lives as teachers. Incorrectly, they view their own mathematical knowledge as static, not as dynamic. The difficulties that many mathematics teachers have in adjusting to content that is also *new* to them is demonstrated, e.g., by their hesitancy to treat themes related to probability. For these reasons, a certain fluctuation of content in the mathematical school curriculum is desirable from a pedagogical perspective. Teachers then have the chance to once again experience and reflect on their own insecurities, doubts, and mental blocks (perhaps guided by sensitive teacher training programs?) – experiences that are quite usual for many students.

103 Cf. the educational literature on project teaching. To list only a few sources: Frey (1982), Gudjons (1989), Emer (1991); on mathematics teaching: Münzinger (1977), Volk (1995). Many of the more interesting suggestions for lesson plans, whether, e.g., from MUED or from the Laboratory School in Bielefeld, proceed from a project type of instruction.

104 An instructional unit on the topic of 'city guides and maps', put to the test for the first time in 1994, was developed for mathematics and social sciences instruction (with the subject geography integrated) at the double grade level of 5 and 6 (Eulenstein 1994).

105 In the fifth book of "Laws", he writes, "For ... no single branch of educational science possesses so great an influence as the study of numbers: its chief advantage is that it wakes up the man who is by nature drowsy and slow of wit, and makes him quick to learn, mindful and

sharp-witted, progressing beyond his natural capacity by art divine" (Plato 1961 [1926], pp. 387, 389).

106 At some point between 1793 and 1796, he wrote, "Mathematics is, indeed, a wonderful science, but the mathematicians are often not even good enough for the hangman. Mathematics is quite similar to theology. For just as those who zealously engage in this latter art, especially if they happen to have an authoritative position, lay claim to a considerable allowance of holiness or to a closer relationship to God, the so-called mathematician demands to be thought of as a profound thinker, even though among them there are the greatest blockheads one could find, unsuitable for any task that calls for deliberation if it cannot immediately be accomplished by simply connecting signs produced more by routine mechanisms than by reflection." (Lichtenberg 1971, p. 433)

107 Cf., e.g., Jungwirth's (1993, p. 214) report on a more recent survey of adults' attitudes toward mathematics: Of 413 persons who participated in a further education program containing mathematics, 76% agreed that mathematics offers cognitive training that promotes logical and precise thinking. By contrast, only 55% of the same group felt that mathematics is useful in everyday life.

108 Reports on subsequent studies which were able to replicate the essential results of the original study are in: Lochhead (1980), Mestre/Lochhead (1983), Cooper (1986), Franke/Wynands (1991). For commentaries and theoretical considerations of the phenomenon, see: Fischer/Malle 81985, pp. 33ff), Philipp (1992), Gardner (1993, pp. 202ff), and Malle (1993, pp. 93ff).

109 For ninth graders who were supposed to solve a similar problem, Franke and Wynands (1991, p. 682) list the following averages for the percentage of correct answers (according to type of school): college-preparatory secondary school (*Gymnasium*): 26% (N=224); secondary modern school (*Realschule*): 17% (N=210); comprehensive school (*Gesamtschule*): 10% (N=52); *Hauptschule*: 8% (N=99).

110 Malle (1993, pp. 93ff) provides an informative overview of various interpretations of the 'Rosnick-Clement-phenomenon', lists further sources, and documents the replication of the phenomenon.

111 Other forms of understanding, such as 'understanding' someone personally or generally understanding a language, can be overlooked here.

112 Cf. Skemp (1976), Backhouse (1978), Haylock (1982), Herscovics/Bergeron (1983), Resnick et al. (1987), Maier (1988, 1991, 1994), MacGregor (1991), Stern (1992), Hiebert/Carpenter (1992), Vollrath (1993), Sierpinska (1994).

113 B. Smith (1975, pp. 10ff); Schoenfeld (1989, pp. 100ff and. 1992); Maier (1991); MacGregor (1991).

114 Both theoretical perspectives involve problems that I would only like to briefly indicate here. Representational theory evades the following issue: the postulated external structures can, in the final analysis, only be communicated by means of internal representations (e.g., of scientifically trained people); empirical studies of this type tend to assume the omniscience of the researcher who compares the reconstructed cognitive structures of the learners with the 'reality' he or she is familiar with (cf. Stern 1992). Moreover, with the representation of knowledge structures in conceptual networks, the dynamic character of understanding, the immediacy and fluid nature of cognitive activities can easily be lost. – The constructivist perspective manages to avoid these problems, but encounters a problem of a different sort when it attributes the 'construction of meaning' to the individual (e.g., Maier 1991, p. 55; Bauersfeld 1993, p. 245 – perhaps the misunderstanding results simply from an abbreviated manner of speech): The experience of meaning (or of understanding) cannot be directly influenced. Speaking of an individual who constructs meaning induces one to conceive of an individual who could arbitrarily construct meaning for himself or herself. But this is contrary to everyday experiences – of learners, e.g., who try in vain to comprehend the meaning of a mathematical statement. – The most promising recent German research project on processes of understanding in the classroom explicitly attempts to relate these theoretical strands to each other as complementary perspectives (Maier 1988, 1991, 1994).

115 Cf. the statement made by an education student cited at the beginning of the introduction.

116 Of course, all of this does not imply that I find it meaningful to simply formally list such rules. M. Winter (1994) forcefully describes a rather depressing scene from everyday classroom

activity that demonstrates how formal mastery of computational principles can become totally independent of meaningful computation.

117 In elementary school teaching, inventing 'stories' for arithmetical expressions is a widespread practice. Herget (1993) offers a number of examples for 'inverted' problems (and for stimulating 'divergent' problems in general) for grade levels up to the upper grades in secondary schools.

118 Wittenberg (1963, p. 68 and passim) can be understood as arguing in favor of continuity.

119 There is no standardized terminology here: Following Kant, Wagner (1959, p. 179), for example, speaks of regulative principles (cited in Müller 1985, p. 205), Polya (e.g., 1962, 1963, 1980) of "(plausible and demonstrable) methods of forming conclusions", Bauer (1978) of "general mathematical processes". The element common to all of these cases is that they deal with intellectual activities that, on the one hand, do not seem immediately connected to particular content and, on the other hand, encompass 'more' than simply forming logical conclusions.

120 With the increasing intellectual demands of the mathematics to be learned, even 'mathematically more talented' persons can encounter situations in which they can no longer make connections to their familiar world of experience (that also encompasses the mathematics they have mastered and 'understood'). Since in conventional mathematics studies, one refrains much more extremely from making explicit connections to prior experience in more elementary mathematics (and neglects non-mathematical prior experiences even more) than in school classrooms, many young people who, up to this point, have considered themselves mathematically talented, begin to doubt their own abilities and to question the value of their major subject, which they originally selected because they were so interested in its content. The autobiographical reports of mathematics students contained in Fischer/Malle (1985, pp. 325ff) offer impressive evidence of such misgivings.

121 Berg's (1990) project "Instruction in the Art of Teaching" testifies to an actual Wagenschein renaissance. See also: Hentig (1989); Vollrath (1989), Ramseger (1991, pp. 162ff); Redeker (1993).

122 This demand is implicit throughout all of Wagenschein's work. In addition to the publications listed in the text above, it is very succinctly expressed in Wagenschein (1965, p. 419 and 1970, pp. 175ff).

123 Cf. Zimmermann (1991). Especially for mathematics teaching in the United States, 'problem solving' is often cited as the guiding principle of mathematics instruction. See, for instance, Polya (1980), Schoenfeld (1989, 1992), and the "Curriculum and Evaluation Standards" of the National Council of Teachers of Mathematics (1989) – a particularly representative and influential publication. On Australia, cf. Stacey (1991).

124 Cf. especially Winter (1989), who, in turn, repeatedly refers to Wagenschein and discusses Socratic teaching (ibid., pp. 12ff, pp. 24-25) and, in certain respects, genetic teaching as prerequisites of discovery learning.

125 A collection and discussion of substantial objections to Wagenschein's (and Wittenberg's) teaching concept, still worth reading today, is contained in Lenné (1969, pp. 62ff). Part of Lenné's argument can now be seen as a reflection of its time. Notable are the high priority that Lenné, a collaborator of Robinsohn, assigns to increasing the utilitarian effectiveness of instruction (cf. his objections to Wagenschein's example of dealing with the topic of infinitely many prime numbers on grounds of the amount of time required, pp. 64-65) and his unbridled optimism in relying on empirical studies as a means of clarifying all currently unresolved teaching problems.

126 For this reason, this topic was treated systematically within the scope of the present text in reference to the problem of encyclopedic knowledge. In his later publications, Wagenschein made the exemplary principle subordinate to the genetic principle (cf. 1975, p. 4 and pp. 55ff).

127 This main idea is so often hidden behind various detailed elaborations of the genetic principle that I consider it worthwhile to expose it and express it with this simplicity. I am disregarding all of the ideas on whether and to what extent genetic teaching should be oriented to the historical development of specific scientific disciplines, to what extent it might be 'didactically simplified', etc. On all of these and similar issues, consult the relevant specialized publications as listed and discussed by Schubring (1978).

128 Roth (1969, p. 116; [1949]) is not explicitly discussing the 'genetic' principle here, but the principle of 'original encounter'. The differences between the two principles can be neglected with reference to the aspect illuminated by the citation above.

129 Perhaps Wagenschein's success as a teacher is, in part, to be credited to the fact that he intuitively understood how to make his teaching stimulating at the informal level. This would also explain why his suggestions on classroom instruction seem so 'unrepeatable': What he explicitly reports is only one particular segment of the entire instructional activity!

130 Lenné (1969, p. 66) made early mention of this point.

131 This distinction stems from Jürgen Mittelstraß, who contends that in addition to the technical knowledge grounded in 'technical rationality' and necessary for technological civilizations such as ours, both older and more modern societies require background knowledge providing orientation (1982, pp. 61-62). But, according to Mittelstraß, this type of knowledge is on the decline: "Modern industrial societies increasingly offer only (potential) technical knowledge of nature and society, no longer background knowledge providing orientation in natural surroundings and in society. Under the paradigm of technical rationality imposed on them by society, the sciences allow the critical potential of orientation to degenerate" (ibid., p. 7).

132 The organizing principle of the new syllabus for comprehensive schools in North Rhine-Westphalia is no longer the conventional systematization of school mathematics, but a (relatively loose) sequence of interconnected themes that allows for a spiral curricular involvement with important mathematical concepts.

133 Intriguing examples are listed in Krämer (1991), Köhler (1992), and Dewdney (1994).

134 Discussions with students at the upper grade levels of secondary schools could be supported, e.g., by Davis/Hersh (1986).

135 "In principle" is, of course, not to be misunderstood as a recommendation to limit oneself to such simple examples. There is certainly nothing wrong with complicated and mathematically challenging projects as long as the preconditions for their realization are given. But if this is questionable, it would seem more important to detect or produce more simple occasions for promoting critical thinking and to make consistent use of such occasions.

136 Cited in: "*Der Spiegel*" 41, 1991, p. 352. This example is also mentioned in the new syllabus for college-preparatory secondary schools in North Rhine-Westphalia – as one of many examples of examination problems in which not simply the mastery of a technique is assessed, but, instead, the emphasis is on deliberate and critical treatment of the mathematics learned (Kultusministerium des Landes Nordrhein-Westfalen 1993, p. 75).

137 "Neue Westfälische" (Bielefeld), issue of Sept. 5, 1994, p. 1 and 2.

138 This tension can also be observed in natural science subjects, which – like mathematics – contend to be primarily committed to a claim of "objectivity". Since there is a relatively high number of lesson hours involved in mathematics, perhaps the consequences here are even more serious.

139 Since the beginning of the 1980s, this concept, which can be traced to Herbart (1965 [1806], p. 22 [A2]), has often found its way as a guiding principle into syllabi and pedagogical guidelines. As such, it has at times been subject to considerable misconception – as Ramseger (1991) demonstrates.

140 Maier (1991, p. 118; 1994, pp. 7ff) has a somewhat different emphasis and speaks of "collectively working out mathematics instruction".

141 For many other subjects, the situation is very similar. In a survey of 181 lessons involving 81 teachers of German, social studies, and natural sciences at the lower grade levels of college-preparatory secondary schools, Hage et al. (1985) determined the rate of teacher-centered instruction as more than 75%; the rate of dialog-oriented instruction was nearly 50% (cited in Meyer 1987, pp. 60-61). Lukesch/Kischkel (1987) summarize the results of empirical studies from 1965 to 1985, in which the observed rate of teacher-centered instruction was between 60% and nearly 90%. For their own survey (German, English, mathematics, and physics at college-preparatory secondary schools), they estimate the time spent on teacher-centered instruction at more than 75%.

142 Moreover, on the basis of my experience with teachers, I suspect that those who design questioning instruction with didactical skill and sensitivity toward their students are often precisely the teachers who frequently alter the form of interaction and who plan phases of instruction that are not teacher-centered. If this is the case, then there is little reason to hope

that the quantitative predominance of teacher-centered instruction is often compensated by high quality teaching.

143 The concept 'culture of instruction' was not used at the time, but it represents what is meant very aptly, for it thematizes both the interaction structure and what is indirectly and, to a certain extent, unintentionally learned through the collective treatment of the mathematical subject matter.

144 Liebau/Huber (1985) and Huber (1990, 1991a, 1991b) offer informative overviews.

145 Nolte-Fischer (1989), who examines the comparable postsecondary socialization of science teachers, also refers to the somewhat dated studies by Bürmann, Reiss, and Libau/Huber. There are more recent studies on attitudes, subjective theories, and occupational knowledge, although these are more concerned with quantifying existing conditions than with socialization theory and analysis of causes. One example is Tietze's (1990) study of teachers at college-preparatory secondary schools, a study that supports some of the statements on trends reported below.

146 It should at least be mentioned that within postsecondary mathematics there are efforts to change this situation. With his concept of "general mathematics", Rudolf Wille attempts to break down the social isolation of mathematics and to establish new emphases in postsecondary mathematics teaching. Cf. Wille (1995).

147 I will never forget a conversation I had during work on my thesis for state certification as a mathematics teacher. I had wanted to discuss a didactic issue with my adviser, who remarked, "Don't fool yourself! The students don't want your pedagogical concern; they want your knowledge."

148 The existing research on the topic 'women and mathematics' has not yet sufficiently concerned itself with this issue (cf. Jungwirth 1994b).

149 Cf. note 120 above.

150 This statement applies largely independently of how mathematics is otherwise characterized: as a "device with a use value for the individual", as an "esoteric teaching", as a "computational devise in the service of the community" or as a "logically sound structure of thought" (Jungwirth 1994a, pp. 77-78, survey of adults in which factor analysis was implemented). – The results of a survey on "mathematical conceptions of the world" among novice mathematics and chemistry students might signal a change in this trend (Törner/Grigutsch (1994)): Contrary to the researchers' expectations, about 70% of the students declared that they did not view mathematics simply as a system, but also as a process involving ideas, concepts, and interconnections between concepts.

151 54% of the students attending college-preparatory secondary schools surveyed by Brückner et al. view mathematics as a "subject as unsuitable as possible for students who like to engage in discussions" (Brückner et al. 1983, p. 229).

152 For a report on teacher experience, see Meisner (1995). For more general information on the topic, cf. Nickson 1992, pp. 109-110).

153 Since these examples are from France, not all of the aspects can be related to the conditions in German schools; in many instances, mathematics instruction in France exhibits, in comparison, even more pronounced features of traditional instruction geared to rote memorization.

154 Hartmut Köhler's work (e.g., 1993) manifests a similar approach, although, to my knowledge, there has been no significant exchange of ideas or, much less, collaboration between him and Andelfinger.

155 Minsky (1990 [1985], p. 193).

REFERENCES

Adorno, Theodor W. *Kritik. Kleine Schriften zur Gesellschaft.* Frankfurt/M.: Suhrkamp, 1971.

- *Studien zum autoritären Charakter.* Frankfurt/M.: Suhrkamp, 1973.
- "Theorie der Halbbildung." In *Bildungstheorien,* edited by Jürgen-Eckhardt Pleines, 89-99. Freiburg: Herder, 1978 [1962].

Alexy, Robert. *Theorie der juristischen Argumentation.* Frankfurt/M.: Suhrkamp, 1978.

Alheit, Peter. *Zivile Kultur.* Frankfurt/M.: Campus, 1994.

Andelfinger, Bernhard. "Sanfter Mathematikunterricht." In *Sanfter Mathematikunterricht - Bildung in der ökologischen Krise,* edited by Bernhard Andelfinger and Hans Schmitt, 23-43. Papenburg: Tagungsbericht, werkstatt: schule + mathe-gesprächskreis aurich, 1989.

Sanfter Mathematikunterricht. Bildung in der EINEN WELT. Ulm: werkstatt: schule, 1993.

Auernheimer, Georg. "Zur Bedeutung der Perspektive für einen demokratischen Bildungsbegriff." *Demokratische Erziehung* 5 (1979): 190-200.

- *Einführung in die interkulturelle Erziehung.* Darmstadt: Wiss. Buchgesellschaft, 1990.

Austin, Joe D., ed. *Applications of Secondary School Mathematics.* Reston, VA.: National Council of Teachers of Mathematics, 1991.

Backhouse, John K. "Understanding School Mathematics - A Comment." *Mathematics Teaching* 82 (1978): 39-41.

Baethge, Martin. "Qualifikation - Qualifikationsstruktur." In *Wörterbuch der Erziehung,* edited by Christoph Wulf, 478-484. München : Piper, 1974.

Baireuther, Peter. *Konkreter Mathematikunterricht.* Bad Salzdetfurth: Franzbecker, 1990.

Ballauff, Theodor. *Funktionen der Schule.* Köln/Wien: Böhlau 2, 1984.

- *Pädagogik als Bildungslehre.* Frankfurt: Haag + Herchen, 1986.

Bardy, Peter. "Mathematische Anforderungen in Ausbildungsberufen." In *Mathematik in der Berufsschule,* edited by Peter Bardy et al, 37-48. Essen: Girardet, 1985.

Baruk, Stella. *Wie alt ist der Kapitän? Über den Irrtum in der Mathematik.* Basel: Birkhäuser, 1989.

Bauer, Ludwig. *Mathematische Fähigkeiten.* Paderborn: Schöningh, 1978.

Bauersfeld, Heinrich. "Kommunikationsmuster im Mathematikunterricht." In *Fallstudien und Analysen zum Mathematikunterricht,* edited by Heinrich Bauersfeld, 158-170. Hannover: Schroedel, 1978.

- "Hidden dimensions in the so-called reality of a mathematics classroom." *Educational Studies in Mathematics* 11 (1980): 23-41.
- "Subjektive Erfahrungsbereiche als Grundlage einer Interaktionstheorie des Mathematiklernens und -lehrens." In *Lernen und Lehren im Mathematikunterricht,* edited by Heinrich Bauersfeld et al, 1-56. Vol VI of *Untersuchungen zum Mathematikunterricht.* Köln: Aulis, 1983.
- "Warum Interaktionsanalysen in didaktischer Forschung und Lehrerfortbildung?" In *"Habt ihr das immer noch nicht kapiert?" - Fachspezifische Interaktionsanalysen für Schule und Unterricht,* edited by Heinrich Bauersfeld, et al, 10-19. Kassel: Hess. Instit. f. Lehrerfortbildung, 1986.
- "Mathematische Lehr-Lern-Prozesse bei Hochbegabten - Bemerkungen zu Theorie und möglicher Förderung." *Journal für Mathematik-Didaktik* 14 (1993): 243-267.
- /Krummheuer, Götz, and Jörg Voigt. "Interactional Theory of Learning and Teaching Mathematics and related Microethnographical Studies." In *Foundations and Methology of the Discipline Mathematics Education,* edited by Hans-Georg Steiner and A. Vermandel, 174-188. Antwerpen: University of Antwerp, 1988.

Beck, Ulrich. *Risikogesellschaft.* Frankfurt/M.: Suhrkamp, 1986.

- *Gegengifte.* Frankfurt/M.: Suhrkamp, 1988.

Becker, Egon, and Wolfgang Ruppert, ed. *Ökologische Pädagogik – pädagogische Ökologie.* Frankfurt/M.: Verl. f. Interkulturelle Kommunikation, 1987.

Becker, Gerhard, et al. *Anwendungsorientierter Mathematikunterricht in der Sekundarstufe I.* Bad Heilbrunn: Klinkhardt, 1979.

- *Neue Beispiele zum anwendungsorientierten Mathematikunterricht in der Sekundarstufe I.* Bad Heilbrunn: Klinkhardt, 1983.

Becker, Hellmut. "Bildung für die Welt von morgen." In *Auf dem Weg zur lernenden Gesellschaft,* edited by Hellmut Becker, 313-332. Stuttgart: Klett-Cotta, 1980.

Belmont, John M., et al. "To Secure Transfer of Training Instruct Self-Management Skills." In *How and How Much Can Intelligence Be Increased?* Edited by Douglas K. Detterman and Robert J. Sternberg, 147-154. Norwood, N.J.: Ablex, 1982.

Benden, Magdalene, ed. *Ziele der Erziehung und Bildung.* Bad Heilbrunn: Klinkhardt 2, 1982.

Bender, Peter. "Umwelterschließung im Mathematikunterricht durch operative Begriffsbildung." *Der Mathematikunterricht* 24 (1978): 25ff.

Berg, Hans C. "Lehrkunst." *Neue Sammlung* 30 (1990).

Bernhard, Arnim, and Dieter Sinhart-Pallin, ed. *Bildung für Emanzipation und Überleben.* Weinheim: Deutscher Studien Verlag, 1989.

Biehler, Rolf, et al, ed. *Mathematik allgemeinbildend unterrichten: Impulse für Lehrerbildung und Schule.* Köln: Aulis, 1995.

Bishop, Alan J. *Mathematical Enculturation.* Dordrecht: Kluwer, 1988.

Blankertz, Herwig. *Geschichte der Pädagogik. Von der Aufklärung bis zur Gegenwart.* Wetzlar: Büchse der Pandora, 1982.

- /Born, Wolfgang. "Auf dem Wege zu einem (Minimal-)Konsens in der Didaktik." In *Didaktische Trends,* edited by Wolfgang Born and Gunter Otto, 26-47. München: Urban & Schwarzenberg, 1978.

Blum, Werner, ed. *Anwendungen und Modellbildung im Mathematikunterricht. Beiträge aus dem ISTRON-Wettbewerb.* Hildesheim: Franzbecker, 1993.

- et al, ed. *Applications and Modelling in Learning and Teaching Mathematics.* Chichester: Ellis Horwood, 1989a.
- et al, ed. *Modelling, Applications and Applied Problem Solving. Teaching Mathematics in a Real Context.* Chichester: Ellis Horwood, 1989b.
- /Sträßer, Rudolf. Mathematics *Teaching in Technical and Vocational Colleges - Professional Training versus General Education.* Occasional Paper 132 Bielefeld: IDM, 1992.

Bodenheimer, Aron R. *Warum? Von der Obszönität des Fragens.* Stuttgart: Reclam, 1984.

Borovcnik, Manfred et al. "Mathematik in der beruflichen Praxis." Vol. V of *Schriftenreihe Didaktik der Mathematik. Univ. Klagenfurt.* Wien/Stuttgart: Hölder-Pichler-Tempsky/Teubner, 1981.

Bracht, Ulla, and H. Zimmer. "Die neokonservative Allgemeinbildungs-diskussion und die Erziehungswissenschaft." In *Rezeption und Verwendung erziehungswissenschaftlichen Wissens in pädagogischen Entscheidungsfeldern,* edited by Eckard König and Peter Zedler, 229- 250. Weinheim: 1989.

Brückner, P., et al. "Motivation und Einstellung zum Beruf des Gymnasiallehrers im Fach Mathematik und in den naturwissenschaftlichen Fächern." In *Die heile Welt der Wissenschaft,* edited by Rainer Brämer and Georg Nolte, 209-236. Marburg: Redaktionsgemeinschaft Soznat, 1983.

Brumlik, Micha. "Interaktionismus, Symbolischer." In Vol I of *Pädagogische Grundbegriffe,* edited by Dieter Lenzen, 764-781. Reinbek: Rowohlt, 1989.

Bruner, Jerome S. *The Process of Education.* Cambridge, Mass.: Harvard University Press, 1965 (1960).

- *Entwurf einer Unterrichtstheorie.* Berlin: Berlin-Verlag, 1974.

Buber, Martin: *Reden über Erziehung.* Heidelberg: Schneider, 1969 [1935].

Buddrus, Volker, and Gerhard W. Schnaitmann, ed. *Friedenspädagogik im Paradigmenwechsel. Allgemeinbildung im Atomzeitalter.* Weinheim: Deutscher Studien-Verlag, 1991.

Bundesminister für Bildung und Wissenschaft, ed. *Schlüsselqualifikationen und Weiterbildung.* Bad Honnef: Bock, 1986a.

- ed. "Bildung heute. Bedeutung und Anerkennung in der Gesellschaft." In Vol. XXIX of *Schriftenreihe Studien zur Bildung und Wissenschaft,* Bad Honnef: Bock, 1986c.

Bürmann, J. "Der "typische Naturwissenschaftler", ein intelligenter Versager?" In *Fachsozialisation im mathematisch-naturwissenschaftlichen Unterricht,* edited by R. Brämer, 33-61. Marburg: 1977.

Burscheid, Hans J. "Beiträge zur Anwendung der Mathematik im Unterricht" *Zentralblatt für Didaktik der Mathematik* 12 (1980): 63-69.

Bussmann, Hans, and Heymann, Hans W. "Bildung mit dem Computer – LOGO kritisch hinterfragt." *Zeitschr. f. Sozialisationsforschung und Erziehungssoziologie* 5 (1985): 239-254.

- /Heymann, Hans W. "Computer und Allgemeinbildung." *Neue Sammlung* 27 (1987): 2-39.

Calchera, Franco, and Johannes C. Weber. "Entwicklung und Förderung von Basiskompetenzen/ Schlüsselqualifikationen." *Berichte zur beruflichen Bildung* 116. Berlin/Bonn: Bundesinstitut für Berufsbildung, 1990.

Closs, Michael P., ed. *Native American Mathematics.* Austin: Univ. of Texas, 1986.

Cobb, Paul, et al. "A Constructivist Alternative to the Representational View of Mind in Mathematics Education." *Journal for Research in Mathematics Education* 23 (1992): 2-33.

Cockroft, W. H., et al. Mathematics counts ("Cockroft-Report"). In *Report of the Committee of Inquiry into Teaching of Mathematics in School under the Chairmanship of Dr W H Cockroft.* London: Her Majesty's Stationery Office, 1982.

Cohen, St., and Taylor, L. *Ausbruchsversuche. Identität und Widerstand in der modernen Lebenswelt.* Frankfurt/M.: 1977.

College de France. "Vorschläge für das Bildungswesen der Zukunft." In *Das Bildungswesen der Zukunft,* edited by Sebastian Müller-Rolli, Stuttgart: Klett-Cotta, 1987.

Cooper, M. "The Dependance of Multiplicative Reversal on Equation Format." *Journal of Mathematical Behavior* 5 (1986): 115-120.

Czerwenka, Kurt, et al. *Schülerurteile über die Schule.* Frankfurt/M.: Lang, 1990.

Dahmer, Ilse, and Wolfgang Klafki. *Geisteswissenschaftliche Pädagogik am Ausgang ihrer Epoche - Erich Weniger.* Weinheim: Beltz, 1968.

Damerow, Peter: *Die Reform des Mathematikunterrichts in der Sekundarstufe I. Eine Fallstudie zum Einfluß gesellschaftlicher Rahmenbedingungen auf den Prozeß der Curriculum-Reform.* Stuttgart: Klett-Cotta, 1977.

- "Wieviel Mathematik braucht ein Hauptschüler?" *mathematica didactica* 3 (1980): 69-86.
- "Mathematics for all - ideas, problems, implications." *Zentralblatt für Didaktik der Mathematik* 3 (1984): 81-85.
- et al. *Lernen für die Praxis? Ein exemplarischer Versuch zur Bestimmung fachüberschreitender Curriculumziele.* Stuttgart: Klett, 1974.

Danner, Helmut. *Verantwortung und Pädagogik.* Königstein: Hain 2, 1985.

Davidson, Neil, ed. *Cooperative Learning in Mathematics. A Handbook for Teachers.* Menlo Park et al: Addison-Wesley, 1990.

Davis, Philip J., and Reuben Hersh. *Erfahrung Mathematik.* Basel/Boston/Stuttgart: Birkhäuser, 1986.

Derbolav, Josef. *Pädagogik und Politik.* Stuttgart et al: Kohlhammer, 1975.

- ed. *Kritik und Metakritik der Praxeologie.* Kastellaun: Henn, 1976.

Deutscher Bildungsrat. *Empfehlungen der Bildungskommission. Strukturplan für das Bildungswesen.* 4th ed. Stuttgart: Klett, 1972.

Dewdney, Alexander K. *200 Prozent von nichts. Die geheimen Tricks der Statistik und andere Schwindeleien mit Zahlen.* Basel et al: Birkhäuser, 1994.

Dichgans, Johannes. "Die Plastizität des Nervensystems. Konsequenzen für die Pädagogik." *Zeitschrift für Pädagogik* 40 (1994): 229-246.

Dick, Lutz van, ed. *Lernen in der Friedensbewegung.* 2nd ed. Weinheim: Beltz, 1987.

- et al, ed. *Ideen für Grüne Bildungspolitik.* Weinheim: Beltz, 1986.

Döbert, Rainer et al, ed. *Entwicklung des Ichs.* Köln: Kiepenheuer & Witsch, 1977.

Duncker, Karl. *On Problem-Solving.* Washington: American Psych. Ass., 1945.

Dweck, Carol S., and Elaine S.Elliot "Achievement Motivation". In Socialization, Personality and Social Development, edited by E. Mavis Hetherington, 643-691. Vol. IV of *Handbook of Child Psychology.* 4th ed. New York et al: Wiley, 1983.

Emer, Wolfgang, ed. *Wie im richtigen Leben ... . Projektunterricht für die Sekundarstufe II.* Bielefeld: Oberstufenkolleg, 1991.

Erichson, Christa. *Von Lichtjahren, Pyramiden und einem regen Wurm. Erstaunliche Geschichten, mit denen man rechnen muß.* Hamburg: Verl. für päd. Medien, 1992.

- "Sachtexte statt Textaufgaben - Ein fächerübergreifender Ansatz zur Erschließung der Verschrifteten Umwelt." In *Beiträge zum Mathematikunterricht 1993.* 116-119. Hildesheim: Franzbecker, 1993.

Erikson, Erik H. *Identität und Lebenszyklus.* Frankfurt/M.: Suhrkamp, 1966.

Eulenstein, Kerstin. *Stadtpläne und Landkarten. Eine Unterrichtseinheit für den Mathematik- bzw. Sowiunterricht der Jahrgangsstufen 5/6.* Bielefeld: Laborschule Bielefeld - Wissenschaftliche Einrichtung an der Universität Bielefeld, 1994.

Faller, Kurt. "Verfassungsauftrag und demokratische Bildungsinhalte." *Demokratische Erziehung* 5 (1979): 181-188.

Fauser, Peter, et al, ed. *Lernen mit Kopf und Hand. Berichte und Anstöße zum praktischen Lernen in der Schule.* 2nd ed. Weinheim/Basel: Beltz, 1990.

Fend, Helmut. *Gesellschaftliche Bedingungen schulischer Sozialisation.* Weinheim/Basel: Beltz, 1974.

- *Theorie der Schule.* München/Wien/Baltimore: Urban & Schwarzenberg, 1980.

Fetscher, Iring. "Krise der Gesellschaft und Zukunft der Bildung". In *Bildung für eine menschliche Zukunft,* edited by Jochen Schweitzer, 19-27. Weinheim and München: Juventa, 1986.

Fischer, Roland. "Fundamentale Ideen bei den reellen Funktionen." *Zentralblatt für Didaktik der Mathematik* 8 (1976): 185-192.

- /Malle, Günther. *Mensch und Mathematik. Eine Einführung in didaktisches Denken und Handeln.* Mannheim et al.: BI & Wissenschaftsverlag, 1985.

Fitzgerald, A., and K. M. Rich. *Mathematics in Employment* (16 - 18). Bath: University-Report, 1981.

Flitner, Wilhelm. *Das Selbstverständnis der Erziehungswissenschaft in der Gegenwart.* 4th ed. Heidelberg: Quelle & Meyer, 1966.

Floer, Jürgen. "Geometrie und Umwelterschließung im Mathematikunterricht." *Die Grundschule* 19 (1987): 52-55.

Fölsch, G. "Die Stellung des Mathematikunterrichtes an allgemeinbildenden Schulen." *Zentralblatt für Didaktik der Mathematik* 7 (1975): 155-159.

Franke, Marianne, and Alexander Wynands. "Zum Verständnis von Variablen – Testergebnisse in 9. Klassen Deutschlands." *Mathematik in der Schule* 29 (1991): 674-691.

Freudenthal, Hans. *Mathematik als pädagogische Aufgabe.* (2 volumes) Stuttgart: Klett, 1973.

Frey, Gerhard. *Die Mathematisierung unserer Welt.* Stuttgart et al: Kohlhammer, 1967.

Frey, Karl. *Die Projektmethode.* Weinheim/Basel: Beltz, 1982.

- /Isenegger, Urs. "Bildung curricularer Sequenzen und Strukturen." In Vol. II of *Curriculum-Handbuch,* edited by Karl Frey, 158-164. München/Zürich: Piper, 1975.

Friedrich-Verlag, ed. "Bildschirm - Faszination oder Information." *Friedrich Jahresheft* 3. Velber: 1985.

Führer, Lutz. "Ich denke, also irre ich. Anfänge und Grenzen der Fehlerkunde." *mathematik lehren* 5 (1984): 2-9.

Gallin, Pete, and Urs Ruf. *Sprache und Mathematik in der Schule. Auf eigenen Wegen zur Fachkompetenz.* Zürich: Verlag Lehrerinnen und Lehrer Schweiz, 1990.

- "Sprache und Mathematik in der Schule. Ein Bericht aus der Praxis." *Journal für Mathematik-Didaktik* 14 (1993a): 3-33.

- "Lernen durch Schreiben - auch in Mathematik." In *Beiträge zum Mathematikunterricht 1993,* 14-21. Hildesheim: Franzbecker, 1993b.

- "Singuläre Schülertexte als Basis eines allgemeinbildenden Mathematikunterrichts." edited by Rolf Biehler et al, 58-82. (1995).

Gamm, Hans-Jochen. *Das Elend der spätbürgerlichen Pädagogik.* München: List, 1972.

Gardner, Howard. *Der ungeschulte Kopf. Wie Kinder denken.* Stuttgart: Klett-Cotta, 1993.

Gay, J., and M. Cole. *The New Mathematics in an Old Culture.* New York: Holt et al, 1967.

Gehlen, Arnold. *Anthropologische Forschung.* Reinbek: Rowohlt, 1961.

Geissler, Georg. "Die Situation der Schule in der Gegenwart." In *Theorie der Schule,* edited by Hermann Röhrs, 168-182. Frankfurt/M.: Akademische Verlagsgesellschaft, 1968.

Geißler, Erich E. *Allgemeinbildung in einer freien Gesellschaft.* Düsseldorf: 1977.

- *Die Schule - Theorien, Modelle, Kritik.* Stuttgart: 1984.

Gerner, Berthold. *Das exemplarische Prinzip.* 4th ed. Darmstadt: Wiss. Buchgesellschaft, 1970 [1963].

Geulen, Dieter. *Das vergesellschaftete Subjekt.* 2nd ed. Frankfurt/M.: 1989.

Glasersfeld, Ernst v. "Learning as a Constructive Activity." In *Problems of Representation in the Teaching and Learning of Mathematics,* edited by C. Janvier, 3-18. Hillsdale, NJ: Lawrence Erlbaum Ass., 1987a.

- *Wissen, Sprache und Wirklichkeit. Arbeiten zum radikalen Konstruktivismus.* Braunschweig: Vieweg, 1987b.

- *Radical Constructivism in Mathematics Education.* Dordrecht et al: Kluwer, 1991.

Goethe, Johann W. v. Wilhelm Meisters Wanderjahre. Vol. 8 of *Goethes Werke,* Hamburger edition in 14 volumes, München (Beck) 1977 [ca. 1820].

Göppel, Rolf. „Umwelterziehung. Katastrophenpädagogik? Moralerziehung? Ökosystemlehre? Oder ästhetische Bildung?" *Neue Sammlung* 31 (1991), 25-38.

Göstemeyer, Karl Franz. "Pädagogik nach der Moderne? Vom kritischen Umgang mit Pluralismus und Dogmatismus." *Zeitschrift für Pädagogik* 39 (1993): 857-870.

Gudjons, Herbert. *Handlungsorientiert lehren und lernen. Projektunterricht und Schüleraktivität.* Heilbronn: Klinkhardt, 1989.

Hahn, Kurt. *Erziehung zur Verantwortung. Reden und Aufsätze.* Stuttgart: Klett, 1958.

Handschuh, Gerhard. *Die gesellschaftliche Verantwortung der Wissenschaftler.* Frankfurt/M.: Haag + Herchen, 1982.

Hansmann, Otto: *Kritik der sogenannten "theoretischen Äquivalente" von "Bildung",* edited by Otto Hansmann, and Winfried Marotzki, 21-54. 1988.

Hansmann, Otto, and Winfried Marotzki, ed. *Diskurs Bildungstheorie I: Systematische Markierungen.* Weinheim: Deutscher Studien Verlag, 1988.

- ed. *Diskurs Bildungstheorie II: Problemgeschichtliche Orientierungen.* Weinheim: Deutscher Studien Verlag, 1989.

Hardörfer, Ludwig. *Denkenlernen und Gesamtorientierung.* München: Kösel, 1978.

Harris, Pam. *Mathematics in a Cultural Context.* Geelong, Vict.: Deakin University, 1980.

Harten, Gerd v., and Heinz Steinbring. "Lesson transcripts and their role in the inservice training of mathematics teachers." *Zentralblatt für Didaktik der Mathematik* 23 (1991): 169-177.

Haylock, Derek W. "Understanding in Mathematics: Making Connections." *Mathematics Teaching* 98 (1982): 54-56.

Heid, Helmut. „Ökologie als Bildungsfrage?" *Zeitschrift für Pädagogik* 38 (1992): 114-138.

Heitele, D. "An epistemological view on fundamental stochastic ideas." *Educational Studies in Mathematics* 6 (1975): 187-205.

Heitger, Marian. *Pädagogik ohne Bildung - Jugend ohne Perspektive?* Düsseldorf: Bildungswerk der Nordrhein-Westfälischen Wirtschaft e.V., 1979.

Heitkämper, Peter, and Rolf Huschke-Rhein, ed. *Allgemeinbildung im Atomzeitalter.* Weinheim/ Basel: Beltz, 1986.

Helling, F. "Weltverständnis statt Brockenwissen." In *Neue Allgemeinbildung,* edited by F. Helling. Schwelm: 1963.

Hentig, Hartmut v. *Einführung zur 8., ergänzten Auflage von Wagenschein,* (1975): 7-22

- *Magier oder Magister? Über die Einheit der Wissenschaft im Verständigungsprozeß.* Stuttgart: Klett, 1972.
- *Das allmähliche Verschwinden der Wirklichkeit.* München/Wien: Hanser, 1984.
- *Die Menschen stärken, die Sachen klären. Ein Plädoyer für die Wiederherstellung der Aufklärung.* Stuttgart: Reclam, 1985a.
- "Eine Antwort an Theodor Wilhelm." *Neue Sammlung* 25 (1985b): 151-167.
- Das Denken bestimmt die Grenzen des Denkens. *Neue Sammlung* 29 (1989): 315-319.
- *Die Schule neu denken.* 2nd ed. München/Wien: Hanser 1993.

Herbart, Johann Friedrich. Allgemeine Pädagogik, nach dem Zweck der Erziehung abgeleitet." In Vol. II of *Johann Friedrich Herbart: Pädagogische Schriften,* edited by Walter Asmus, 9-155. Düsseldorf: Küpper, 1965 [1806].

Herget, Wilfrid. "Mathe-(Klausur-)Aufgaben – einmal anders?!" In *Wieviel Termumformung braucht der Mensch?,* edited by Horst Hischer, 58-69. Hildesheim: Franzbecker, 1993.

Hering, Hermann. "Spuren der Differentation in der Sekundarstufe I. Genetische Aspekte einer fundamentalen Idee." *Der Mathematikunterricht* 31 (1985): 72-86.

Herscovics, Nicholas, and Jaques C. Bergeron. Models of Understanding. *Zentralblatt für Didaktik der Mathematik* 15 (1983): 75-83.

Heydorn, Heinz-Joachim. "Ziele und Inhalte demokratischer Erziehung und Bildung." *Demokratische Erziehung* 5 (1979): 176-181.

Heymann, Hans W. "Didaktisches Handeln im Mathematikunterricht aus Lehrersicht." In *Analysen zum Unterrichtshandeln,* edited by H. Bauersfeld et al, 141-167. Köln: Aulis, 1982.

- "Mathematikunterricht als schulischer Alltag." In *Mathematikunterricht zwischen Tradition und neuen Impulsen,* edited by Hans W. Heymann 80-115. Köln: Aulis, 1984.
- "Innere Differenzierung im Mathematikunterricht." *mathematik lehren* 49 (1991): 63-66.
- /van Lück, Willi, ed. Vol. XXXVII of *Allgemeinbildung und öffentliche Schule: Klärungsversuche (Materialien und Studien),* Bielefeld: IDM, 1990.
- et al. "Allgemeinbildung als Aufgabe der öffentlichen Schule." In Heymann, H. W., and W. van Lück (1990): 9-20.

Hiebert, James, and Thomas P. Carpenter. "Learning and Teaching with Understanding." In *Handbook of Research on Mathematics Teaching and Learning (NTCM),* edited by Douglas A. Grouws, 65-97. New York: Macmillan, 1992.

Hirsch, E. D., Jr. *What Every American Needs to Know.* Boston: Houghton Mifflin, 1987.

Horkheimer, Max, and Theodor W. Adorno. *Dialektik der Aufklärung.* Frankfurt a.M.: Fischer, 1969.

Howson, Geoffrey: "Motivation and Attainment." *Zentralblatt für Didaktik der Mathematik* 16 (1984): 85-87.

- /Wilson, Bryan, ed. *School Mathematics in the 1990s.* (ICMI Study Series) Cambridge et al.: University Press, 1986.

Huber, Ludwig. "Fachkulturen. Über die Mühen der Verständigung zwischen den Disziplinen." *Neue Sammlung* 31 (1991a): 3-24.

- "Sozialisation in der Hochschule." In *Neues Handbuch der Sozialisationsforschung,* edited by Klaus Hurrelmann and Dieter Ulich, 417-441.Weinheim/Basel: Beltz, 1991b.

Humboldt, Wilhelm v. "Theorie der Bildung des Menschen. Bruchstück." In *Allgemeine Bildung,* edited by Heinz-Elmar Tenorth, 32-38. Weinheim/München: Juventa, 1986 [1790].

Hunger, Edgar. *Mathematik und Bildung.* Braunschweig: 1949.

Hurrelmann, Klaus. *Erziehungssystem und Gesellschaft.* Reinbek: Rowohlt, 1975.

Jahnke, Hans N. "Al-Khwarizmi und Cantor in der Lehrerbildung." In Rolf Biehler et al. (1995), 114-136.

Jahnke, Thomas. "Das Simpsonsche Paradoxon verstehen. Ein Beitrag des Mathematikunterrichts zur Allgemeinbildung." *Journal für Mathematik-Didaktik* 14 (1993). 221-242.

- "Beispiele für Themen in einem allgemeinbildenden Mathematikunterricht an Schule und Hochschule." In Rolf Biehler et al. (1995): 137-151.

Jonas, Hans. *Das Prinzip Verantwortung.* Frankfurt/M.: Suhrkamp, 1979.

Jung, Walter. "Zum Begriff einer mathematischen Bildung." *mathematica didactica* 1 (1978): 161-176.

Jungwirth, Helga. "Zwischen Ehrfurcht und Verdammung - wie Nicht-MathematikerInnen die Mathematik sehen." *Beiträge zum Mathematikunterricht 1993.* 213-216. Hildesheim: Franzbecker, 1993.

- "Erwachsene und Mathematik - eine reife Beziehung?" *mathematica didactica* 17 (1994a): 69-89.

- "Die Forschung zu Frauen und Mathematik: Versuch einer Paradigmenklärung." *Journal für Mathematik-Didaktik* 15 (1994b): 253-276.

- /Steinbring, Heinz, Jörg Voigt, and Bernd Wollring. " Interpretative Unterrichtsforschung in der Lehrerbildung." In *Verstehen und Verständigung,* edited by Hermann Maier and Jörg Voigt, 12-42. Köln: Aulis, 1994.

Kade, Jochen. "Bildung oder Qualifikation? Zur Gesellschaftlichkeit beruflichen Lernens." *Zeitschrift für Pädagogik* 29 (1983): 859-876.

Kahlert, Joachim. "Die mißverstandene Krise. Theoriedefizite in der umweltpädagogischen Diskussion." *Zeitschrift für Pädagogik* 37 (1991): 97-122.

Kaiser-Meßmer, Gabriele. *Anwendungen im Mathematikunterricht,* 2 volumes Bad Salzdetfurth: Franzbecker, 1986.

- "Aktuelle Richtungen innerhalb der Diskussion um Anwendungen im Mathematikunterricht." *Journal für Mathematik-Didaktik* 10 (1989), 309-347.

Kalb, Peter E. et al. *Unterrichten – und was sonst? Zum Berufsverständnis von Lehrerinnen und Lehrern.* Weinheim/Basel: Beltz, 1990.

Kant, Immanuel. "Kritik der reinen Vernunft," Vol. XXXVII a of *Philosophische Bibliothek*, edited by Raymund Schmidt, Hamburg: Meiner, 1956 [1781].

- "Was ist Aufklärung?" In Vol VIII of *Werke,* edited by Königl. Preuß. Akad. der Wissenschaften. Berlin : reprinted 1968 [1784].

Kazis, Cornelia, ed. *Buchstäblich sprachlos. Analphabetismus in der Informationsgesellschaft.* Basel: Lenos, 1991.

Keck, Rudolf W et al, ed. *Fachdidaktik zwischen Allgemeiner Didaktik und Fachwissenschaft.* Bad Heilbrunn: Klinkhardt, 1990.

Kegan, Robert. *Die Entwicklungsstufen des Selbst.* München: Kindt, 1986.

Keitel, Christine. "Mathematik für alle – ein Ziel." *Zentralblatt für Didaktik der Mathematik* 6 (1985): 177-186.

Kern, Peter, and Hans-Georg Wittig. *Pädagogik im Atomzeitalter.* Freiburg/Breisgau: Herder, 1982.

Kindt, Martin, and Jan de Lange Jzn. "Realistic math for (almost) all. The Hewet Project." In *Mathematics for all ... in the computer age (Proceedings of the 37th CIEAEM meeting),* edited by Jan de Lange Jzn, 14-41. Utrecht: University/Researchgroup OW & OC, 1986.

Kirk, Gordon. *The Core Curriculum.* London et al: Hodder and Stoughton, 1986.

Klafki, Wolfgang. *Das pädagogische Problem des Elementaren und die Theorie der kategorialen Bildung.* 3rd/4th ed. Weinheim: Beltz, 1964.

- "Engagement und Reflexion im Bildungsprozeß. Zum Problem der Erziehung zur Verantwortung." In *Studien zur Bildungstheorie und Didaktik,* edited by Wolfgang Klafki, 46-71. 5th/7th ed. Weinheim: Beltz, 1965a.
- "Kategoriale Bildung." In *Studien zur Bildungstheorie und Didaktik,* edited by Wolfgang Klafki, 25-45. 5th/7th ed. Weinheim:Beltz 1965.
- "Konturen eines neuen Allgemeinbildungskonzepts." In *Neue Studien zur Bildungstheorie und Didaktik,* edited by Wolfgang Klafki, 12-30. Weinheim/Basel: Beltz, 1985a. [4th revised edition of 1994; in the present book, citations are from the edition of 1985].
- "Thesen zur "Wissenschaftsorientierung" des Unterrichts." In *Neue Studien zur Bildungstheorie und Didaktik,* edited by Wolfgang Klafki, 108-118. 4th revised edition of 1994, Weinheim/Basel:Beltz, 1985b.
- "Exemplarisches Lehren und Lernen." In *Neue Studien zur Bildungstheorie und Didaktik,* edited by Wolfgang Klafki, 87-107. 4th revised edition of 1994, Weinheim/Basel:Beltz, 1985c.
- "Die Bedeutung der klassischen Bildungstheorien für ein zeitgemäßes Konzept allgemeiner Bildung." *Zeitschrift für Pädagogik* 32 (1986): 455-476.

Kleber, Eduard W., ed. *Grundzüge ökologischer Pädagogik. Eine Einführung in ökologisch-pädagogisches Denken.* Weinheim: Juventa, 1993.

Knauer, Ulrich. *Mathematische Modellierung. Laster, Busse und Schweine im Mathematikstudium.* Braunschweig/Wiesbaden: Vieweg, 1992.

Knox, C. *Numeracy and school leavers. A survey of employer´s needs.* Sheffield: 1977.

Koch, Jens-Jörg. *Lehrer-Studium und Beruf. Einstellungswandel in den beiden Phasen der Ausbildung.* Ulm: Süddeutsche Verlagsgesellschaft, 1972.

Kohlberg, Lawrence. *The Philosophy of Moral Development. Moral Stages and the Idea of Justice.* San Francisco: Harper & Row, 1981.
- *The Psychology of Moral Development. Nature and Validity of Moral Stages.* San Francisco: Harper Row, 1984.
- "Der 'Just Community'-Ansatz der Moralerziehung in Theorie und Praxis." In *Transformation und Entwicklung,* edited by Fritz Oser, Reinhard Fatke, and Otfried Höffe, 21-55. Frankfurt/M.: Suhrkamp, 1986.

Köhler, Hartmut. *Über Relevanz und Grenzen von Mathematisierungen. Anregungen zur Ermöglichung von Bildung im Mathematikunterricht.* Buxheim/ Eichstätt: Polygon, 1992.
- *Bildung und Mathematik in der gefährdeten Welt. Annäherungen an die Wirklichkeit.* Buxheim/ Eichstätt: Polygon, 1993.

Kolakowski, Leszek. "Der Anspruch auf die selbstverschuldete Unmündigkeit." In *Vom Sinn der Tradition,* edited by Leonhard Reinisch, 1-15. München: Beck, 1970.

Krämer, Walter. *So lügt man mit Statistik.* Frankfurt/M.: Campus, 1991.

Krappmann, Lothar. *Soziologische Dimensionen der Identität.* Stuttgart: Klett, 1971.

Krippner, Wolfgang. *Mathematik differenziert unterrichten.* Hannover: Schroedel, 1992.

Krol, Gerd-Jan. "Ökologie als Bildungsfrage? Zum sozialen Vakuum der Umweltbildung." *Zeitschrift für Pädagogik* 39 (1993): 651-672.

Krummheuer, Götz, and Jörg Voigt. "Interaktionsanalysen im Mathematikunterricht." In *Interpretative Unterrichtsforschung,* edited by Hermann Maier and Jörg Voigt, 13-33. Köln: Aulis, 1991.

Kultusminister des Landes Nordrhein-Westfalen, ed. *Richtlinien für den Unterricht in der Höheren Schule. Mathematik.* Ratingen: Henn, 1963.
- ed. *Richtlinien und Lehrpläne für das Gymnasium – Mathematik – Sekundarstufe I.* Frechen: Ritterbach, 1993.

Kütting, Herbert. "Stochastisches Denken in der Schule. Grundlegende Ideen und Methoden." *Der Mathematikunterricht* 31 (1985): 87-106.

Lancy, David F. *Cross-Cultural Studies in Cognition and Mathematics.* New York et al: Academic Press, 1983.

Lange Jzn, Jan van, and Martin Kindt. "The Hewet Project. Report on an Experiment Leading to a New Curriculum for the Pre-University Students." *Zentralblatt für Didaktik der Mathematik* 16 (1984): 74-79.

Lapsley, Daniel K., and Clark F. Power. *Self, Ego, and Identity. Integrative Approaches.* New York et al: Springer, 1988.

Lean, G. A. *Counting Systems of Papua New Guinea.* Lae, Papua New Guinea: Departm. of Math., Papua New Guinea University of Technolgy, 1986.

Lenhardt, Gero. *Berufliche Weiterbildung und Arbeitsteilung in der Industrieproduktion.* Frankfurt/ M.: Suhrkamp, 1974.

Lenné, Helge. *Analyse der Mathematikdidaktik in Deutschland.* Stuttgart: Klett, 1969.

Leschinsky, Achim, and Peter-Martin Roeder. *Schule im historischen Prozeß.* Stuttgart: Klett, 1976.

- /Roeder, Peter-Martin. "Gesellschaftliche Funktionen der Schule." In Vol. III of *Handbuch Schule und Unterricht,* edited by Walter Twellmann, 107-154. Düsseldorf: Schwann, 1981.

Lichtenberg, Georg C. Vol. II of *Schriften und Briefe. Sudelbücher II.* Darmstadt: Wiss. Buchgesellschaft, 1971.

Liebau, Eckart, and Ludwig Huber. "Die Kulturen der Fächer." *Neue Sammlung* 25 (1985): 314-339.

Litt, Theodor. *"Führen" oder "wachsenlassen"?* Leipzig et al: Teubner, 1927.

Lochhead, Jack. "Faculty interpretations of simple algebraic statements: The professor's side of the equation." *Journal of Mathematical Behavior* 3 (1980): 29-37.

Loevinger, Jane. *Ego Development.* San Francisco: Jossey-Bass, 1976.

Lohmann, Knut, ed. *Der Beitrag der Unterrichtsfächer zur Allgemeinbildung.* Rinteln: Merkur, 1990.

Lörcher, Gustav A. "Mathematik als Fremdsprache." *Reflektierte Schulpraxis.* 10th installment. Villingen: Neckar-Verlag, 1976.

- "Allgemeinbildender und berufsbildender Mathematikunterricht. Diskrepanzen und Koordinationsmöglichkeiten." *Zentralblatt für Didaktik der Mathematik* 12 (1980): 129-134.

Löwisch, Dieter-Jürgen. *Kultur und Pädagogik.* Darmstadt: Wiss. Buchges., 1989.

Lukesch, Helmut, and Karl-Heinz Kischkel. "Unterrichtsformen an Gymnasien." *Zeitschrift für erziehungswissenschaftliche Forschung* 21 (1987): 237-256.

MacGregor, Mollie E. *Making Sense of Algebra: Cognitive Processes Influencing Comprehension.* Geelong, Victoria: Deakin Univ., 1991.

Maier, Hans. "Allgemeinbildung in der arbeitsteiligen Industriegesellschaft." In *Allgemeinbildung im Computerzeitalter,* edited by Bundesmin. für Bildung und Wissenschaft, 17-33. Bonn: 1986.

Maier, Hermann: ""Verstehen" im Mathematikunterricht – Explikationsversuch zu einem vielverwendeten Begriff." In *Mathematikdidaktik: Theorie und Praxis,* edited by Peter Bender, 131-142. Berlin: Cornelsen, 1988.

- "Verstehen als individueller Prozeß der Sinnkonstruktion." *mathematik lehren* 49 (1991): 55-60.
- *Bericht über das Forschungsprojekt "Verstehen von Lehrerinstruktionen und -erklärungen durch Schüler im Mathematikunterricht".* Universität Regensburg: temporary copy, 1994.
- and Jörg Voigt, ed. *Interpretative Unterrichtsforschung.* Köln: Aulis, 1991.
- and Jörg Voigt. "Teaching styles in mathematics education." *Zentralblatt für Didaktik der Mathematik* 24 (1992): 249-253.
- and Jörg Voigt, ed. *Verstehen und Verständigung. Arbeiten zur interpretativen Unterrichtsforschung.* Köln: Aulis, 1994.

Malle, Günther. *Didaktische Probleme der elementaren Algebra.* Braunschweig: Vieweg, 1993.

Mason, John, and J. Davis. *Modelling with Mathematics in Primary and Secondary Schools.* Geelong, Australia: Deakon University Press, 1991.

Mathematical Sciences Education Board et al, ed. *Everybody Counts. A Report to the Nation on the Future of Mathematics Education.* Washington, D.C.: Nation. Acad. Press, 1989.

Mead, George H. *Geist, Identität und Gesellschaft.* Frankfurt/M.: Suhrkamp, 1968 [1934].

Mead, Margaret. *Der Konflikt der Generationen. Jugend ohne Vorbild.* 3rd ed. Olten: Walter, 1972.

Meisel, Klaus et al. *Schlüsselqualifikationen in der Diskussion.* Frankfurt/M.:Deutscher Volkshochschulverband, 1989.

Meisner, Andreas. "Erfahrungen mit einer Unterrichtseinheit zum Goldenen Schnitt – Forderungen für ein Ausbildungskonzept. In Rolf Biehler et al, 92-102. 1995.

Menze, Clemens. "Wissenschaftsorientierung als Problem der Schule." *Vierteljahresschrift für wissenschaftliche Pädagogik* 56 (1980): 177-188.

Merten, Roland. "Haben Kinder und Jugendliche keine Werte mehr? Zur moralischen Sozialisation." *Neue Sammlung* 34 (1994): 233-246.

Mertens, Dieter. "Schlüsselqualifikationen. Thesen zur Schulung für eine moderne Gesellschaft." *Mitteilungen aus Arbeitsmarkt und Berufsforschung* (1974): 36-43.

Mestre, José P., and Jack Lochhead. "The Variable-reversal Error among five Cultural Groups." *Proceedings of PME-NA-5.* Vol. I. (1983): 180-188.

Meyer, Hilbert. *UnterrichtsMethoden II: Praxisband.* Frankfurt/M.: Scriptor, 1987.

Meyer, Meinert A., and Wilfried Plöger, ed. *Allgemeine Didaktik, Fachdidaktik und Fachunterricht.* Weinheim/Basel: Beltz, 1994.

Meyer-Abich, Klaus M. *Wege zum Frieden mit der Natur.* München/Wien: Hanser, 1984.

Miller, Alice. *Am Anfang war Erziehung.* Frankfurt/M.: Suhrkamp, 1980.

Minsky, Marvin. *Mentopolis ("The Society of Mind").* Stuttgart: Klett-Cotta, 1990 [1985].

Mitchelmore, Michael C. "Three-dimensional geometrical drawing in three cultures." *Educational Studies in Mathematics* 11 (1980): 205-216.

Mittelstraß, Jürgen. *Wissenschaft als Lebensform.* Frankfurt/M.: Suhrkamp, 1982.

MUED e. V. (Mathematik-Unterrichts-Einheiten-Datei), ed. *Unterrichtsmaterialien. Überblick für die Jahrgangsstufen 5 - 10.* Appelhülsen: MUED, 1994a.

- *Unterrichtsmaterialien. Überblick für die Jahrgangsstufen 11 - 13.* Appelhülsen: MUED, 1994b.

Müller, Manfred. *Mathematisches Denken.* Frankfurt/M.: Lang, 1985.

Müller-Rolli, Sebastian. "Einleitung." In *Das Bildungswesen der Zukunft,* edited by Sebastian Müller-Rolli, 7-29. Stuttgart: Klett-Cotta, 1987.

Münzinger, Wolfgang, ed. *Projektorientierter Mathematikunterricht.* München et al.: Urban & Schwarzenberg, 1977.

National Council of Teachers of Mathematics – Commission on Standards for School Mathematics, ed. *Curriculum and Evaluation Standards for School Mathematics.* Reston/VA.: NCTM, 1989.

Nicklis, Werner S. "Allgemeinbildung heute. Allgemeine wissenschaftsorientierte Grundbildung für alle? Konsequenzen für die Schulstruktur." In *Allgemeinbildung und Schulstruktur,* edited by Franz-Wilhelm Babilon and Heinz-Jürgen Ipfling, 45-57. Bochum: Kamp, 1980.

Nickson, Marilyn. "The Culture of the Mathematics Classroom: An Unknown Quantity?" In *Handbook of Research on Mathematics Teaching and Learning (NCTM),* edited by Douglas A. Grouws, 101-114. New York: Macmillan, 1992.

Nipkow, Karl E. *Die Individualität als pädagogisches Problem bei Pestalozzi, Humboldt und Schleiermacher.* Weinheim: Beltz, 1960.

- "Bildung und Entfremdung. Überlegungen zur Rekonstruktion der Bildungstheorie." *Zeitschrift für Pädagogik* 14 (1977): 205-229.

Niss, Mogens et al., ed. *Teaching of Mathematical Modelling and Applications.* New York: Ellis Horwood, 1991.

Nitzschke, Volker, ed. *Multikulturelle Gesellschaft - multikulturelle Erziehung.* Stuttgart: J.B. Metzler, 1982.

Nolte-Fischer, Georg. *Bildung zum Laien.* Weinheim: Deutscher Studien Verlag, 1989.

Nüse, Ralf et al. *Über die Erfindungen des Radikalen Konstruktivismus. Kritische Gegenargumente aus psychologischer Sicht.* Weinheim: Deutscher Studien Verlag, 1991.

Osthoff, Ralf. *Grundlagen einer ökologischen Pädagogik.* Frankfurt/M.: dipa, 1986.

Otto, Gunter, and Wolfgang Schulz. "Der Beitrag der Curriculumforschung. Ziele und Inhalte der Erziehung und des Unterrichts." In *Enzyklopädie Erziehungswissenschaft,* edited by Hans-Dieter Haller, and Hilbert Meyer, 49-105.Vol III. Stuttgart: Klett, 1986.

Parsons, Talcott. "The School Class as a Social System: Some of its Functions in American Society." In *Social Structure and Personality.* London and New York: Free Press, 1965 (1959), pp. 129-154.

Pascal, Blaise. *Gedanken. Eine Auswahl.* Stuttgart: Reclam, 1987. [ca. 1660].

Paulos, John A. *Zahlenblind. Mathematisches Analphabetentum und seine Konsequenzen.* (with a preface by Douglas R. Hofstadter.) München: Heyne, 1990.

Paulsen, Friedrich. "Geschichte des gelehrten Unterrichts." Vol. II of *Der gelehrte Unterricht im Zeichen des Neuhumanismus.* Leipzig: 1885.

- "Die Aufgabe des Unterrichts überhaupt." In *Theorie der Schule,* edited by Hermann Röhrs, 43-59. Frankfurt/M.: Akad. Verlagsges., 1968. [1911]

Peddiwell, J. Abner. *Das Säbelzahn-Curriculum.* Stuttgart: Klett, 1974. [1938].

Peschek, Werner. "Mathematikunterricht und Qualifizierung." *Journal für Mathematik-Didaktik* 2 (1981): 249-279.

Philipp, Randolph A. "A Study of Algebraic Variables: Beyond the Student-Professor Problem." *Journal of Mathematical Behavior* 11 (1992): 161-176.

Picht, Georg. "Der Begriff der Verantwortung." In *Wahrheit, Vernunft Verantwortung,* edited by Georg Picht, Stuttgart: 1969.

Picker, Bernold. "Mathematikunterricht als Vermittlung von grundlegenden Ideen." *Der Mathematikunterricht* 31 (1985a): 6-9.

- "Intuitives Erfassen und Gebrauchen von grundlegenden Ideen der Analysis im mathematischen Anfangsunterricht." *Der Mathematikunterricht* 31 (1985b): 46-71.

Pinxten, Rik et al. *Anthropology of Space.* Philadelphia, Pa.: Univ. of Pennsylvania Press, 1983.

Plato. *Laws.* With an English Translation by R. G. Bury. Vol. I London: Heinemann 1961 (1926).

Pleines, Jürgen-Eckardt. "Das Problem des Allgemeinen in der Bildungstheorie." *Zeitschrift für Pädagogik* (1987): 35-40.

Plöger, Ursula. "Was kein Mann mehr sagen kann. Kritische Anmerkungen zu einer feministischen Okkupation von Sprache und Bewußtsein." In Vol. XXVIII (Politik II) of *Philosophie. Beiträge zur Unterrichtspraxis,* edited by Jürgen Hengelbrock, 79-88. Berlin: Cornelsen, 1994.

Plöger, Wilfried. *Naturwissenschaftlich-technischer Unterricht unter dem Anspruch der Allgemeinbildung.* Frankfurt/M.: Lang, 1989.

Polya, Georg. "Induktion und Analogie in der Mathematik." Vol. I of *Mathematik und plausibles Schließen,* Basel: Birkhäuser, 1962.

- "Typen und Strukturen plausibler Folgerung." Vol. II of *Mathematik und plausibles Schließen*, Basel: Birkhäuser, 1963.

- *Schule des Denkens. Vom Lösen mathematischer Probleme.* 3rd ed. Bern/München: Francke, 1980.

Prior, Harm ed. *Soziales Lernen.* Düsseldorf: Pädagogischer Verlag Schwann, 1976.

Projektgruppe Schlüsselqualifikationen in der beruflichen Bildung. *Wege zur beruflichen Mündigkeit* (part I). Weinheim: Deutscher Studien Verlag, 1992.

Raatz, U. *Mathematik am Arbeitsplatz - zwei empirische Untersuchungen.* (Materialien zu VHS-Zertifikaten Nr. 15). Frankfurt/M.: 1974.

Rademacher, Hans, and Otto Toeplitz. *Von Zahlen und Figuren. Proben mathematischen Denkens für Liebhaber der Mathematik.* Berlin: Springer, 1968. [1933].

Ramseger, Jörg. *Was heißt "durch Unterricht erziehen"? Erziehender Unterricht und Schulreform.* Weinheim/Basel: Beltz, 1991.

Reble, Albert. *Geschichte der Pädagogik.* 11th ed. Stuttgart: Klett, 1971a.

- *Geschichte der Pädagogik.* Dokumentationsband II. Stuttgart: Klett, 1971b.

Redeker, Bruno. "Martin Wagenschein – Feiertagsdidaktik oder Notwendigkeit einer Renaissance?" *Neue Sammlung* 33 (1993): 15-30.

Redl, Fritz. *Erziehung schwieriger Kinder.* München: Piper & Co., 1971.

- and David Wineman. *Kinder, die hassen. Auflösung und Zusammenbruch der Selbstkontrolle.* München: Piper & Co., 1979. [1951].

Reichel, Hans-Christian, and J. Zöchling. "Tausend Gleichungen, und was nun? Computertomographie als Einstieg in ein aktuelles Thema im Mathematikunterricht." *Didaktik der Mathematik* 18 (1990): 245-270.

Reiß, Veronika. "Fachspezifische Sozialisation in der Ausbildung von Gymnasiallehrern." *Neue Sammlung* 15 (1975): 298-314.

- "Zur theoretischen Einordnung von Sozialisationsphänomenen im Mathematikunterricht." *Zeitschrift für Pädagogik* 25 (1979): 275-289.

Rentz, Winrich. "Die optimale Dosenform." *Mathematik in der Schule* 29 (1991): 841-844.

Resnick, Lauren B. et al. "Understanding Algebra." In *Cognitive Processes in Mathematics,* edited by John A. Sloboda, and Don Rogers, 169-203. Oxford: Clarendon, 1987b.

- and Leopold E. Klopfer. "Toward the Thinking Curriculum: An Overview." In *Toward the Thinking Curriculum: Current Cognitive Research,* edited by Resnick, Lauren B. and Leopold E. Klopfer, 1-18. Washington D. C.: ASCD, 1989.

- and Leopold E. Klopfer. "Toward the Thinking Curriculum: Concluding Remarks." In *Toward the Thinking Curriculum: Current Cognitive Research,* edited by Resnick, Lauren B. and Leopold E. Klopfer, 206-211. Washington D.C.: ASCD, 1989.

Rhyn, Heinz. "Allgemeinbildung, Staat und Politik. Zur aktuellen Diskussion um die angelsächsische "liberal education"." *Zeitschrift für Pädagogik* 40 (1994): 607-625.

Richter, Ingo. "Entscheidungsstrukturen für Bildungsfragen in offenen Gesellschaften." *Zeitschrift für Pädagogik* 40 (1994). 181-191.

Riedel, Christoph. *Subjekt und Individuum. Zur Geschichte des philosophischen Ich-Begriffes.* Darmstadt: Wissenschaftliche Buchgesellschaft, 1989.

Robinsohn, Saul B. *Bildungsreform als Revision des Curriculum und ein Strukturkonzept für Curriculumentwicklung.* 4th ed. Neuwied/Berlin: Luchterhand, 1973. [1967/1969].

Röhrs, Hermann. Frieden – eine pädagogische Aufgabe. Braunschweig: Pedersen, 1983.

Rolff, Hans-Günter, and Peter Zimmermann, ed. *Neue Medien und Lernen.* Weinheim/Basel: Beltz, 1985.

Rosnick, Peter, and John Clement. "Learning without understanding: The effect of tutoring strategies on algebra misconceptions." *Journal of Mathematical Behavior* 3 (1980): 3-27.

Roth, Heinrich. "Die realistische Wendung in der pädagogischen Forschung." *Neue Sammlung* 2 (1962): 486 ff.

- *Pädagogische Psychologie des Lehrens und Lernens.* 9th ed. Hannover: Schroedel, 1966.

Rousseau, Jean-Jacques. *Emile oder Über die Erziehung.* 4th ed. Paderborn: Schöningh, 1978. [1762].

Rülcker, Tobias. *Bildung, Gesellschaft und Wissenschaft.* Heidelberg: Quelle & Meyer, 1976.

Rumpf, Horst. *Unterricht und Identität.* München: Juventa, 1976.

- *Die übergangene Sinnlichkeit.* München: Juventa, 1981.
- "Die Schule, der Körper und das handgreifliche Tun." *Neue Sammlung* 23 (1983): 333-346.
- *Belebungsversuche. Ausgrabungen wider die Verödung der Lernkultur.* Weinheim/München: Juventa, 1987.

Russell, Bertrand. *Erziehung ohne Dogma. Pädagogische Schriften.* München: Nymphenburger, 1974. [1961].

- "Education for a Difficult World." In *Fact and Fiction.* London and New York: Routledge, 1994 (1961), pp. 144-149.

Savigny, Eike v. *Grundkurs im wissenschaftlichen Definieren.* 4th ed. München: dtv, 1976.

Scheibe, Wolfgang. *Die Reformpädagogische Bewegung 1900-1932.* Weinheim: Beltz, 1969.

Scheilke, Christoph, ed. *Bildung durch Schlüsselqualifikationen? Zum Verhältnis von Bildung und Beruf.* Münster: Comenius-Institut, 1991.

Schleiermacher, Friedrich. *Pädagogische Schriften 1. Die Vorlesungen aus dem Jahre 1826,* edited by von E. Weniger, and Th. Schulze. Frankfurt a.M./Berlin/Wien: Klett-Cotta/Ullstein, 1983.

Schoenfeld, Alan H. "Teaching Mathematical Thinking and Problem Solving." In *Towards the Thinking Curriculum: Current Cognitive Research,* edited by Lauren B. Resnick, and Leopold E. Klopfer, 83-103. Washington D.C.: ASCD, 1989.

- "Learning to Think Mathematically: Problem Solving, Metacognition, and Sense Making in Mathematics." In *Handbook of Research on Mathematics Teaching and Learning (NTCM),* edited by Douglas A. Grouws, 334-370. New York; Macmillan, 1992.

Schreiber, Alfred. "Universelle Ideen im mathematischen Denken - ein Forschungsgegenstand der Fachdidaktik." *mathematica didactica* 2 (1979): 165-171.

- "Bemerkungen zur Rolle universeller Ideen im mathematischen Denken." *mathematica didactica* 6 (1983): 65-76.

Schubring, Gert. *Das genetische Prinzip in der Mathematik-Didaktik.* Stuttgart: Klett-Cotta, 1978.

Schulze, Theodor. *Das Allgemeine der Bildung und das Spezielle der Fächer*, edited by Knut Lohmann, 16-38. 1990.

Schupp, Hans. "Optimieren als Leitlinie im Mathematikunterricht." *Mathematische Semesterberichte* 31 (1984): 59-76.

- "Anwendungsorientierter Mathematikunterricht in der Sekundarstufe I zwischen Tradition und neuen Impulsen." *Der Mathematikunterricht* 34 (1988): 5-16.
- *Optimieren: Extremwertbestimmung im Mathematikunterricht.* Mannheim et al: BI & Wissenschaftsverlag, 1992.

Schweiger, Fritz. "Fundamentale Ideen." *Journal für Mathematik-Didaktik* 13 (1992): 199- 214.

Schweitzer, Friedrich. *Identität und Erziehung. Was kann der Identitätsbegriff für die Pädagogik leisten?* Weinheim/Basel: Beltz, 1985.

Seibert, Herbert, and Helmut J. Serve, ed. *Bildung und Erziehung an der Schwelle zum dritten Jahrtausend.* München: PimS, 1994.

Sierpinska, Anna. *Understanding in Mathematics.* London/Washington D.C.: Falmer Press, 1994.

Skemp, Richard R. "Relational Understanding and Instrumental Understanding. *Mathematics Teaching* 77 (1976): 20-26.

Smith, Frank. *Comprehension and Learning.* New York et al.: Holt, Rinehart & Winston, 1975.

Spiegel, Hartmut. *Ergebnisse einer Umfrage unter Studienanfängern. Unveröffentlichtes Manuskript.* Paderborn: 1988.

- "Sokratische Gespräche über mathematische Themen mit Erwachsenen – Absichten und Erfahrungen." *mathematik lehren* 33 (1989): 54-59.

Spies, Werner. "Zerfallende Selbstverständlichkeit. Auflösung des Bildungskonzepts – Bemühen um neue Bezugspunkte." *Die Deutsche Schule* 68 (1976):7-16.

Stacey, Kaye. "Linking application and acquisition of mathematical ideas through problem solving." *Zentralblatt für Didaktik der Mathematik* 23 (1991): 8-14.
Stachowiak, Herbert. *Allgemeine Modelltheorie.* Wien/New York: Springer, 1973.
Steen, Lynn A., ed . *On the Shoulders of Giants. New Approaches to Numeracy.* Washington, D.C.: National Academy Press, 1990.
Stegmaier, Werner. "Allgemeinbildung und Weltorientierung." *Universitas* 39 (1984): 619-628.
Stern, Elsbeth. "Warum werden Kapitänsaufgaben gelöst? Das Verstehen von Textaufgaben aus psychologischer Sicht." *Der Mathematikunterricht* 38 (1992): 7-29.
Sträßer, Rudolf. "Mathematik als Element beruflicher Qualifikation." In *Mathematikunterricht zwischen Tradition und neuen Impulsen,* edited by H. W. Heymann, 49-79. Köln: Aulis, 1984.
- et al. "Skills versus Understanding." *Zentralblatt für Didaktik der Mathematik* 21 (1989): 197-201.
Struve, Rolf, and Jörg Voigt. Die Unterrichtsszene im Menon-Dialog." *Journal für Mathematik-Didaktik* 9 (1988): 259-285.
Sühl-Strohmenger, Wilfried. *Horizonte von Bildung und Allgemeinbildung.* Frankfurt/M.: Lang, 1984.
Tenorth, Heinz-Elmar. *Geschichte der Erziehung.* Weinheim/München: Juventa, 1988.
- "Neue Konzepte der Allgemeinbildung." edited by Hans W. Heymann , and Willi van Lück, (1990): 111-130.
- *"Alle alles zu lehren". Möglichkeiten und Perspektiven allgemeiner Bildung.* Darmstadt: Wiss. Buchges., 1994.
Terhart, Ewald. "Moralerziehung in der Schule. Positionen und Probleme eines schulpädagogischen Programms." *Neue Sammlung* 29 (1989): 376-394.
Tietze, Uwe P. "Der Mathematiklehrer an der gymnasialen Oberstufe. Zur Erfassung berufsbezogener Kognitionen." *Journal für Mathematik-Didaktik* 11 (1990): 177-243.
- et al. *Didaktik des Mathematikunterrichts in der Sekundarstufe II.* Braunschweig: Vieweg, 1982.
Tillmann, Klaus-Jürgen. "Kooperationsbereitschaft - Flexibilität – Kundenorientierung. Ein neuer Reformdialog zwischen Wirtschaft und Schule?" *Neue Sammlung* 34 (1994): 135-148.
Törner, Günter, and Stefan Grigutsch. ""Mathematische Weltbilder" bei Studienanfängern - eine Erhebung." *Journal für Mathematik-Didaktik* 15 (1994): 211-251.
Tylor, Edward B. *Primitive Culture.* London: 1871.
Voigt, Jörg. "Der kurztaktige, fragend-entwickelnde Mathematikunterricht. Szenen und Analysen." *mathematica didactica* 7 (1984a): 161-186.
- *Interaktionsmuster und Routinen im Mathematikunterricht.* Weinheim/Basel: Beltz, 1984b.
- "Entwicklung mathematischer Themen und Normen im Unterricht." In *Verstehen und Verständigung,* edited by Hermann Maier, and Jörg Voigt, 77-111. Köln, Aulis, 1994.
Volk, Dieter. *Handlungsorientierende Unterrichtslehre am Beispiel Mathematikunterricht.* Bensheim: päd-extra-Buchverlag, 1979.
- *Zur Wissenschaftstheorie der Mathematik. Orientierungen für emanzipatorischen Mathematikunterricht.* Bensheim: päd-extra-Buchverlag, 1980.
- *Mathematik für's tägliche Leben.* Soest: LSW, 1994.
Vollrath, Hans-Joachim. "Rettet die Ideen!" *Der mathematische und naturwissenschaftliche Unterricht* 31 (1978): 449-455.
- "Anstöße - Gedanken zu Martin Wagenschein." *Journal für Mathematik-Didaktik* 10 (1989): 349-363.
- "Paradoxien des Verstehens von Mathematik." *Journal für Mathematik-Didaktik* 14 (1993): 35-58.
Wagenschein, Martin. *Ursprüngliches Verstehen und exaktes Denken.* Stuttgart: Klett, 1965.
- *Ursprüngliches Verstehen und exaktes Denken II.* Stuttgart: Klett, 1970.
- "Der Vorrang des Verstehens. Pädagogische Anmerkungen zum mathematisierenden Unterricht." *Neue Sammlung* 14 (1974): 144-160.
- *Verstehen lehren. Genetisch - Sokratisch - Exemplarisch.* 5th ed. Weinheim/Basel: Beltz, 1975.
- *Naturphänomene sehen und verstehen. Genetische Lehrgänge.* Stuttgart: Klett, 1980.
Wagner, Hans. *Philosophie und Reflexion.* München et al.: Reinhardt, 1959.
Weber, Hellmar. *Grundlagen einer Didaktik des Mathematisierens.* Frankfurt: Lang, 1980.
Weinert, Franz E., and Bernhard Treiber. "Einleitung." In *Lehr-Lern-Forschung,* edited by Franz E.Weinert, and Bernhard Treiber, 7-11. München et al.: Urban & Schwarzenberg, 1982.
Weizsäcker, Carl F. v. *Die Zeit drängt. Eine Weltversammlung der Christen für Gerechtigkeit, Frieden und die Bewahrung der Schöpfung.* München/Wien: Hanser, 1986.
Welsch, Wolfgang. *Unsere postmoderne Moderne.* Weinheim: VCH, 1987.

Weniger, Erich. *Die Eigenständigkeit der Erziehung in Theorie und Praxis.* Weinheim: Beltz, 1953.

- *Didaktik als Bildungslehre. Teil 1. Theorie der Bildungsinhalte und des Lehrplans.* Weinheim: Beltz, 31960.
- "Die Autonomie der Pädagogik." In *Erich Weniger - Ausgewählte Schriften zur geisteswissenschaftlichen Pädagogik,* edited by Bruno Schonig, 11-27. Weinheim/Basel: Beltz, 1975 [1929].

Werge, Christian. *Anwendungsaufgaben im Mathematikunterricht.* Diss. Universität Leipzig: 1987.

Westbury, Ian, and Alan C. Purves, ed. *Cultural Literacy and the Idea of General Education.* Chicago: NSSE, 1988.

Westphalen, Klaus. "Was heißt heute Allgemeinbildung?" *Anregung* 28 (1982): 72-81.

Whitehead, Alfred N. "Die Gegenstände des mathematischen Unterrichts." *Neue Sammlung* 2 (1962 [1913]): 257-266.

- "The Mathematical Curriculum." In *The Aims of Education and Other Essays.* New York: Free Press 1967 (1913), pp. 77-89.

Wilhelm, Theodor. *Theorie der Schule.* 2nd ed. Stuttgart: Metzler, 1969. [1967]

- *Pflegefall Staatsschule – Nachtrag zur "Theorie der Schule".* Stuttgart: Metzler, 1982.
- "Die Allgemeinbildung ist tot – Es lebe die Allgemeinbildung!" *Neue Sammlung* 25 (1985): 120-150.

Wille, Rudolf. *"Allgemeine Mathematik" als Bildungskonzept für die Schule,* edited by Rolf Biehler, et al. (1995): 41-55.

Wilsdorf, Dieter. *Schlüsselqualifikationen.* München: Lexika, 1991.

Winter, Heinrich. *Sachrechnen in der Grundschule.* Bielefeld: CVK, 1985.

- *Entdeckendes Lernen im Mathematikunterricht.* Braunschweig/Wiesbaden: Vieweg, 1989.
- "Bürger und Mathematik." *Zentralblatt für Didaktik der Mathematik* 22 (1990): 131-147.

Winter, Martin. "Läßt sich allgemeinbildender Mathematikunterricht im Grundkurs realisieren?" *mathematik lehren* 33 (1989): 43-49.

- "Unterrichtskultur bestimmt mathematische Bildung!" *Mathematik in der Schule* 32 (1994): 65-70.
- "Zuhören können - auch im Mathematikunterricht." In Rolf Biehler, et al. 103-111. (1995).

Wittenberg, Alexander I. *Bildung und Mathematik.* Stuttgart: Klett, 1963.

Wittmann, Erich. *Grundfragen des Mathematikunterrichts.* Braunschweig: Vieweg, 1974.

Zabeck, Jürgen. "'Schlüsselqualifikationen' – Zur Kritik einer didaktischen Zielformel." *Wirtschaft und Erziehung* 3 (1989): 77-86.

Zimmermann, Bernd, ed. *Problemorientierter Mathematikunterricht.* Bad Salzdetfurth: Franzbecker, 1991.

INDEX

Mathematics Education Library

Managing Editor: A.J. Bishop, Melbourne, Australia

1. H. Freudenthal: *Didactical Phenomenology of Mathematical Structures.* 1983
   ISBN 90-277-1535-1; Pb 90-277-2261-7
2. B. Christiansen, A. G. Howson and M. Otte (eds.): *Perspectives on Mathematics Education.* Papers submitted by Members of the Bacomet Group. 1986.
   ISBN 90-277-1929-2; Pb 90-277-2118-1
3. A. Treffers: *Three Dimensions.* A Model of Goal and Theory Description in Mathematics Instruction The Wiskobas Project. 1987 ISBN 90-277-2165-3
4. S. Mellin-Olsen: *The Politics of Mathematics Education.* 1987
   ISBN 90-277-2350-8
5. E. Fischbein: *Intuition in Science and Mathematics.* An Educational Approach. 1987
   ISBN 90-277-2506-3
6. A.J. Bishop: *Mathematical Enculturation.* A Cultural Perspective on Mathematics Education. 1988 ISBN 90-277-2646-9; Pb (1991) 0-7923-1270-8
7. E. von Glasersfeld (ed.): *Radical Constructivism in Mathematics Education.* 1991
   ISBN 0-7923-1257-0
8. L. Streefland: *Fractions in Realistic Mathematics Education.* A Paradigm of Developmental Research. 1991 ISBN 0-7923-1282-1
9. H. Freudenthal: *Revisiting Mathematics Education.* China Lectures. 1991
   ISBN 0-7923-1299-6
10. A.J. Bishop, S. Mellin-Olsen and J. van Dormolen (eds.): *Mathematical Knowledge: Its Growth Through Teaching.* 1991 ISBN 0-7923-1344-5
11. D. Tall (ed.): *Advanced Mathematical Thinking.* 1991 ISBN 0-7923-1456-5
12. R. Kapadia and M. Borovcnik (eds.): *Chance Encounters: Probability in Education.* 1991 ISBN 0-7923-1474-3
13. R. Biehler, R.W. Scholz, R. Sträßer and B. Winkelmann (eds.): *Didactics of Mathematics as a Scientific Discipline.* 1994 ISBN 0-7923-2613-X
14. S. Lerman (ed.): *Cultural Perspectives on the Mathematics Classroom.* 1994
   ISBN 0-7923-2931-7
15. O. Skovsmose: *Towards a Philosophy of Critical Mathematics Education.* 1994
   ISBN 0-7923-2932-5
16. H. Mansfield, N.A. Pateman and N. Bednarz (eds.): *Mathematics for Tomorrow's Young Children.* International Perspectives on Curriculum. 1996
   ISBN 0-7923-3998-3
17. R. Noss and C. Hoyles: *Windows on Mathematical Meanings.* Learning Cultures and Computers. 1996 ISBN 0-7923-4073-6; Pb 0-7923-4074-4

Mathematics Education Library

18. N. Bednarz, C. Kieran and L. Lee (eds.): *Approaches to Algebra.* Perspectives for Research and Teaching. 1996 ISBN 0-7923-4145-7; Pb ISBN 0-7923-4168-6
19. G. Brousseau: *Theory of Didactical Situations in Mathematics.* Didactique des Mathématiques 19701990. Edited and translated by N. Balacheff, M. Cooper, R. Sutherland and V. Warfield. 1997 ISBN 0-7923-4526-6
20. T. Brown: *Mathematics Education and Language.* Interpreting Hermeneutics and Post-Structuralism. 1997 ISBN 0-7923-4554-1
    Second Revised Edition. 2001 Pb ISBN 0-7923-6969-6
21. D. Coben, J. O'Donoghue and G.E. FitzSimons (eds.): *Perspectives on Adults Learning Mathematics.* Research and Practice. 2000 ISBN 0-7923-6415-5
22. R. Sutherland, T. Rojano, A. Bell and R. Lins (eds.): *Perspectives on School Algebra.* 2000 ISBN 0-7923-6462-7
23. J.-L. Dorier (ed.): *On the Teaching of Linear Algebra.* 2000 ISBN 0-7923-6539-9
24. A. Bessot and J. Ridgway (eds.): *Education for Mathematics in the Workplace.* 2000 ISBN 0-7923-6663-8
25. D. Clarke (ed.): *Perspectives on Practice and Meaning in Mathematics and Science Classrooms.* 2001 ISBN 0-7923-6938-6; Pb ISBN 0-7923-6939-4
26. J. Adler: *Teaching Mathematics in Multilingual Classrooms.* 2001 ISBN 0-7923-7079-1; Pb ISBN 0-7923-7080-5
27. G. de Abreu, A.J. Bishop and N.C. Presmeg (eds.): *Transitions Between Contexts of Mathematical Practices.* 2001 ISBN 0-7923-7185-2
28. G.E. FitzSimons: *What Counts as Mathematics?* Technologies of Power in Adult and Vocational Education. 2002 ISBN 1-4020-0668-3
29. H. Alrø and O. Skovsmose: *Dialogue and Learning in Mathematics Education.* Intention, Reflection, Critique. 2002 ISBN 1-4020-0998-4
30. K. Gravemeijer, R. Lehrer, B. van Oers and L. Verschaffel (eds.): *Symbolizing, Modeling and Tool Use in Mathematics Education.* 2002 ISBN 1-4020-1032-X
31. G.C. Leder, E. Pehkonen and G. Törner (eds.): *Beliefs: A Hidden Variable in Mathematics Education?* 2002 ISBN 1-4020-1057-5; Pb ISBN 1-4020-1058-3
32. R. Vithal: *In Search of a Pedagogy of Conflict and Dialogue for Mathematics Education.* 2003 ISBN 1-4020-1504-6
33. H.W. Heymann: *Why Teach Mathematics?* A Focus on General Education. 2003 ISBN 1-4020-1786-3

KLUWER ACADEMIC PUBLISHERS – DORDRECHT / BOSTON / LONDON